Interpretive Solutions for Dynamic Structures Through ABAQUS Finite Element Packages

Interpretive Solutions for Dynamic Structures Through ABAQUS Finite Element Packages

Farzad Hejazi
Hojjat Mohammadi Esfahani

CRC Press
Taylor & Francis Group
Boca Raton London New York

CRC Press is an imprint of the
Taylor & Francis Group, an **informa** business

First edition published 2022
by CRC Press
6000 Broken Sound Parkway NW, Suite 300, Boca Raton, FL 33487-2742

and by CRC Press
2 Park Square, Milton Park, Abingdon, Oxon, OX14 4RN

Library of Congress Cataloging-in-Publication Data
Names: Hejazi, Farzad, author. | Esfahani, Hojjat Mohammadi, author.
Title: Interpretive solutions for dynamic structures through ABAQUS finite
    element packages / Farzad Hejazi, Hojjat Mohammadi Esfahani.
Description: Boca Raton, FL : CRC Press, 2022. | Includes bibliographical
    references and index.
Identifiers: LCCN 2021035409 (print) | LCCN 2021035410 (ebook) |
    ISBN 9781032113517 (hbk) | ISBN 9781032113524 (pbk) |
    ISBN 9781003219491 (ebk)
Subjects: LCSH: Structural dynamics—Data processing. | Finite element
    method. | Abaqus (Electronic resource)
Classification: LCC TA347.F5 H45 2022 (print) | LCC TA347.F5 (ebook) |
    DDC 620.001/51535—dc23
LC record available at https://lccn.loc.gov/2021035409
LC ebook record available at https://lccn.loc.gov/2021035410

ISBN: 978-1-032-11351-7 (hbk)
ISBN: 978-1-032-11352-4 (pbk)
ISBN: 978-1-003-21949-1 (ebk)

DOI: 10.1201/9781003219491

Typeset in Times
by codeMantra

Contents

List of Figures xi
List of Tables xxvii
Preface xxix
Authors xxxi

Chapter 1 Development of Subroutines for ABAQUS Software 1

    1.1   Introduction 1
    1.2   Problem Description 2
    1.3   Objectives 2
    1.4   Developing a UMAT Subroutine 2
    1.5   Modeling 8
         1.5.1   Parts Module 9
         1.5.2   Property Module 11
              1.5.2.1   Material Properties 11
              1.5.2.2   Section Properties 11
              1.5.2.3   Section Assignment 11
         1.5.3   Assembly Module 11
         1.5.4   Step Module 13
              1.5.4.1   Implicit Dynamics 14
              1.5.4.2   Explicit Dynamics 17
         1.5.5   Required Output 17
         1.5.6   Interaction Module 17
         1.5.7   Load Module 19
         1.5.8   Mesh Module 22
    1.6   Analysis: Job Module 22
    1.7   Analysis Results 22

Chapter 2 Evaluate Performance of Steel Wall in Structures Subjected
to Cyclic Load 27

    2.1   Introduction 27
    2.2   Problem Description 27
    2.3   Objectives 27
    2.4   Modeling 28
         2.4.1   Part Module 28
              2.4.1.1   Create a New Model Database 28
              2.4.1.2   Create Parts 28
              2.4.1.3   Define an I Shape with Dimensions 30
              2.4.1.4   Create Partition 31
         2.4.2   Property Module 31

2.4.2.1   Material Properties 32
2.4.2.2   Section Properties (Type, Thickness,
          and Material Assignment)..................... 34
2.4.2.3   Section Assignment 36
2.4.3   Assembly Module.. 37
2.4.3.1   Assemble Part Instances into the Model ... 37
2.4.4   Step Module 37
2.4.4.1   Create an Analysis Step: Step 1 39
2.4.5   Interaction Module 42
2.4.5.1   Tie Constraint 42
2.4.6   Load Module 42
2.4.7   Mesh Module.. 46
2.4.7.1   Mesh: Seed the Part (50 mm Elements) ... 46
2.4.7.2   Assign an ABAQUS Element Type 46
2.4.7.3   Seed and Mesh the Model 48
2.4.7.4   Verify Mesh 50
2.5   Analysis: Job Module 52
2.5.1   Create an Analysis Job: Job-1............................... 52
2.5.2   Monitor Solution in Progress 54
2.6   Visualization Module 54
2.6.1   View the Results of the Analysis........................... 54
2.6.2   Visualization/Results Module 55

Chapter 3   Performance of Reinforced-Concrete Frame with Embedded
CFRP Rod under Cyclic Load... 59

3.1   Introduction 59
3.2   Problem Description.. 59
3.2.1   Problem Statement 59
3.2.1.1   Material Properties 60
3.3   Objectives 61
3.4   Modeling... 62
3.4.1   Part Module.. 62
3.4.1.1   Create a New Model Database.............. 62
3.4.1.2   Create a New Model Database and
          a New Part 63
3.4.1.3   Define a Rectangle with Dimensions..... 63
3.4.1.4   Define Section of the Rebars
          Section with Dimensions 64
3.4.2   Property Module 65
3.4.2.1   Material Properties 65
3.4.2.2   Section Properties................................ 68
3.4.2.3   Section Assignment 70
3.4.3   Mesh Module... 70
3.4.3.1   Mesh... 70

        3.4.4   Assembly Module 72

                3.4.4.1   Assemble Part Instances into the Model ... 73

        3.4.5   Interaction Module 73

                3.4.5.1   Tie Constraint 73

                3.4.5.2   Embedded Region Constraint 75

                3.4.5.3   Create an Embedded Region Constraint ... 75

        3.4.6   Step Module 77

                3.4.6.1   Create an Analysis Step: Apply Load 78

        3.4.7   Load Condition Module 80

                3.4.7.1   Apply Cyclic Loading to the Frame 84

   3.5   Analysis: Job Module 85

        3.5.1   Create an Analysis Job: Job-1 85

        3.5.2   Monitor the Solution in Progress 87

   3.6   Visualization Module 87

        3.6.1   View the Results of the Analysis 87

   3.7   Analysis Result 88

        3.7.1   Hysteresis Results 88

        3.7.2   Stresses in Frame 90

        3.7.3   Total Strain in Frame 90

        3.7.4   Plasticity Contour Plots in Frame 91

   3.8   Discussions 92

Chapter 4   Behavior of Precast Beam–Column Dowel Connection
under Cyclic Loads 101

   4.1   Introduction 101

   4.2   Problem Description 101

        4.2.1   Geometric Properties 101

        4.2.2   Material Properties 101

   4.3   Objectives 102

   4.4   Modeling 105

        4.4.1   Part Module 105

                4.4.1.1   Create a New Model Database 105

                4.4.1.2   Create a New Model Database and
                                a New Part 105

                4.4.1.3   Define Section of the Precast Beam
                                with Dimension 106

                4.4.1.4   Create a Circular Hole 108

                4.4.1.5   Create Partition 109

                4.4.1.6   Create a New Part 115

                4.4.1.7   Define Bottom Bar Section with
                                Dimension 115

        4.4.2   Property Module 117

                  4.4.2.1   Material Properties 117

                4.4.2.2   Section Properties 119

4.4.2.3   Section Assignments............................ 121
        4.4.3   Assembly Module.............................. 122
        4.4.4   Step Module 125
        4.4.5   Interaction Module 128
                4.4.5.1   Surface-to-Surface Contact Interaction... 129
                4.4.5.2   Tie Constraint 135
        4.4.6   Load Condition Module 136
        4.4.7   Mesh Module.................................143
        4.4.8   History Output Definition 145
                4.4.8.1   Create Sets of Nodes........... 146
                4.4.8.2   History Output for the Sets.......... 146
    4.5   Analysis: Job Module 148
        4.5.1   Create an Analysis Job: Job-1............ 148
    4.6   Visualization Module 149
        4.6.1   View the Results of the Analysis........ 149
    4.7   Contour plots 151
    4.8   Conclusion 158

Chapter 5   Simulation of the Preloaded Bolt Connection under
            Cyclic Loading 159

    5.1   Introduction 159
    5.2   Problem Description................................ 159
    5.3   Objectives 161
    5.4   Modeling... 161
        5.4.1   Part Module 162
                5.4.1.1   Modifying Parts for Use in the
                          Connection......................... 168
        5.4.2   Material Properties.......................... 170
        5.4.3   Section Properties 173
        5.4.4   Section Assignment........................ 174
        5.4.5   Assembly Module.......................... 175
        5.4.6   Meshing Module 181
        5.4.7   Step Module 189
        5.4.8   Interaction Module 190
                5.4.8.1   Coupling Constraint............ 192
        5.4.9   Load Module 194
        5.4.10   Boundary Condition 194
    5.5   Analysis: Job Module 199
    5.6   Results 199

Chapter 6   Beam–Column Connection Retrofitted with CFRP Sheets
            Subjected to Pushover Loading...................... 205

    6.1   Introduction 205

6.2   Problem Description..205
6.3   Objectives 206
6.4   Modeling..206
       6.4.1   Part Module..206
               6.4.1.1   Create a New Model Database............207
               6.4.1.2   Create a New Model Database and
                         a New Part 207
               6.4.1.3   Define a Rectangle with Dimensions...208
               6.4.1.4   Extrude Solid to Create a Beam 208
       6.4.2   Property Module 212
               6.4.2.1   Material Properties.............................212
               6.4.2.2   Section Properties...............................216
               6.4.2.3   Assign Sections to Parts 217
       6.4.3   Mesh Module...217
       6.4.4   Assembly Module..221
       6.4.5   Step Module 222
       6.4.6   Interaction Module 226
       6.4.7   Load Module 227
6.5   Analysis: Job Module 230
6.6   Visualization Module 232

Chapter 7   Hollow Circular Ultra-High-Performance Concrete
            (UHPFRC) Section under Lateral Cyclic Load 235

7.1   Introduction 235
7.2   Problem Description..235
       7.2.1   Geometric Properties 236
7.3   Objectives 236
7.4   Modeling..236
       7.4.1   Part Module..236
       7.4.2   Property Module 242
               7.4.2.1   Material Properties.............................244
               7.4.2.2   Section Properties...............................246
               7.4.2.3   Assigning the Defined Section to
                         the Parts 250
       7.4.3   Assembly Module..251
       7.4.4   Step Module 253
               7.4.4.1   Create an Analysis Step: Cyclic...........253
               7.4.4.2   Create an Amplitude............................253
               7.4.4.3   Create Set for Request History Output... 253
               7.4.4.4   Create Partitions for the Hollow
                         Circular Section 256
       7.4.5   Interaction Module 259
               7.4.5.1   General Contact Interaction..................259
               7.4.5.2   Embedded Region Constraint..............259

        7.4.6   Load Condition Module 260
                 7.4.6.1   Apply the Cyclic Displacement as a
                           Boundary Condition 260
                 7.4.6.2   Apply Boundary Condition to the
                           Column Base 264
        7.4.7   Mesh Module 265
   7.5 Analysis: Job Module 266
   7.6 Visualization Module 267
   7.7 Analysis Result 271

Chapter 8   Modal Analysis of a Three-Story Building 275
   8.1 Introduction 275
   8.2 Problem Description 275
   8.3 Objectives 276
   8.4 Modeling 276
        8.4.1   Part Module 276
                 8.4.1.1   Create a New Model Database 276
                 8.4.1.2   Create Part 276
        8.4.2   Property Module 277
                 8.4.2.1   Material Properties 277
                 8.4.2.2   Section Properties 278
                 8.4.2.3   Section Assignment 280
        8.4.3   Assembly Module 280
        8.4.4   Step Module 281
        8.4.5   Interaction Module 287
                 8.4.5.1   Tie Constraint 287
        8.4.6   Load Module 289
        8.4.7   Mesh Module 290
   8.5 Analysis: Job Module 294
   8.6 Visualization Module 295

Index 301

List of Figures

FIGURE 1.1 Dataflow and actions of ABAQUS/Standard software.............. 3

FIGURE 1.2 UMAT flowchart 4

FIGURE 1.3 Overall dimensions of the considered frame (meters)............... 4

FIGURE 1.4 IPE200 dimensions.. 5

FIGURE 1.5 "El Centro" acceleration time history 5

FIGURE 1.6 UMAT subroutine interface 7

FIGURE 1.7 ABAQUS/CAE... 9

FIGURE 1.8 Creating the IPE 200 beam 10

FIGURE 1.9 Defining the material.. 12

FIGURE 1.10 Creating a solid homogenous section 12

FIGURE 1.11 Assigning the section.. 13

FIGURE 1.12 Rotating the beam.. 14

FIGURE 1.13 Defining a pattern of the instance as columns 15

FIGURE 1.14 Adding and positioning new instances as beams..................... 16

FIGURE 1.15 Defining the analysis step... 18

FIGURE 1.16 Changing the output frequency 18

FIGURE 1.17 Defining the tie constraint.. 20

FIGURE 1.18 Defining the boundary conditions 21

FIGURE 1.19 Mesh generation 22

FIGURE 1.20 Defining the job.. 23

FIGURE 1.21 Plot displacement time history 24

FIGURE 1.22 Extracting Von Mises stress contour at the end and during an earthquake 25

FIGURE 2.1 Create a new model database 29

FIGURE 2.2 Create a new part.. 29

FIGURE 2.3 Create Beam as 3D solid part... 30

FIGURE 2.4 Define I shape section and dimensions..................................... 30

FIGURE 2.5   Modeling part 31

FIGURE 2.6   Create partition by using the face sketching method 32

FIGURE 2.7   Create partition by using the extrusion method 33

FIGURE 2.8   Partition completed.. 34

FIGURE 2.9   Identify the material density for steel...................................... 34

FIGURE 2.10   Define elasticity for steel 35

FIGURE 2.11   Identify the material elasticity for steel.................................. 35

FIGURE 2.12   Define section 36

FIGURE 2.13   Assign the beam section 36

FIGURE 2.14   Assembly of the entire individual parts into a single model 38

FIGURE 2.15   Define analysis step 39

FIGURE 2.16   Time period, nonlinear geometry, and incrementation 40

FIGURE 2.17   Create History output 40

FIGURE 2.18   Create Set and History output 41

FIGURE 2.19   Interaction module... 42

FIGURE 2.20   Create Tie connection between all parts by Find Contact
              Pair option 43

FIGURE 2.21   Create boundary condition 44

FIGURE 2.22   Select the base surface... 44

FIGURE 2.23   Define the boundary condition for the base surface................ 45

FIGURE 2.24   Create pressure loading 46

FIGURE 2.25   Input the load value and amplitude 47

FIGURE 2.26   Selecting the element type... 48

FIGURE 2.27   Assign the approximate global size... 49

FIGURE 2.28   Choose element shape and meshing technique and mesh part 50

FIGURE 2.29   Verify mesh 51

FIGURE 2.30   Select part for mesh verification... 51

FIGURE 2.31   Checking mesh part for analysis warnings and errors 52

FIGURE 2.32   Define a job.. 52

FIGURE 2.33   Create job by default.. 53

FIGURE 2.34 Job submission 53

FIGURE 2.35 Monitoring 54

FIGURE 2.36 Stress contour 55

FIGURE 2.37 Select the reaction force history outputs 56

FIGURE 2.38 Save the reaction force with SUM operation............................ 56

FIGURE 2.39 Save the displacement records with "as is" operation 57

FIGURE 2.40 Create XY data 57

FIGURE 2.41 Operate on XY data 58

FIGURE 2.42 Hysteresis graph of the reaction force vs. displacement 58

FIGURE 3.1 Frame structure formed by Lu et al. (2008) 60

FIGURE 3.2 The dimensions for the beam and column of the frame model ... 60

FIGURE 3.3 The position of the steel and FRP rod 61

FIGURE 3.4 Getting started 62

FIGURE 3.5 Create new parts: solid base feature and wire base feature...... 63

FIGURE 3.6 Define a rectangle and dimensions of the geometry to the
rectangle 64

FIGURE 3.7 Depth of the beam 64

FIGURE 3.8 Beam 3D model 64

FIGURE 3.9 Define bar dimension 65

FIGURE 3.10 Material properties for CFRP 66

FIGURE 3.11 Material properties for concrete 67

FIGURE 3.12 Define concrete section properties 68

FIGURE 3.13 Edit concrete section properties 69

FIGURE 3.14 Define steel and FRP section type 69

FIGURE 3.15 Edit steel and FRP section properties 69

FIGURE 3.16 Section assignment to the beam 70

FIGURE 3.17 Select the element type 71

FIGURE 3.18 Assign the approximate global size for the mesh 72

FIGURE 3.19 Choose the element shape and meshing technique option 72

FIGURE 3.20 Meshed frame 73

FIGURE 3.21 Assemble part instances into the model 74

FIGURE 3.22 Complete assembly of the model... 74

FIGURE 3.23 Find contact pairs 75

FIGURE 3.24 Tie constraint by find contact pairs definition........................ 76

FIGURE 3.25 Create embedded region constraint... 76

FIGURE 3.26 Concrete to bars embedded region constraint 77

FIGURE 3.27 Initial step... 78

FIGURE 3.28 Analysis step... 79

FIGURE 3.29 Step configurations and time incrementation.......................... 79

FIGURE 3.30 Create amplitude... 80

FIGURE 3.31 Create sets... 81

FIGURE 3.32 Edit history output request 82

FIGURE 3.33 Create boundary condition 83

FIGURE 3.34 Edit boundary condition... 83

FIGURE 3.35 Create boundary condition 84

FIGURE 3.36 Edit the boundary condition to apply the cyclic load.............. 85

FIGURE 3.37 Define a job... 86

FIGURE 3.38 Create job.. 86

FIGURE 3.39 Submit Job-1 for analysis... 86

FIGURE 3.40 View the results of the analysis 87

FIGURE 3.41 Create XY Data 88

FIGURE 3.42 Select the displacement results... 88

FIGURE 3.43 Save displacement diagram 89

FIGURE 3.44 Select reaction forces results 89

FIGURE 3.45 Save reaction forces as "Force"..90

FIGURE 3.46 Create the Force–Displacement plot 91

FIGURE 3.47 Hysteresis loops in the Force–Displacement plot.................... 91

FIGURE 3.48 Von Mises stress contour plot in the whole model.................. 92

FIGURE 3.49 Von Mises stress contour plot in concrete parts...................... 93

FIGURE 3.50 Von Mises stress contour plot in rods...................................... 94

FIGURE 3.51   Set Maximum principal strain as the primary variable in contour plots 94

FIGURE 3.52   Maximum principal strain in the whole model 95

FIGURE 3.53   Maximum principal strain in concrete parts 96

FIGURE 3.54   Maximum principal strain in rods 97

FIGURE 3.55   Set Plastic strain contour plot as the primary variable in contour plots 97

FIGURE 3.56   Plastic strain contour plot in concrete parts 98

FIGURE 3.57   Plastic strain contour plot in rods 99

FIGURE 4.1   Three-dimensional model of the precast frame 102

FIGURE 4.2   Details of the precast reinforced concrete frame 103

FIGURE 4.3   Dimension of the precast beam 105

FIGURE 4.4   Create the precast beam 106

FIGURE 4.5   ABAQUS software sketcher to draw precast beam section ... 107

FIGURE 4.6   Edit base extrusion 107

FIGURE 4.7   Precast beam 107

FIGURE 4.8   Create a circular cut hole 108

FIGURE 4.9   Select the top surface as the plane of the hole 108

FIGURE 4.10   A circular hole is created on the precast beam 109

FIGURE 4.11   Create a datum plane 109

FIGURE 4.12   Select the left side plane to be offset 110

FIGURE 4.13   Create a partition using a defined datum plane 110

FIGURE 4.14   Complete definition of partition 111

FIGURE 4.15   Define face sketch partition 111

FIGURE 4.16   Select the face to be partitioned 112

FIGURE 4.17   Draw a circle in the sketcher 112

FIGURE 4.18   Define extrude edge partition 113

FIGURE 4.19   Use extrude along direction 113

FIGURE 4.20   Wireframe view of the partitioned precast beam 114

FIGURE 4.21   (a) Dimension of the precast column; (b) shaded and wireframe views of the precast column 114

FIGURE 4.22   (a) Dimension of the rubber bearing pad; (b) shaded and
              wireframe views of the rubber bearing pad 115

FIGURE 4.23   (a) Dimension of the dowel bar; (b) shaded view of the
              dowel bar 115

FIGURE 4.24   Create bottom bar part...116

FIGURE 4.25   Draw and add dimensions of the bottom bar sketch 116

FIGURE 4.26   Dimensions of the top bar...116

FIGURE 4.27   Dimension of the shear link 117

FIGURE 4.28   Create concrete material properties...118

FIGURE 4.29   Enter the parameters for the concrete damaged plasticity
              model 119

FIGURE 4.30   Create section property for the precast beam......................... 120

FIGURE 4.31   Solid section of concrete... 120

FIGURE 4.32   Create the section property for shear link 121

FIGURE 4.33   Section assignment to the concrete beam.............................. 122

FIGURE 4.34   The beam assigned section.. 122

FIGURE 4.35   Insert all parts to assembly... 123

FIGURE 4.36   Translate, rotate, and linear pattern tools.............................. 123

FIGURE 4.37   Define the steel reinforcement of the beam........................... 124

FIGURE 4.38   Define the steel reinforcement for the beam and columns..... 124

FIGURE 4.39   The precast frame model (translucency mode activated)....... 125

FIGURE 4.40   Create Gravity step... 126

FIGURE 4.41   Gravity step settings 126

FIGURE 4.42   Create amplitude... 128

FIGURE 4.43   Create and define the interaction property 129

FIGURE 4.44   Edit concrete–rubber contact properties 130

FIGURE 4.45   Create surface-to-surface contact interaction.........................131

FIGURE 4.46   Edit the contact interaction.. 133

FIGURE 4.47   Embedded region constraint... 134

FIGURE 4.48   Edit constraint for the embedded region 134

FIGURE 4.49   Create tie constraint for the concrete–dowel connection 135

FIGURE 4.50 Create the fixed-end boundary condition 137

FIGURE 4.51 Select boundary regions and edit the boundary condition 137

FIGURE 4.52 Fixed-end boundary condition 138

FIGURE 4.53 Create gravity loading 139

FIGURE 4.54 Input the load value for gravity load 139

FIGURE 4.55 Create a permanent load on the precast beam 140

FIGURE 4.56 Select the surface to apply pressure 141

FIGURE 4.57 Input the load value 141

FIGURE 4.58 Create a cyclic load on the frame 142

FIGURE 4.59 Select the surfaces to apply the cyclic load and edit the
boundary condition 142

FIGURE 4.60 Select the element type for beam reinforcement 143

FIGURE 4.61 Assign the approximate global size for the mesh elements 144

FIGURE 4.62 Meshing on beam reinforcement is done 144

FIGURE 4.63 Verify the mesh 145

FIGURE 4.64 Whole meshed model 145

FIGURE 4.65 Create a displacement node set 146

FIGURE 4.66 Select the displacement node on beam 146

FIGURE 4.67 Create history output for displacement 147

FIGURE 4.68 Create history output for reaction force 147

FIGURE 4.69 Create job 148

FIGURE 4.70 Job submission 148

FIGURE 4.71 Select result to enter the Visualization module 149

FIGURE 4.72 Extracting displacement history output 149

FIGURE 4.73 Extracting reaction force history output summation 150

FIGURE 4.74 Combine the displacement and force graphs into a single
graph 150

FIGURE 4.75 Von Mises stresses contour distribution for precast
concrete frame 151

FIGURE 4.76 Von Mises stresses contour distribution for steel
reinforcement 152

FIGURE 4.77   Von Mises stresses contour distribution for dowel bars 152

FIGURE 4.78   Von Mises stresses contour distribution for rubber
bearing pads 153

FIGURE 4.79   Plastic strain contour distribution for precast concrete
frame... 153

FIGURE 4.80   Plastic strain contour distribution for steel reinforcement 154

FIGURE 4.81   Plastic strain contour distribution for dowel bars................... 154

FIGURE 4.82   Concrete tensile damage contour distribution........................ 155

FIGURE 4.83   Concrete compressive damage contour distribution............... 155

FIGURE 4.84   Total maximum principal strain in the whole model 156

FIGURE 4.85   Displacement vs. time graph 157

FIGURE 4.86   Force vs. displacement hysteresis and envelop curves 157

FIGURE 5.1   Bolt connection assembly 159

FIGURE 5.2   IPE400 dimensions.. 160

FIGURE 5.3   Bolt, Nut, and L-plate dimensions.. 160

FIGURE 5.4   Run ABAQUS/CAE 161

FIGURE 5.5   Creating the beam 162

FIGURE 5.6   Draw the IPE section sketch.. 163

FIGURE 5.7   Equal length constraints 163

FIGURE 5.8   Add dimensions to drawing... 164

FIGURE 5.9   Extrusion of the drawing 164

FIGURE 5.10   Copy of the beam as a column 164

FIGURE 5.11   Create bolt and nut.. 165

FIGURE 5.12   Draw the bolt head.. 165

FIGURE 5.13   Add dimensions to the bolt head 166

FIGURE 5.14   Extruding bolt head 166

FIGURE 5.15   Bolt body: Open sketcher 166

FIGURE 5.16   Drawing bolt body.. 167

FIGURE 5.17   Open sketcher to draw the nut.. 167

FIGURE 5.18   Modeling the nut... 168

FIGURE 5.19   Drawing the L-plate.. 169

FIGURE 5.20   Modeling the L-plate 169

FIGURE 5.21   Creating the connection holes for the beam 170

FIGURE 5.22   Creating the connection holes for the column 171

FIGURE 5.23   Creating the connection holes in the L-plate 172

FIGURE 5.24   Defining the material model 173

FIGURE 5.25   Defining the section 173

FIGURE 5.26   Assigning the section 174

FIGURE 5.27   Adding a beam in assembly 175

FIGURE 5.28   Recoloring instances in the assembly to display better 175

FIGURE 5.29   Adding the column in the assembly 176

FIGURE 5.30   Defining the datum point in the middle of the beam flange
                 and column top face 177

FIGURE 5.31   Defining a datum point of distance 1 cm from the beam to
                 the column 178

FIGURE 5.32   Rotating column 178

FIGURE 5.33   Translating column to beam 179

FIGURE 5.34   Adding and positioning the first L-plate in the assembly 179

FIGURE 5.35   Adding and positioning the second L-plate in the assembly 180

FIGURE 5.36   Adding and positioning the bolt 181

FIGURE 5.37   Linear pattern a bolt 182

FIGURE 5.38   Adding and positioning other bolts 183

FIGURE 5.39   Importing the beam into the mesh module 183

FIGURE 5.40   Defining type and method of partition 184

FIGURE 5.41   Partitioning the beam 184

FIGURE 5.42   Defining the element size by seed 185

FIGURE 5.43   Meshing of the beam 185

FIGURE 5.44   Partitioning the column 186

FIGURE 5.45   Seeding and meshing column 186

FIGURE 5.46   Partitioning the bolt by point and normal method 187

FIGURE 5.47   Defining a datum plane 188

FIGURE 5.48   Partitioning the bolt by the datum and plane 188

FIGURE 5.49   Seeding and meshing the bolt.. 189

FIGURE 5.50   Partitioning the L-plate 189

FIGURE 5.51   Seeding and meshing the L-plate 190

FIGURE 5.52   Defining the Bolt_Load step as the first step 191

FIGURE 5.53   Defining the Cyclic analysis step as the second step 192

FIGURE 5.54   Defining general contact and contact interaction properties 193

FIGURE 5.55   Defining datum point as the reference point 194

FIGURE 5.56   Coupling constraint 195

FIGURE 5.57   Hiding all parts except bolts... 196

FIGURE 5.58   Defining the axis for the bolts 196

FIGURE 5.59   Defining the bolt loads 197

FIGURE 5.60   Modifying bolt loads in the Cyclic step 198

FIGURE 5.61   Defining the fixed boundary condition on the column........... 199

FIGURE 5.62   Defining the boundary condition for the reference point of
               the beam 199

FIGURE 5.63   Defining and submitting the job..200

FIGURE 5.64   Opening the visualization module...200

FIGURE 5.65   PEEQ contour plot for the whole model................................ 201

FIGURE 5.66   PEEQ contour plot for the bolts..202

FIGURE 5.67   Displacement contour for the model......................................202

FIGURE 5.68   Extracting force and displacement diagrams 203

FIGURE 5.69   Extracting the force–displacement diagram...........................204

FIGURE 6.1   3D model of concrete column reinforced by steel bars and
               CFRP...206

FIGURE 6.2   Getting started 207

FIGURE 6.3   Create a new part...208

FIGURE 6.4   Draw a rectangle as the column section sketch......................209

FIGURE 6.5   Depth of the column...209

FIGURE 6.6   Define the CFRP 210

FIGURE 6.7   Define steel parts 211

FIGURE 6.8   Solid extrude on the column surface.....................................211

FIGURE 6.9 Sketch the geometry of a beam section 212

FIGURE 6.10 Define the beam depth.. 212

FIGURE 6.11 Complete the column–beam definition 213

FIGURE 6.12 Identify the material of concrete 214

FIGURE 6.13 Identify the material of CFRP and steel............................. 215

FIGURE 6.14 Define concrete section 216

FIGURE 6.15 Define CFRP section 216

FIGURE 6.16 Define steel bar section..217

FIGURE 6.17 Concrete section assignment 217

FIGURE 6.18 CFRP section assignment..218

FIGURE 6.19 Steel bars section assignment..219

FIGURE 6.20 Define the element type 220

FIGURE 6.21 Partitioning the column 221

FIGURE 6.22 Define seeds for the parts 222

FIGURE 6.23 Meshing the parts 223

FIGURE 6.24 Define a column instance and rotate it 223

FIGURE 6.25 Define CFRP instances, rotate and translate it to the
proper position.. 224

FIGURE 6.26 Define steel part instances and create their assembly by a
linear pattern 225

FIGURE 6.27 Final assembly 225

FIGURE 6.28 Create an analysis step 226

FIGURE 6.29 Analysis step configurations... 227

FIGURE 6.30 Edit field output request.. 228

FIGURE 6.31 Tie constraints between CFRPs and Column and Embedded
constraint between steel reinforcement and column 229

FIGURE 6.32 The boundary condition applied at the base of the column ... 229

FIGURE 6.33 The load applied to CFRPs...230

FIGURE 6.34 Create job... 231

FIGURE 6.35 Job definitions.. 231

FIGURE 6.36 Submit the job for analysis 232

FIGURE 6.37 Activating the Visualization module 232

FIGURE 6.38 Von Mises stress report 233

FIGURE 6.39 Displacement contour plot 234

FIGURE 6.40 Equivalent plasticity contour plot 234

FIGURE 7.1 The complete 3D model 236

FIGURE 7.2 Rebars and stirrups 237

FIGURE 7.3 Segment height 238

FIGURE 7.4 Segment diameter and thickness 238

FIGURE 7.5 Steel sleeve 238

FIGURE 7.6 Reinforcement details 239

FIGURE 7.7 Create a hollow circular section 240

FIGURE 7.8 Draw the circular sketch of the column 240

FIGURE 7.9 Edit base extrusion 241

FIGURE 7.10 Hollow circular column 241

FIGURE 7.11 Steel sleeve definition 242

FIGURE 7.12 Longitudinal bars sketch 243

FIGURE 7.13 Radial pattern 243

FIGURE 7.14 Final geometry of longitudinal bars 244

FIGURE 7.15 Define stirrup 245

FIGURE 7.16 Material definition for steel 246

FIGURE 7.17 Plasticity flow for concrete damage plasticity 247

FIGURE 7.18 Concrete tensile behavior 247

FIGURE 7.19 Concrete compressive behavior 248

FIGURE 7.20 Tension damage parameters 248

FIGURE 7.21 Compression damage parameters 249

FIGURE 7.22 Create the solid section for steel 249

FIGURE 7.23 Steel section definition 250

FIGURE 7.24 Concrete section definition 250

FIGURE 7.25 Section assignment for a concrete hollow cylinder 250

FIGURE 7.26 Section assignments to steel parts 251

FIGURE 7.27   Insert the parts into the assemble platform 252

FIGURE 7.28   Using rotate, translate, and linear pattern tools to define the final assembly 252

FIGURE 7.29   Static step definition 254

FIGURE 7.30   Create amplitude 255

FIGURE 7.31   Define sets for history outputs 255

FIGURE 7.32   History output definition 256

FIGURE 7.33   Create datum planes 257

FIGURE 7.34   Datum plane definition 257

FIGURE 7.35   Partition definition 258

FIGURE 7.36   General contact definition 260

FIGURE 7.37   Create embedded region constraint 261

FIGURE 7.38   Embedded region constraint definition 262

FIGURE 7.39   Create cyclic displacement on the model 262

FIGURE 7.40   Input the displacement value and amplitude 263

FIGURE 7.41   Applied displacement 263

FIGURE 7.42   Create fixed-end boundary conditions 264

FIGURE 7.43   Edit boundary condition for the fixed end 264

FIGURE 7.44   Boundary conditions 265

FIGURE 7.45   Select element type for the hollow circular section 266

FIGURE 7.46   Assign the approximate global size for the hollow circular section 266

FIGURE 7.47   Assign the approximate global size for steel parts 267

FIGURE 7.48   Meshing parts 268

FIGURE 7.49   Job definition and submission 268

FIGURE 7.50   Visualization module 269

FIGURE 7.51   Save all reaction force graph 269

FIGURE 7.52   Save displacement graph 270

FIGURE 7.53   Operate on XY data 270

FIGURE 7.54   Combining displacement and force graphs 271

FIGURE 7.55   PEEQ contour plot 272

FIGURE 7.56   DAMAGEC contour plot...273

FIGURE 7.57   DAMAGET contour plot...273

FIGURE 8.1   The three-story building. 275

FIGURE 8.2   Create a new model database 277

FIGURE 8.3   Create a new part...278

FIGURE 8.4   Draw the section sketch of the model.....................................279

FIGURE 8.5   Define the depth of the beam 279

FIGURE 8.6   Identify the steel density...280

FIGURE 8.7   Identify the steel elasticity..281

FIGURE 8.8   Identify the material plasticity 282

FIGURE 8.9   Create solid section...283

FIGURE 8.10   Solid, homogenous section definition.....................................283

FIGURE 8.11   Section assignment 283

FIGURE 8.12   Beam section assigned..284

FIGURE 8.13   Create instance and linear pattern...284

FIGURE 8.14   Create instance and linear pattern as beams 285

FIGURE 8.15   Create instance and linear pattern as columns........................285

FIGURE 8.16   Finalizing the assembly 285

FIGURE 8.17   Step module 286

FIGURE 8.18   Define frequency step..286

FIGURE 8.19   Edit step for frequency extraction 287

FIGURE 8.20   Checking preselected default variable as field output 288

FIGURE 8.21   Interaction module...288

FIGURE 8.22   Create tie constraint using the find contact pair tool..............289

FIGURE 8.23   Edit contact pair to tie constraint 289

FIGURE 8.24   Finalizing tie constraints definition..290

FIGURE 8.25   Create boundary condition 291

FIGURE 8.26   Define encastre boundary condition for bases........................291

FIGURE 8.27   Encastred boundary conditions in the viewport......................292

FIGURE 8.28   Selecting the element type in the mesh module 292

FIGURE 8.29 Assign the approximate global size as seed 293

FIGURE 8.30 Meshing the part..293

FIGURE 8.31 Verify meshing 294

FIGURE 8.32 Checking finite element mesh for error and warning.............294

FIGURE 8.33 Create job..295

FIGURE 8.34 Job definition 296

FIGURE 8.35 Job submission..296

FIGURE 8.36 Monitor the job 297

FIGURE 8.37 Visualization module...297

FIGURE 8.38 Accessing the extracted natural frequencies 297

FIGURE 8.39 Result of natural frequency 298

FIGURE 8.40 The first natural frequency and the corresponding mode
shape of the structure 298

FIGURE 8.41 The fifth natural frequency and the corresponding mode
shape of the structure 299

FIGURE 8.42 The ninth natural frequency and the corresponding mode
shape of the structure 299

List of Tables

Table 1.1 Subroutines Provided in ABAQUS Solvers.......................................2

Table 1.2 Stiffness Matrix Components...6

Table 1.3 UMAT Essential Variables...7

Table 1.4 Steel Material Properties as Linear Elastic and Isotropic..............11

Table 2.1 Dimension of Parts...28

Table 3.1 Properties of the FRP Rod..61

Table 4.1 Geometric Properties of the Components.....................................102

Table 4.2 Properties of Concrete C25/30 for the CDP Model......................103

Table 4.3 Properties of the Rubber Bearing Pad...104

Table 4.4 Properties of Steel Reinforcement and Dowel Bar......................105

Table 4.5 Amplitude of Displacement-Based Cyclic Load...........................127

Table 4.6 Von Mises Stresses and Magnitude of Plastic Strain...................157

Table 5.1 Mechanical Properties for All Components..................................172

Preface

By the growing populations in modern cities, the demand for high-rise buildings and infrastructures has incredibly increased. The advanced methods of design and construction of structures lead to build comfortable and safe structures, which are not viable without considering the dynamic loads and vibration effect on structures. In the developed modern cities, there are many dynamic loads and vibration generator sources such as vehicles, trains, mechanical equipment, and machineries; therefore, considering the effect of dynamic loads on complex structures is vital to ensure the stability of buildings. However, the process for evaluation response of complex structures subjected to dynamic loads is very challenging and complicated and using the classic methods requires noticeable efforts and calculations.

The best solution to address this challenge is through the implementation of Finite Element Method (FEM), which is a numerical method that can be employed to solve a complex problem in which no classical solutions can be applied. The ABAQUS software is one successful finite element package, which was developed in 1978 and has been used for numerical simulation of complex problems in various fields such as civil and especially structural engineering. ABAQUS is a general-purpose finite element simulation package mainly used to numerically solve a wide variety of design engineering problems. Although the implementation of ABAQUS in solving many issues in various fields of Civil Engineering has been considered by many experts; however, its application to simulate dynamic structures is highly complicated. Therefore, this book aims to present specific complicated and puzzling challenges encountered in the application of the FEM for solving problems related to Structural Dynamics using ABAQUS software, which can fully utilize this method in complex simulation and analysis.

Therefore, the attempt in this book has been to demonstrate all processes for modeling and analysis of impenetrable problems through simplified step-by-step illustration by presenting screenshots from software in each part and showing graphs.

After reading this book, the skill and knowledge of readers for modeling and analysis of complicated and convoluted problems in structural dynamics will be incredibly enhanced and they will be capable to do the following:

- Solving general and complicated problems in the Structural Dynamics field using ABAQUS software in a reasonable time frame and with reasonable effort.
- The ability to analyze and model different types of structures with various dynamic and cyclic loads.
- Easily applying the ABAQUS software to simulate irregularly shaped objects comprising several different materials with multipart boundary conditions.

- The ability to complete various tutorials from a broad array of applications such as bridges, offshores, dams, and seismic resistant systems.
- Implementing the available tools in ABAQUS software such as scripting mesh and writing subroutines.
- Implementing various types of new materials such as Ultra-High-Performance Fiber Concrete in special structures such as high-rise and towers by considering damage plasticity models.
- Extracting the modal and buckling modes to estimate the tolerance limit of structures.

Authors

Farzad Hejazi is Associate Professor at Department of Civil Engineering, Faculty of Engineering, University of Putra Malaysia (UPM) and has been Senior Visiting Academic at the University of Sheffield (UK) since 2020. He is also innovation champion in UPM since 2013 and a member of the management committee for the Housing Research Center in Faculty of Engineering (UPM). He was appointed as Innovation Coordinator for Faculty of Engineering in 2014 by Deputy Vice Chancellor for Research and Innovation and appointed as the Research Coordinator in Department of Civil Engineering in 2017. He received his Ph.D. in Structural Engineering from University of Putra Malaysia in 2011 and worked as a postdoctoral fellow until 2012 and thereafter employed as a member of Department of Civil Engineering, UPM.

Since 2011, he has been teaching postgraduate courses for master and Ph.D. students in structural engineering fields such as the finite element method, structural dynamics, advanced solid mechanics, advanced structural analysis, earthquake resistance structure, and research methodology.

Prof. Farzad is managing and supervising a research team consisting of 20 Ph.D. students and 8 master students and is involved in many high impact research and industry projects funded by the Ministry of Higher Education Malaysia, Ministry of Science, Technology, and Innovation, PlaTCOM Venture Malaysian Government Agency, University of Putra Malaysia, and industrial companies which led to the filing of more than 15 patents in USA, Japan, Germany, Canada, New Zealand, and Malaysia. Four of his patents related to vibration dissipation devices are already licensed to the industry for mass production and implementation in construction projects for bridges and structures.

Prof. Farzad published six books and more than 100 scientific research papers in high impact journals and presented research papers in several international conferences. He won more than 35 awards in national and international innovation and invention exhibitions such as "Gold Award" in Invention and New Product Exposition in Pittsburgh, Pennsylvania (USA-2010), "Second Prize" of Taiwan National Center for Research on Earthquake Engineering (NCREE) for Introducing and Demonstrating Earthquake Engineering Research in Schools (IDEERS-2017), "Very Best Award" in Malaysian Technology Expo (2015) and Best Innovation Award from Ministry of Science, Technology, and Innovation in Malaysia Technology Expo (2021).

His specific fields of expertise include Structural Engineering, Structural Dynamics, Reinforced concrete Structures, Vibration, Finite Element Method, Inelastic Analysis, Earthquake, Damper Device, Vibration Dissipation Systems, Active and Passive Structural Control Systems, Optimization, Computer Program Coding, and Structure Simulation.

Hojjat Mohammadi Esfahani received his Bachelor of Mechanical Engineering in 2007 and his Master of Science in Mechanical Engineering in 2017. His main expert and focus area is Finite Element Simulation. He has more than 10 years of experience in teaching and training Finite Element Packages such as ABAQUS. Currently, he is involved in many research and industry projects related to the simulation of complex and infrastructures using ABAQUS Finite Element Software.

1 Development of Subroutines for ABAQUS Software

1.1 INTRODUCTION

The development of subroutines for ABAQUS software is one of the prominent features of this software, which allows the program to be customized for solving a specific problem.

The extensive range of solutions for addressing the specific problems and challenges using subroutines, along with a simple and understandable interface, has made the ABAQUS package an extremely useful tool for complex numerical analysis since the default options available in the ABAQUS software are not capable to meet all the needs of users.

The ABAQUS package is capable to provide various customized options and tools to meet the requirements of the users with the aid of the FORTRAN programming code (which is the foundation of several numbers of computing software such as ABAQUS, ANSYS, and NASTRAN). Therefore, the FORTRAN program provided the opportunity for the users to overcome any limitations of developing and implementing numerical solutions to solve the special complex problems.

Both ABAQUS/Standard and ABAQUS/Explicit software consist of subroutines as listed in Table 1.1.

Figure 1.1 illustrates the basic dataflow and actions of ABAQUS/Standard software from the first step of the analysis to the last step. The figure also shows the way and where the steps are implemented during analysis.

Each subroutine in the ABAQUS software consists of a set of variables and parameters that must be written in a predetermined and specific format.

UMAT (User Material) is the name of the ABAQUS sets used in material models developed by users (UMAT in the implicit analysis mode and VUMAT in the explicit analysis mode).

In UMAT, the user must introduce the behavior of materials in the form of FORTRAN code (*.for). Through this program, force–displacement relation, stress–strain relation, and also the yielding criteria for failure or fatigue condition are defined.

Figure 1.2 represents how a UMAT subroutine works.

DOI: 10.1201/9781003219491-1

TABLE 1.1

Subroutines Provided in ABAQUS Solvers

Solver	Subroutines
ABAQUS/ standard	CREEP,DFLOW,DFLUX,DISP,DLOAD,FILM,FLOW,FRIC,FRIC_COEF,GAPCON, GAPELECTR,HARDINI,HETVAL,MPC,ORIENT,RSURFU,SDVINI,SIGINI,UA MP,UANISOHYPER_INV,UANISOHYPER_STRAIN,UCORR,UDECURRENT, UDEMPOTENTIAL,UDMGINI,UEL,UELMAT,UDSECURRENT,UEXPAN,UEX TERNALDB,UFIELD,UFLUID,UFLUIDLEAKOFF,UGENS,UHARD,UHYPEL, UHYPER,UINTER,UMASFL,UMAT,UMATHT,UMESHMOTION,UMOTION, UMULLINS,UPOREP,UPRESS,UPSD,URDFIL,USDFLD,UTEMP,UTRACLOAD, UTRS,UVARM,UWAVE,VOIDRI,
ABAQUS/ explicit	VDISP,VDLOAD,VFABRIC,VFRIC,VFRIC_COEF,VFRICTION,VUAMP,VUANI SOHYPER_INV,VUANISOHYPER_STRAIN,VUEL,VUFIELD,VUFLUIDE XCH, VUFLUIDEXCHEFFAREA,VUHARD,VUINTER,VUINTERACTION,VU MAT,VUMULLINS,VUSDFLD,VUTRS,VUVISCOSITY,VWAVE

1.2 PROBLEM DESCRIPTION

In this chapter, writing and implementing UMAT subroutines in ABAQUS program is demonstrated in detail. For this purpose, one-floor frame structure is considered and subjected to earthquake excitation (Figure 1.3).

The frame is made of IPE200 beams (Figure 1.4) in which all the structural members are welded together.

The frame vibrates by acceleration history using the *El Centro*[1] earthquake record. The acceleration is shown in Figure 1.5.

The material considered for the beam is steel and it is considered as a linear isotropic elastic material. The material is defined in the UMAT subroutine.

The isotropic linear elastic includes "isotropic," which means all frame material properties are the same in all directions.

1.3 OBJECTIVES

1. To evaluate the seismic response of the steel frame subjected to an earthquake acceleration record.
2. To demonstrate the preparation and implementation of a UMAT subroutine for ABAQUS.

1.4 DEVELOPING A UMAT SUBROUTINE

Hooke's law is the basic equation in the structural engineering field that describes the relation between load and displacement. In the finite element method, many differential equations were obtained from Hooke's law such as the stress–strain relation, which should be solved to determine the displacement of the structure

[1] USA - 1940.

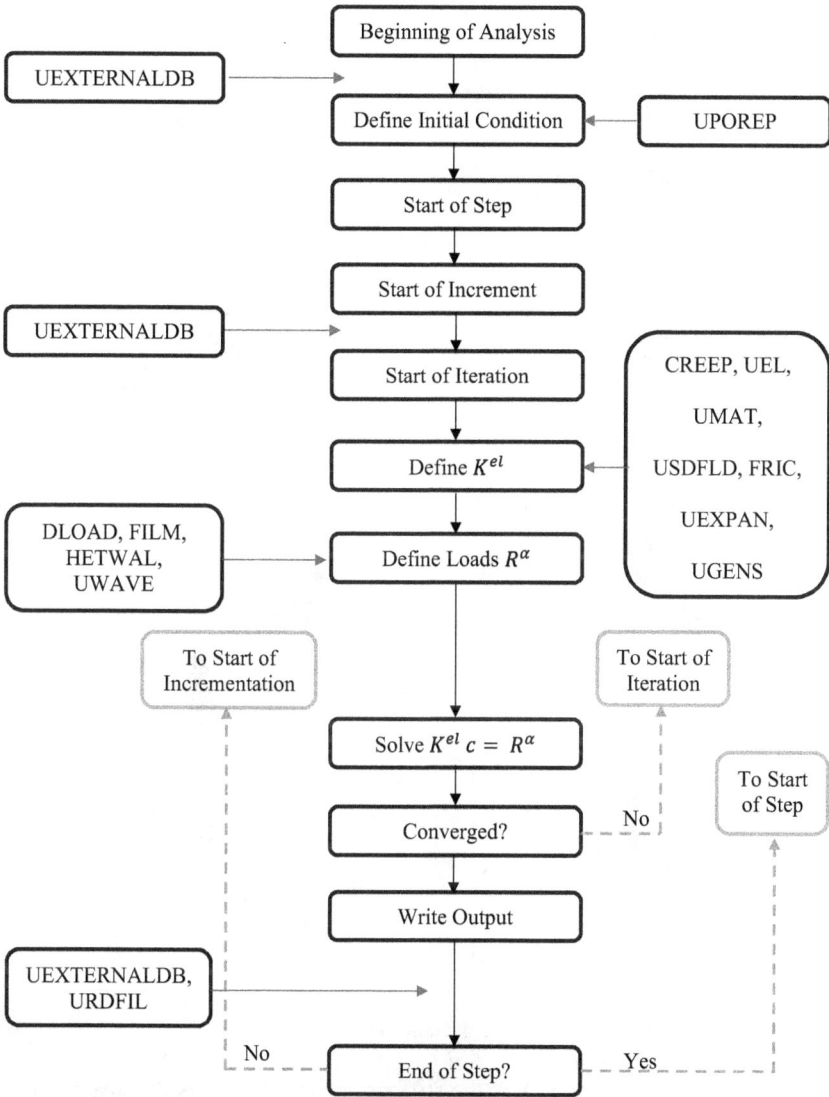

FIGURE 1.1   Dataflow and actions of ABAQUS/Standard software.

under the applied load and then calculate the resulting forces and stress generated in the structural members. Hooke's law can be written in two forms as given in Equations 1.1 and 1.2:

$$\{F\} = [K] \times \{X\} \tag{1.1}$$

$$\{\sigma\} = [D] \times \{\varepsilon\} \tag{1.2}$$

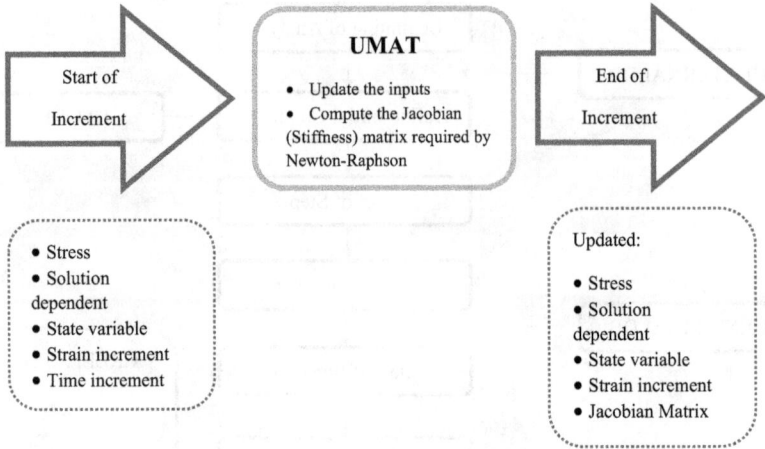

UMAT
- Update the inputs
- Compute the Jacobian (Stiffness) matrix required by Newton-Raphson

Start of Increment

End of Increment

- Stress
- Solution dependent
- State variable
- Strain increment
- Time increment

Updated:
- Stress
- Solution dependent
- State variable
- Strain increment
- Jacobian Matrix

FIGURE 1.2   UMAT flowchart.

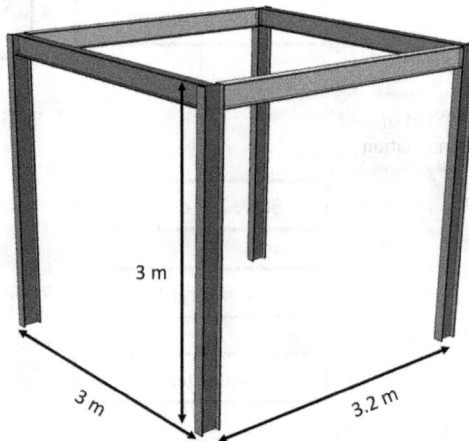

3 m

3 m

3.2 m

FIGURE 1.3   Overall dimensions of the considered frame (meters).

where F is the force vector, K is the stiffness matrix, and X is the displacement vector. Also, σ is the stress vector, D is the elasticity matrix (or module of elasticity in one-dimensional problems), and ε is the strain vector.

Finite element codes such as ABAQUS software use these equations to compute the stresses corresponding to the deformations made in the structure due to the applied load.

The main purpose of this example is to define a stiffness matrix for an element to the ABAQUS software using UMAT subroutines. In UMAT, the user must define all components of the stiffness matrix. Therefore, for this example, at the first step, all normal and shear components of the stiffness matrix have been calculated manually in three corresponding directions.

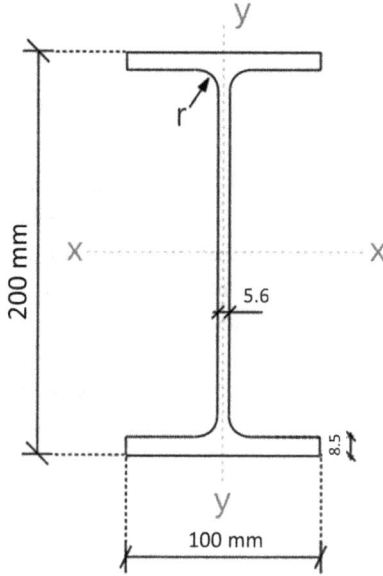

FIGURE 1.4   IPE200 dimensions.

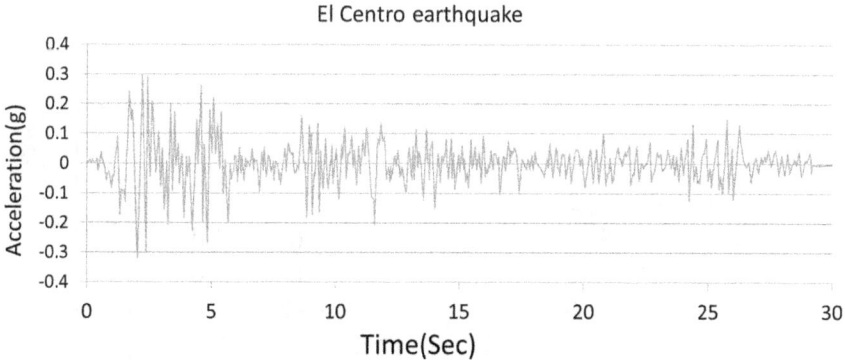

FIGURE 1.5   "El Centro" acceleration time history.

Therefore, the stiffness matrix [K] should include nine components as demonstrated in Table 1.2.

For the isotropic linear elastic material, two independent constant parameters for considered materials are needed to define the elasticity matrix [D] for three-dimensional problems as shown in Equation 1.3. The two parameters are Young's modulus (E), which represents the resistance of the material in the direction of the applied load, and Poisson's ratio (ν), which reflects the deformation of the material in the direction perpendicular to the applied load.

TABLE 1.2
Stiffness Matrix Components

Stiffness Component	Definition
K_{xx}	Stiffness in the x-direction due to loads in the x-direction
K_{yy}	Stiffness in the y-direction due to loads in the y-direction
K_{zz}	Stiffness in the z-direction due to loads in the z-direction
K_{xy}	Stiffness in the x-direction due to loads in the y-direction
K_{yx}	Stiffness in the y-direction due to loads in the x-direction
K_{zy}	Stiffness in the z-direction due to loads in the y-direction
K_{yz}	Stiffness in the y-direction due to loads in the z-direction
K_{xz}	Stiffness in the x-direction due to loads in the z-direction
K_{zx}	Stiffness in the z-direction due to loads in the x-direction

$$
\begin{bmatrix} \sigma_{xx} \\ \sigma_{yy} \\ \sigma_{zz} \\ \sigma_{yz} \\ \sigma_{zx} \\ \sigma_{xy} \end{bmatrix} = \frac{E}{(1+v)(1-2v)}
\begin{bmatrix}
1-v & v & v & 0 & 0 & 0 \\
v & 1-v & v & 0 & 0 & 0 \\
v & v & 1-v & 0 & 0 & 0 \\
0 & 0 & 0 & 1-2v & 0 & 0 \\
0 & 0 & 0 & 0 & 1-2v & 0 \\
0 & 0 & 0 & 0 & 0 & 1-2v
\end{bmatrix}
\begin{bmatrix} \epsilon_{xx} \\ \epsilon_{yy} \\ \epsilon_{zz} \\ \epsilon_{yz} \\ \epsilon_{zx} \\ \epsilon_{xy} \end{bmatrix} \quad (1.3)
$$

However, as mentioned before, the D matrix is equal to E (Young's modulus) in one-dimensional problems.

In the next step, the equation must be defined in the UMAT subroutine. The interface of these subroutines is shown in Figure 1.6.

There are three essential variables for the subroutine, including DDSDDE, STRESS, and STRAIN, which are shown in Table 1.3.

In every load increment, the STRAIN tensor is provided as input to the UMAT. Then, UMAT using DDSDDE and STRAIN matrices updates the STRESS matrix and returns it as an output for the current increment before moving to the next increment. The final UMAT for this example has been shown below. In

```
      SUBROUTINE UMAT(STRESS,STATEV,DDSDDE,SSE,SPD,SCD,
     1 RPL,DDSDDT,DRPLDE,DRPLDT,
     2 STRAN,DSTRAN,TIME,DTIME,TEMP,DTEMP,PREDEF,DPRED,CMNAME,
     3 NDI,NSHR,NTENS,NSTATV,PROPS,NPROPS,COORDS,DROT,PNEWDT,
     4 CELENT,DFGRD0,DFGRD1,NOEL,NPT,LAYER,KSPT,KSTEP,KINC)
C
      INCLUDE 'ABA_PARAM.INC'
C
      CHARACTER*80 CMNAME
      DIMENSION STRESS(NTENS),STATEV(NSTATV),
     1 DDSDDE(NTENS,NTENS),DDSDDT(NTENS),DRPLDE(NTENS),
     2 STRAN(NTENS),DSTRAN(NTENS),TIME(2),PREDEF(1),DPRED(1),
     3 PROPS(NPROPS),COORDS(3),DROT(3,3),DFGRD0(3,3),DFGRD1(3,3)


      user coding to define DDSDDE, STRESS, STATEV, SSE, SPD, SCD
      and, if necessary, RPL, DDSDDT, DRPLDE, DRPLDT, PNEWDT


      RETURN
      END
```

FIGURE 1.6 UMAT subroutine interface.

TABLE 1.3
UMAT Essential Variables

Variable in ABAQUS Subroutines	Definition
DDSDDE	Stiffness matrix[2] of the material
STRESS	Stress tensor matrix
STRAIN	Strain tensor matrix

the subroutine, PROPS (1) and PROPS (2) are mechanical constants that should be defined in the "Create Material" dialog box in the software, which will be explained in the next section.

```
SUBROUTINE UMAT(STRESS,STATEV,DDSDDE,SSE,SPD,SCD,
     1 RPL,DDSDDT,DRPLDE,DRPLDT,
     2 STRAN,DSTRAN,TIME,DTIME,TEMP,DTEMP,PREDEF,DPRED,CMNAME,
     3 NDI,NSHR,NTENS,NSTATV,PROPS,NPROPS,COORDS,DROT,PNEWDT,
     4 CELENT,DFGRD0,DFGRD1,NOEL,NPT,LAYER,KSPT,JSTEP,KINC)
C
      INCLUDE 'ABA_PARAM.INC'
C
      CHARACTER*80 CMNAME
      DIMENSION STRESS(NTENS),STATEV(NSTATV),
     1 DDSDDE(NTENS,NTENS),DDSDDT(NTENS),DRPLDE(NTENS),
     2 STRAN(NTENS),DSTRAN(NTENS),TIME(2),PREDEF(1),DPRED(1),
     3 PROPS(NPROPS),COORDS(3),DROT(3,3),DFGRD0(3,3),
       DFGRD1(3,3),
     4 JSTEP(4)
```

[2] In ABAQUS documentation, stiffness matrix called Jacobian matrix.

```
C     ELASTIC USER SUBROUTINE
      PARAMETER (ONE=1.0D0, TWO=2.0D0)
        E=PROPS(1)
        NU=PROPS(2)
        A=E/(ONE+NU)/(ONE-TWO*NU)
        B=(ONE-NU)
      C=(ONE-TWO*NU)
            DO I=1,NTENS
              DO J=1,NTENS
      DDSDDE(I,J)=0.0D0
              ENDDO
            ENDDO
              DDSDDE(1,1)=(A*B)
              DDSDDE(2,2)=(A*B)
              DDSDDE(3,3)=(A*B)
              DDSDDE(4,4)=(A*C)
              DDSDDE(5,5)=(A*C)
              DDSDDE(6,6)=(A*C)
              DDSDDE(1,2)=(A*NU)
              DDSDDE(1,3)=(A*NU)
              DDSDDE(2,3)=(A*NU)
              DDSDDE(2,1)=(A*NU)
              DDSDDE(3,1)=(A*ANU)
              DDSDDE(3,2)=(A*NU)
          DO I=1,NTENS
              DO J=1,NTENS
      STRESS(I)=STRESS(I)+DDSDDE(I,J)*DSTRAN(J)
              ENDDO
            ENDDO
      RETURN
      END
```

For implementing this subroutine in the software, it should be saved with (*.for) as the extension.

As mentioned earlier, the subroutines for ABAQUS software are programmed in FORTRAN language. Therefore, it is necessary to create them in the FORTRAN programming software, as the developed programs can be checked and debugged by the software. Therefore, the program can be installed and placed in the user's computer and link to the ABAQUS software, as described in the ABAQUS documentation. The professional users need not write the code in the FORTRAN program, as they could write the subroutine in Notepad.exe and save it with an (*.for) extension if they are sure that the developed subroutine is functioning correctly. In any case, the FORTRAN program and the ABAQUS software should be linked.

Save the subroutine as "UMAT.for."

1.5 MODELING

In the next step, the FEA model should be prepared in ABAQUS/CAE software. In this example, a 3D model of a one-floor frame containing one beam is

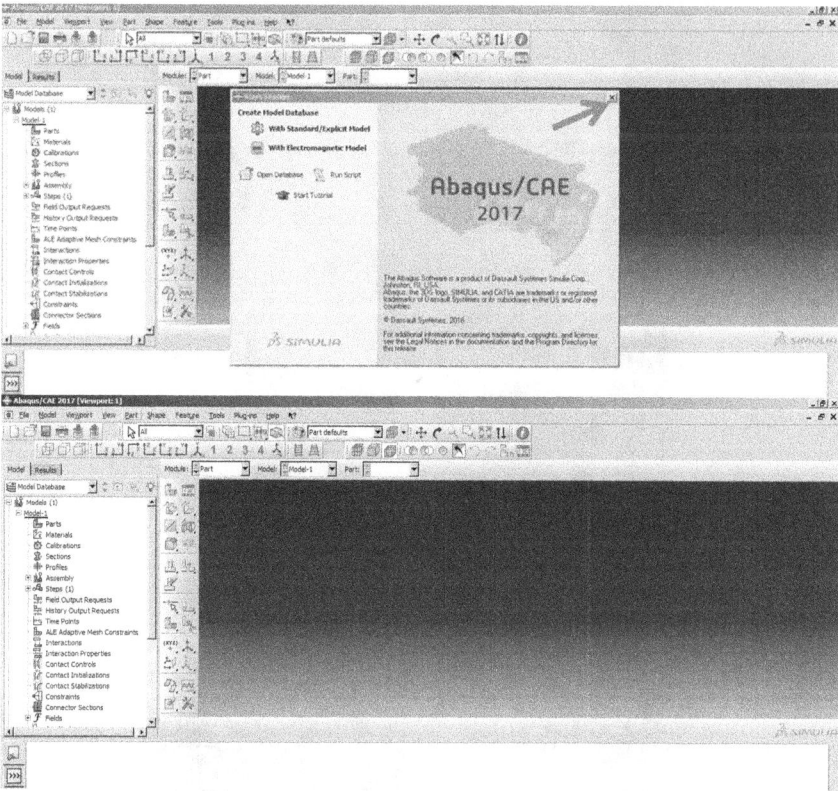

FIGURE 1.7   ABAQUS/CAE.

considered. The frame and section dimensions are shown in Figures 1.3 and 1.4. To simplify modeling, welding fillets in section (*r*) have been ignored.

From the start menu, open ABAQUS/CAE software and close the "Start Session" dialog box (Figure 1.7).

1.5.1   PARTS MODULE

The first step is to define an IPE beam section geometry and its constraints. For this purpose, the dimensions of the section should be defined first.

Double click on the "Parts" in the "Model Tree" to open the "Create Part" dialog box and name the part. Set the approximate size to 2 and click "Continue" to open sketcher.

Select "Create Lines: Connected" and draw an IPE section.
Select "Add Constraint" and make corresponding lines of equal length.
Select "Add Dimension" and determine the dimensions according to Figure 1.4.

Click the middle mouse button twice to open the "Edit Base Extrusion" dialog box. Enter 3 for the "Depth" and click "Ok" to apply and close the dialog box (Figure 1.8).

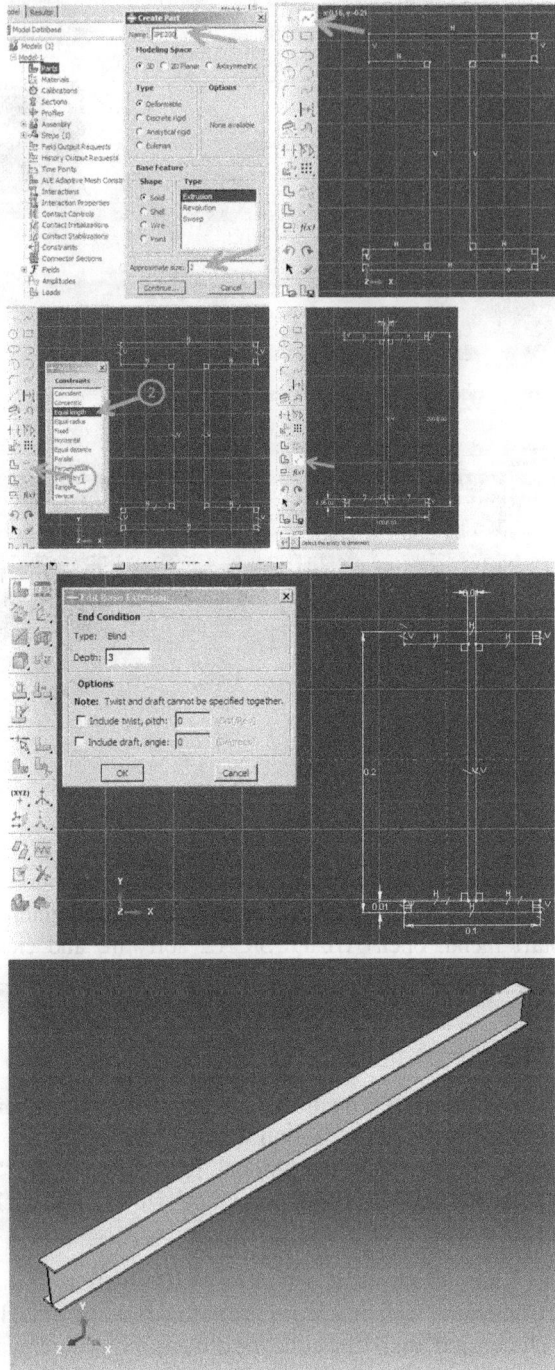

FIGURE 1.8   Creating the IPE 200 beam.

TABLE 1.4

Steel Material Properties as Linear Elastic and Isotropic

Properties	Young's Modulus (GPa)	Poisson's Ratio	Density (kg/m³)
Quantity	200	0.3	7850

1.5.2 PROPERTY MODULE

1.5.2.1 Material Properties

In the considered example, all structural members are made of steel with the material properties listed in Table 1.4.

Double click on "Material" in the "Model Tree" to open the "Create Material" dialog box and name the material.

Select "General → User Material" to define the UMAT subroutine Mechanical Constants; Young's modulus and Poisson's ratios are defined in Table 1.3. After entering Young's modulus, in the next step, create the second row to input Poisson's ratio. They are named *ROPS* (1) and *PROPS* (2) in the UMAT subroutine.

Select "General → Density" and enter 7850 as "Mass Density" and click "Ok" to save the material and close the dialog box (Figure 1.9).

1.5.2.2 Section Properties

Since steel is a homogenous material, the material properties are similar in all locations. Then, double click on "Sections" in the "Model Tree" and then open the "Create Section" dialog box. Name the section and keep the defaults by selecting "Continue" and clicking "Ok" to define a solid homogenous section (Figure 1.10).

1.5.2.3 Section Assignment

In the next step, the section that was defined before should be assigned to the beam.

Double click on "Section Assignments" underneath IPE200 in the "Model Tree". Then, select "Beam" and click the middle mouse button to open the "Edit Section Assignment" dialog box. To assign the section, click "Ok", and close the dialog box (Figure 1.11).

1.5.3 ASSEMBLY MODULE

In the next step, the assembly should be proceeded by defining the dependent part instances for IPE200, which includes eight instances.

Double click on "Instances" underneath "Assembly" in the "Model Tree" to open the "Create Instance" dialog box. Click "Ok" to all and add the IPE200 instance to assembly.

Select "instance → Rotate", then choose beam instance as the instance to rotate and click the middle mouse button. Then choose points, as shown in Figure 1.12, that define the axis of rotation. Click the middle mouse button to accept 90° as the angle of rotation. Then click the middle mouse button to accept the new position (Figure 1.12).

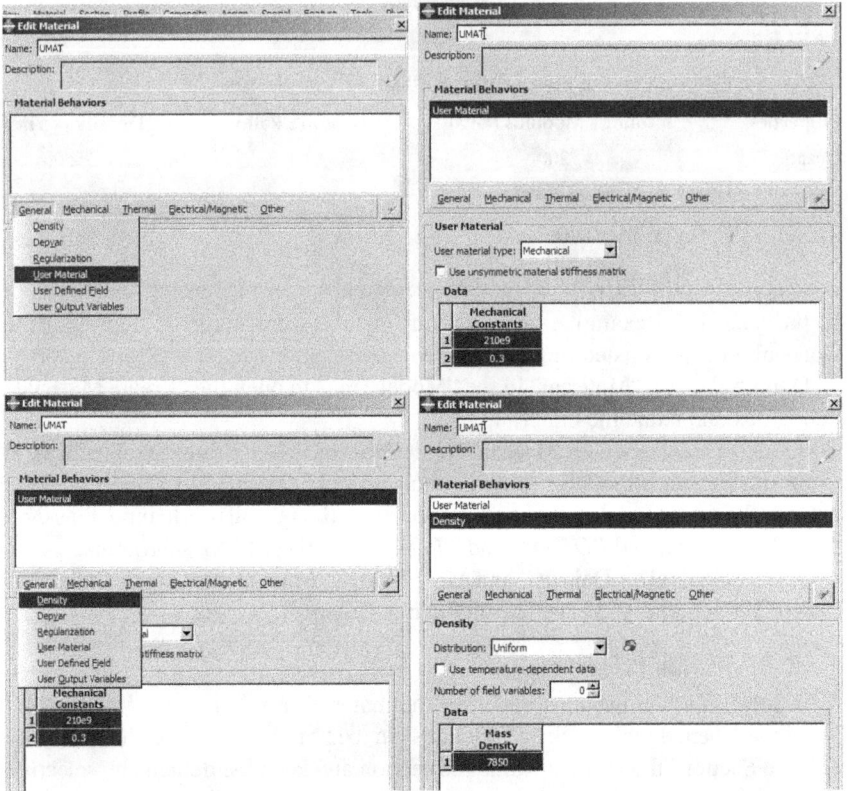

FIGURE 1.9   Defining the material.

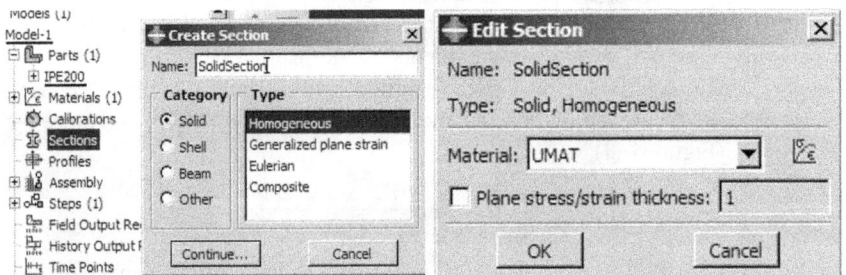

FIGURE 1.10   Creating a solid homogenous section.

Select "Instance → Linear Pattern" to make an array of the frame columns. Click on the "Instance" and click the middle mouse bottom to open the "Linear Pattern" dialog box. Click on the "Edit" icon in the dialog box and select "Axis z" and click the middle mouse button. Then enter 3.2 and 3.0056 as offset along the z- and x-directions, respectively, and click "Ok" to apply and close the dialog box (Figure 1.13).

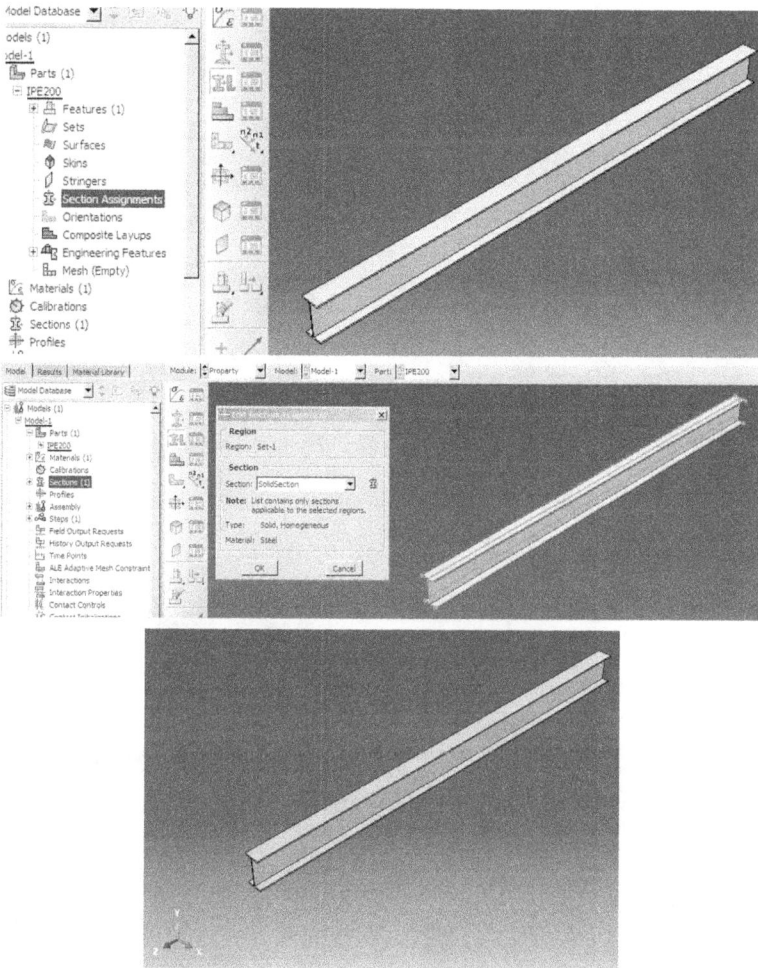

FIGURE 1.11   Assigning the section.

Double click on "Instances" and add a new instance into the assembly. Then select "Instance → Translate" and choose the earlier instance and click the middle mouse button. Then select point, as shown in Figure 1.14, to reposition it properly. Do the same for other instances in the corresponding position.

Add a new instance and rotate it by 90° so that it is in the y-direction. Then translate it to the right position. Perform the same step for other beams (Figure 1.14).

1.5.4   Step Module

In this example, dynamic analysis is considered and two methods for solving dynamic problems are available in the ABAQUS software as demonstrated as follows:

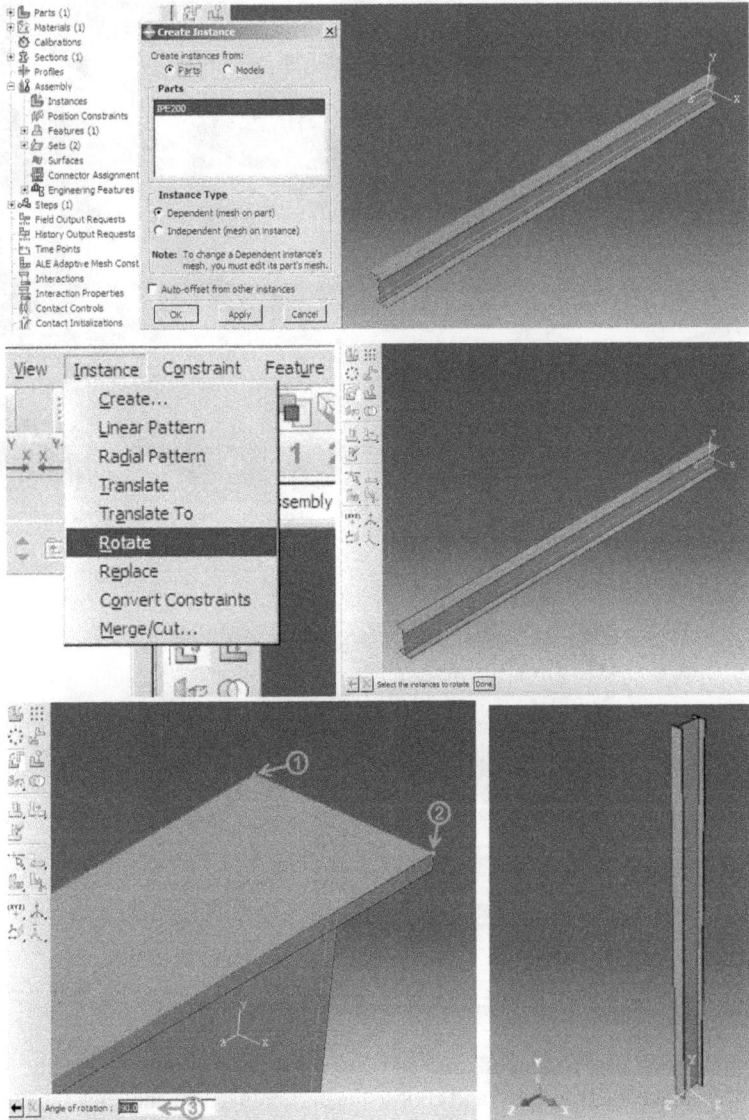

FIGURE 1.12 Rotating the beam.

1.5.4.1 Implicit Dynamics

In this method, as used in the ABAQUS/Standard solver software, the Newton-Raphson method and simultaneous solving equations in each load incremental step are used to analyze the problem. The response of each load increment is calculated based on the matrix methods and on the number of degrees of freedom. The required condition for solving the problem is to satisfy equilibrium equations in each increment. In fact, in this process, due to the nonlinear behavior of the

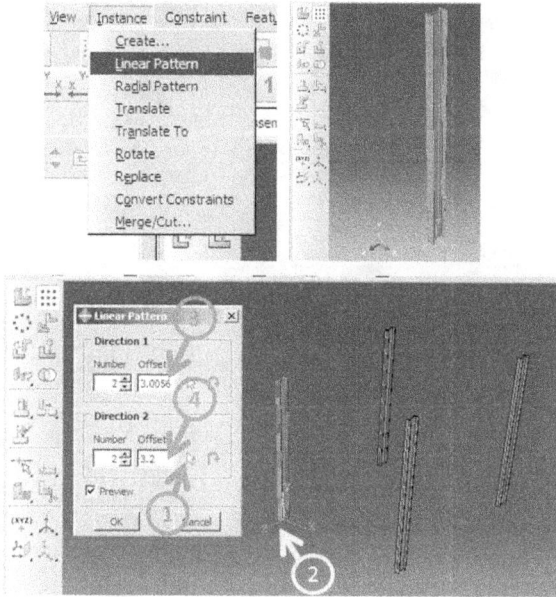

FIGURE 1.13 Defining a pattern of the instance as columns.

structure, the load is applied in an incremental process; so that in each increment, the linear behavior of the structure is considered. The equilibrium equation for each increment and each degree of freedom is defined as given in Equation 1.4

$$F - f = m\ddot{x} \tag{1.4}$$

where F is the external nodal force, f is the internal force, m is the mass of the structure, and $\ddot{x}$ is the acceleration applied to the structure. For example, for a model with 120 degrees of freedom (includes 20 nodes and 6 degrees of freedom per node), the equation of equilibrium is formed. It is necessary to use matrix methods to solve these equations simultaneously to calculate the variables.

Depending on the complexity and smoothness of nonlinear behaviors, the number of load increments may be either high or low. In complex nonlinear problems, many increments need to be formed to obtain accurate results, and if the complex model led to a large matrix for analysis, the computational process will be long and expensive. The speed of computation and analysis highly depends on the computational hardware system, although the high accuracy results involve a lot of computational processes. The implicit method can also be used for static problems as the right side of Equation 1.4 is considered to be zero for static loads (as no acceleration involved in the static analysis). Linear problems are solving one increment using this method. Since the implicit solution method is solving equations simultaneously, then this method has been considered as a powerful method for analyzing a wide range of problems.

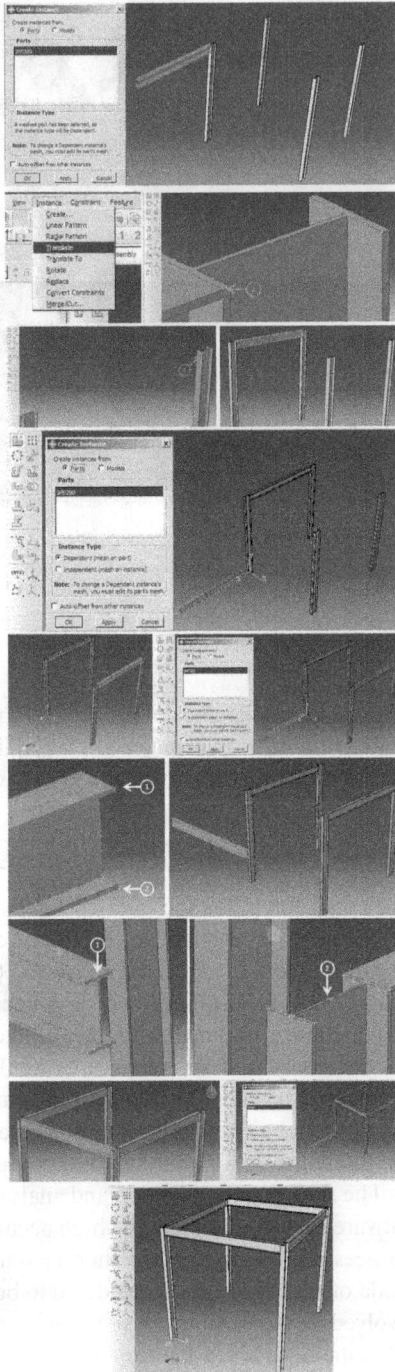

FIGURE 1.14   Adding and positioning new instances as beams.

1.5.4.2 Explicit Dynamics

In the explicit method, as used in the ABAQUS/Explicit solver software, the equations are solved sequentially based on the central difference method, in which the results for the variables in time step of $t + \Delta t$ are obtained based on the variables at a time step of t. In the explicit method, the central difference solution is conditionally stable, and this condition depends on time step stability and the lower time step leads to higher stability conduction. Equations in this method form a diagonal matrix due to the cost-effective equations to solve simultaneously, which are extremely fast. Also, in this method, the size of the elements affects the time ability. This is because the longitudinal decoupled matrix is solving problems using the method, which does not depend on the computational hardware system. This method can be implemented for nonlinear problems with large displacements, while there is no complexity in analysis.

Since the considered example includes earthquake effects, the dynamic analysis procedure should be defined. Although the Implicit and Explicit methods could be considered in the analysis, the implicit method is used to conduct dynamic analysis for the training purpose of the users.

For this purpose, double click on "Steps" in the "Model Tree" to open the "Create Step" dialog box and select "Dynamic, Implicit" and click "Continue" to open the "Edit Step" dialog box.

In the "Basic" tabbed page, change the time period to 30 for the acceleration time history.

In the "Incrementation" tabbed page, consider 0.01 for the "Initial increment size"; 1e-5 for "Minimum increment size", specify 0.1 for the "Maximum increment size", and 10,000 for the "Maximum number of increments". All other options in the dialog box should remain unchanged. Then, click "Ok" to create the step and close the dialog box (Figure 1.15).

1.5.5 Required Output

In the next step, the output variables should be defined in "Field Output Requests". In the sample, the preselected default variables are sufficient. However, they should be specified properly to extract a proper diagram.

Double click on "F-Output-1" underneath "Field Output Request" in the "Model Tree" to open the "Edit field Output Request" dialog box. Change the output frequency to 300 intervals, and the output variables should be changed, as shown in Figure 1.16. This will imply the results that will be achieved in 300-time points (every 0.1 seconds, one frame will be generated that contains the results). Click "Ok" to apply these changes, while all other options remained unchanged (Figure 1.16).

1.5.6 Interaction Module

In this example, all elements are assumed to be welded together. The best way to define welding in ABAQUS software is via the "Tie constraint" that is presented in the "Interaction module". The "Tie constraint" contains pair surfaces (whether Node surface or Element surface) as the "Master" and "Slave". The "Master

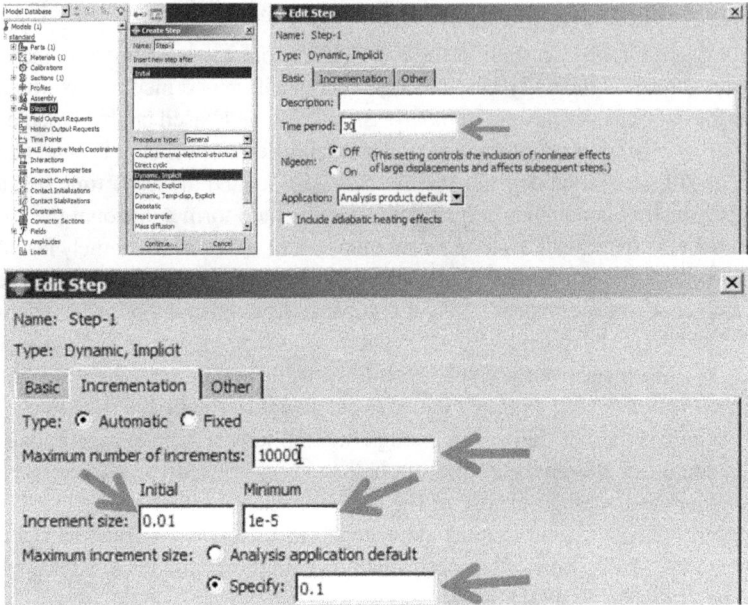

FIGURE 1.15   Defining the analysis step.

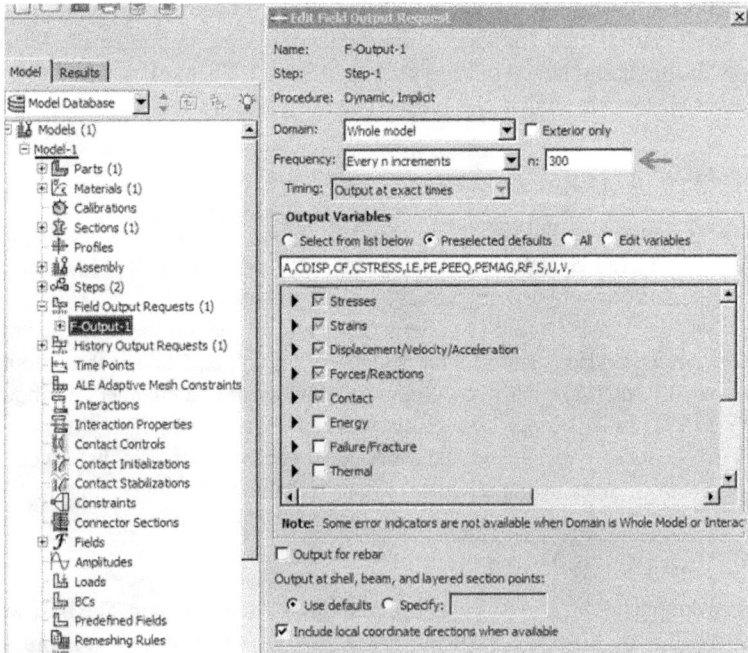

FIGURE 1.16   Changing the output frequency.

surface" is a surface that can control and affect another surface (slave). In other words, the master surface is a more stiffened surface (that usually includes a higher Young's modulus) than a slave surface. Although, in some models (such as in this example), the master and slave materials are the same, and there is no difference between the "Master surface" and the "Slave surface". For this example, all beam cross sections are assumed as "Master surface", and the corresponding column wings and flange (that interact with beams) are considered as "Slave surface".

Double click on "Constraint" in the "Model Tree" to open the "Create Constraint" dialog box and select "Tie" as the type of constraint.

Hide all columns by right-clicking on them in the "Model Tree" and select "Hide." Then, click the middle mouse button and select "Surface" from the "Master type". Hold the shift key on the keyboard and select the beams cross sections and finally click the middle mouse button and then the "Master surfaces" will be colored red.

Show the columns by right-clicking on them and hide the beams by right-clicking and select "Hide". Then, click the middle mouse button and after choosing "Surface" as the "Slave type", select the corresponding wings and flanges of columns while holding down the shift key. The "Slave surfaces" will be colored purple.

Click the middle mouse button to open the "Edit Constraint" dialog box. As a default, the software uses 5% of the element length to constrain the parts that seem to be suitable for this example. The nodes that are generated in the slave surfaces should be placed in the initial position at the beginning of the analysis. In this example, because of the clamping of columns and beams, rotational DOF in addition to the translational DOF should be considered as a constraint. Therefore, any changes in the dialog box are not required. Click "Ok" to define the constraint (Figure 1.17).

1.5.7 LOAD MODULE

In this module, the boundary conditions should be determined. In the example, two boundary conditions should be defined. The first one is about limiting displacement of the frame basis (all four columns) that move along the x-direction, and another one is the acceleration time history that is considered along the x-direction. The earthquake acceleration time history should be defined as an amplitude that the acceleration boundary condition refers to it.

Double click on "BCs" and open the "Create Boundary Condition" dialog box. Then, select "Displacement/Rotation" as the type and click "Continue".

Choose all four column bottom faces and click the middle mouse button to open the "Edit Boundary Condition" dialog box. Check "U2" and "U3" as y and z directions, where the frame is not moved along these directions, and click "Ok" to apply and close the dialog box.

Again, double click on "BCs" and select "Acceleration/Angular acceleration" as the boundary condition type and select "Continue". Select all four column bottom faces and click the middle mouse button to open the "Edit Boundary Condition" dialog box. Check "A1" and consider 9.81 as its magnitude.

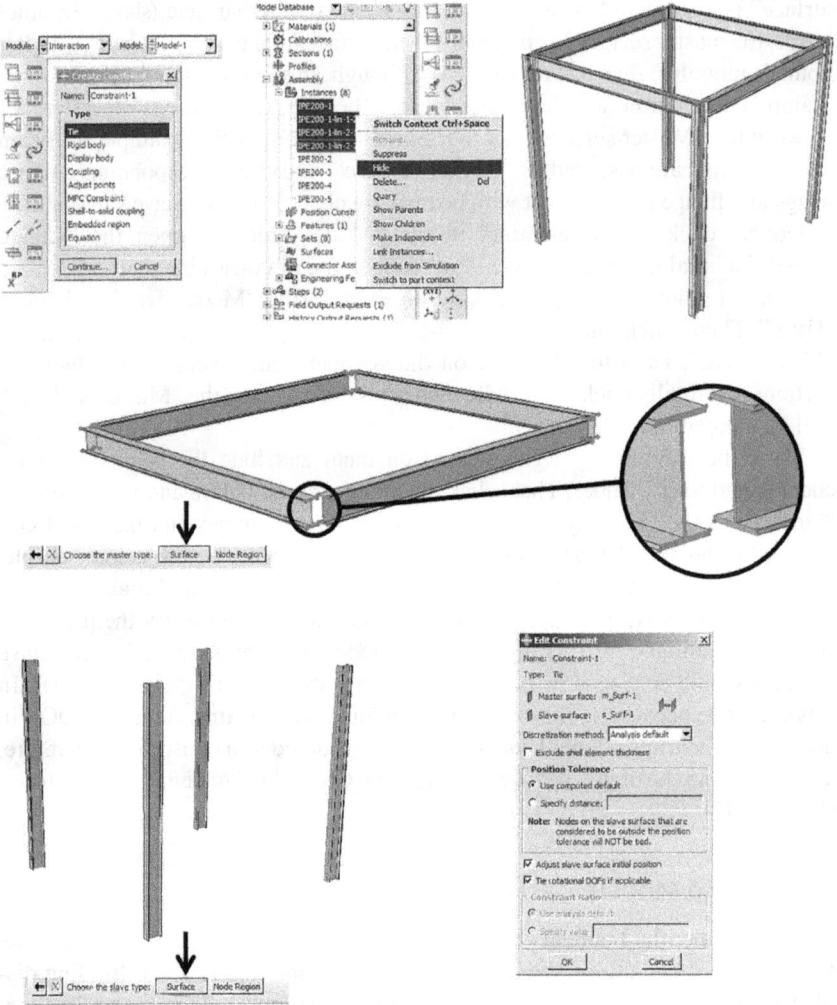

FIGURE 1.17   Defining the tie constraint.

Click the "Create Amplitude" icon to open the "Create Amplitude" dialog box. Click "Continue" and input time/acceleration data provided in the attached *.xlsx file to copy the values. Click "Ok" to create the amplitude and close the dialog box.

Select "Amp-1" in "Edit Boundary Condition" and click "Ok" to apply the boundary condition and close the dialog box (Figure 1.18).

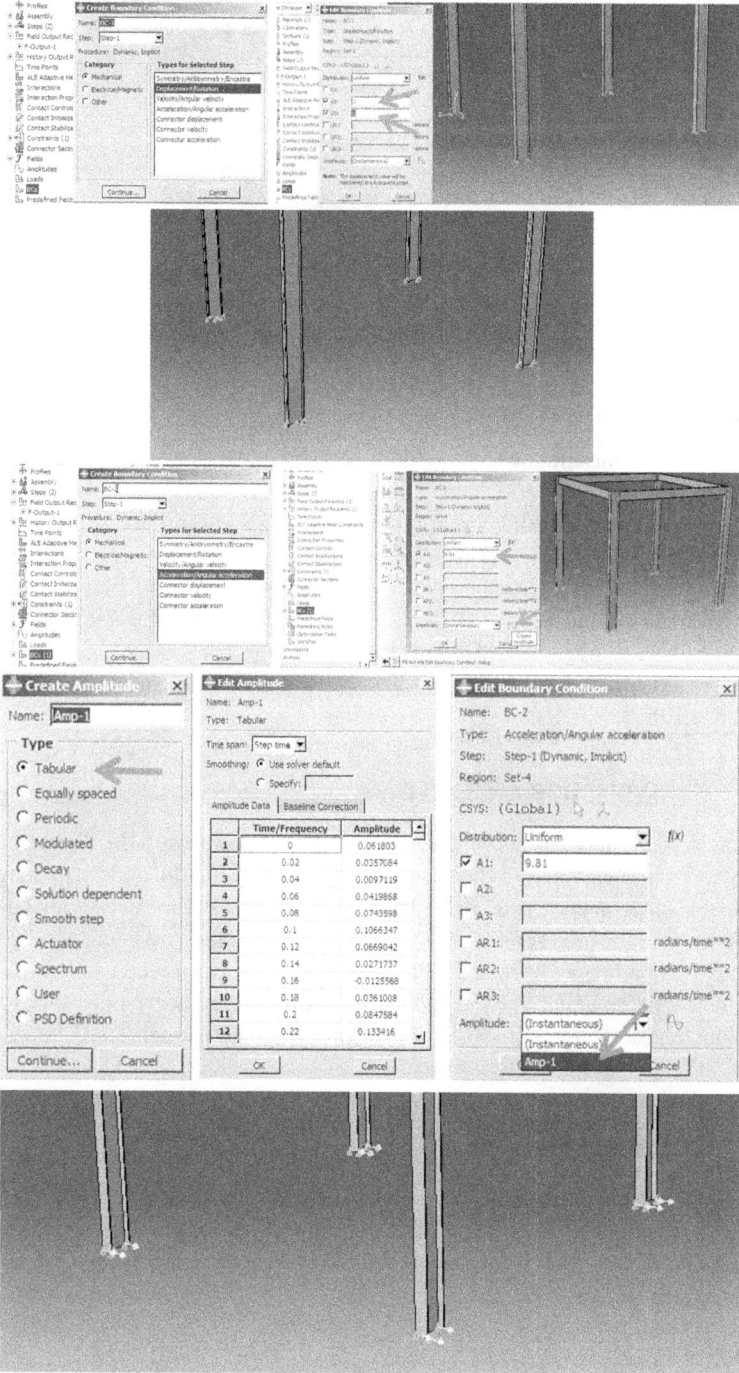

FIGURE 1.18 Defining the boundary conditions.

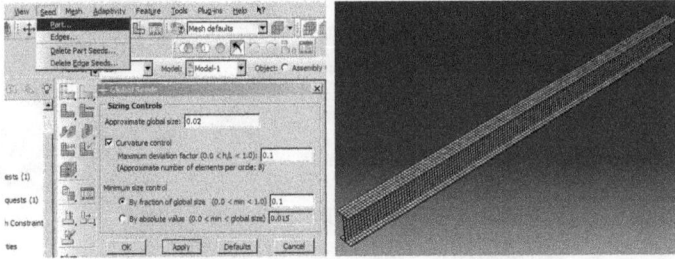

FIGURE 1.19   Mesh generation.

1.5.8 MESH MODULE

The next step is meshing, which includes specifying the element size and generating the mesh.

Double click on "Mesh" (Empty) underneath IPE200 in the "Model Tree" to open the "Mesh module".

Select "Seed → Part" to open the "Global Seeds" dialog box. Enter 0.02 as the "Approximate Global Size" and click "Apply" to check the approximate position for the nodes and then click "Ok" to close the dialog box.

Select "Mesh → Part" and click "Yes" in the prompt area for mesh generation (Figure 1.19).

1.6 ANALYSIS: JOB MODULE

In this step, the job should be defined. The job contains the UMAT subroutine that will be implemented in the procedure and computing the material behavior.

Double click on the "Jobs" in the "Model Tree" to open the "Create Job" dialog box and name the job. Then, click "Continue" to open the "Edit Job" dialog box.

In the tabbed page, browse the subroutine file that was saved as "UMAT.for" and click "Ok" to define the job and close the dialog box.

Right click on the job underneath "Jobs" in the "Model Tree" and select "Submit" to begin the analysis (Figure 1.20).

1.7 ANALYSIS RESULTS

After running the analysis, the results can be extracted. The results are saved in "*.Odb" files and can be checked in the "Visualization module". In this example, displacement at the top of the frame as a time history diagram and Von Mises stress in the frame will be investigated.

For this purpose, right click on the "Completed job" and select "Results" to open the "Visualization module".

Double click on "XYData" to open the "Create XY Data" box. Select "ODB Field Output" and click "Continue" to open "XYData" from the "ODB Field

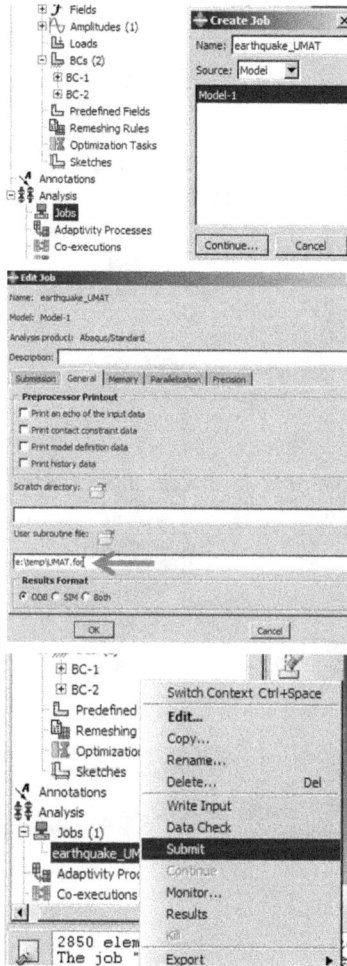

FIGURE 1.20 Defining the job.

Output" dialog box. In the "Variables" tabbed page, select "Unique Nodal" from the "Position of the Output Variables" and choose "U1".

Then, in the "Elements/Nodes" tabbed page, click "Edit Selection" and select a top node and click the middle mouse button to note "1 Node selected" in the box, then click "Plot" to plot displacement time history. The maximum displacement shown in the plot is about 240 mm (Figure 1.21).

Click "Plot Contours" on the "Deformed Shape" icon to extract the "Von Mises stress contour".

Click "Animate Time History" to check the "Maximum Von Mises stress" during the earthquake. The maximum stress in the graph is about 90 MPa (Figure 1.22).

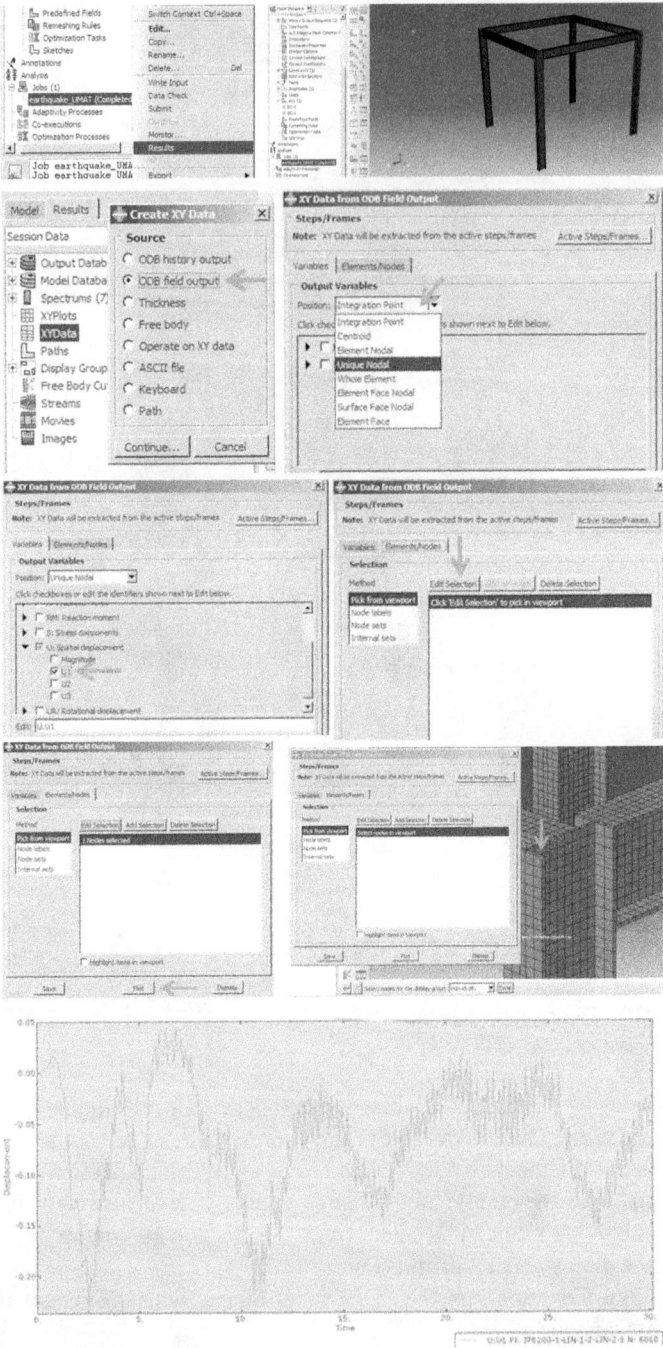

FIGURE 1.21 Plot displacement time history.

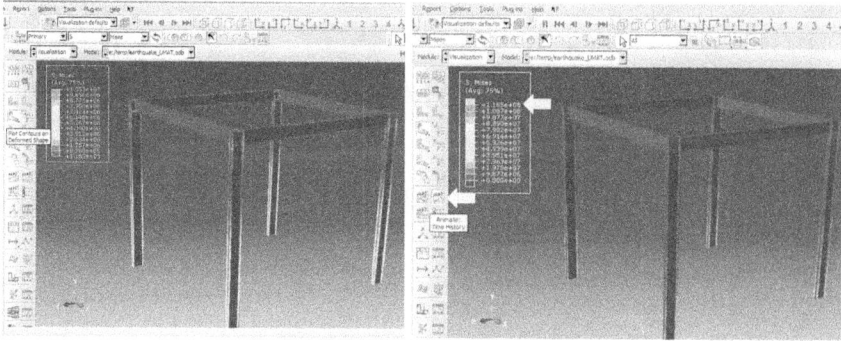

FIGURE 1.22 Extracting Von Mises stress contour at the end and during an earthquake.

2 Evaluate Performance of Steel Wall in Structures Subjected to Cyclic Load

2.1 INTRODUCTION

This chapter presents an example for performing a nonlinear static cyclic analysis using the ABAQUS/Standard finite element program. The evaluation of the seismic response of the structure is vital for engineers in the design of high-strength and stable buildings in earthquake-prone areas.

Steel shear walls are mainly used in various types of structures to provide resistance against seismic vibration; however, the modeling and simulation of steel frames for nonlinear analysis is a challenging procedure. Therefore, this chapter is devoted to demonstrate the development of a three-dimensional model of the frame structure retrofitted with a steel shear wall as the energy dissipation system is subjected to the lateral load. For this purpose, the steel frame structure is subjected to monotonic and cyclic loadings and pushover analysis and nonlinear dynamic analyses have been conducted, respectively. Therefore, this exercise aims to demonstrate the steps, which should be followed by users in creating and analyzing complex 3D frame structure models using the ABAQUS/CAE program.

Researcher: Mohammadhannan Azari (Hannan717@gmail.com)

2.2 PROBLEM DESCRIPTION

A steel frame structure furnished with a steel wall is considered and the response of this frame under the applied incremental load and cyclic load is evaluated through finite element simulation using the ABAQUS software.

The dimensions of all parts in the frame structure are shown in Table 2.1.

2.3 OBJECTIVES

- To develop a finite element model of the steel frame structure equipped with a steel shear wall subjected to earthquake excitations.
- To perform a nonlinear analysis of the steel frame structure with a steel shear wall and evaluate the seismic response of the structure.
- To study the performance of the steel shear wall in retrofitting a bare frame structure.

DOI: 10.1201/9781003219491-2

TABLE 2.1
Dimension of Parts

Name	Property/Dimension (mm)	Quantity/Length (mm)
Beam	H $150 \times 150 \times 10$	1/2000
Column	H $150 \times 150 \times 10$	2/2000
Connection	$300 \times 300 \times 150 \times 10$	4
Jack plate	$150 \times 150 \times 20$	2
Inner plate	$1300 \times 1300 \times 10$	1
Upper and down plate	$1300 \times 150 \times 10$	2
Steel box	$1500 \times 1500 \times 10$	1

2.4 MODELING

2.4.1 PART MODULE

This module processes the creation of the geometry required for the considered problem. To create a 3D geometry, first a 2D model should be created and then manipulated to obtain the solid geometry.

2.4.1.1 Create a New Model Database

Run the ABAQUS software from programs in the "Start" menu, then close the "Start Session" dialog box (see Figure 2.1).

Click on "With Standard/Explicit Model". This step allows the user to start modeling, where the user can create a new file and save it under any name in a new folder.

When the Part module is loaded, it displays the Part module toolbox on the left side of the ABAQUS/CAE main window. Each module displays its own set of tools in the module toolbox.

2.4.1.2 Create Parts

From the main menu bar, select "Part → Create" in order to create a new part. Then the "Create Part" dialog box appears (see Figure 2.2).

Use the "Create Part" dialog box to name the part and to choose its modeling space, type, and base feature and to set the approximate size. The name of the part may be edited once it has been created, but the modeling space, type, or base feature cannot be changed.

Name the part "Beam." Then next, choose a three-dimensional planar; a deformable body and then choose "Solid and Extrusion" from base feature (see Figure 2.3).

In the "Approximate size text" field, type 2000. The value entered in the approximate size text field at the bottom of the dialog box sets the approximate size of the new part. Click "Continue" to exit the "Create Part" dialog box.

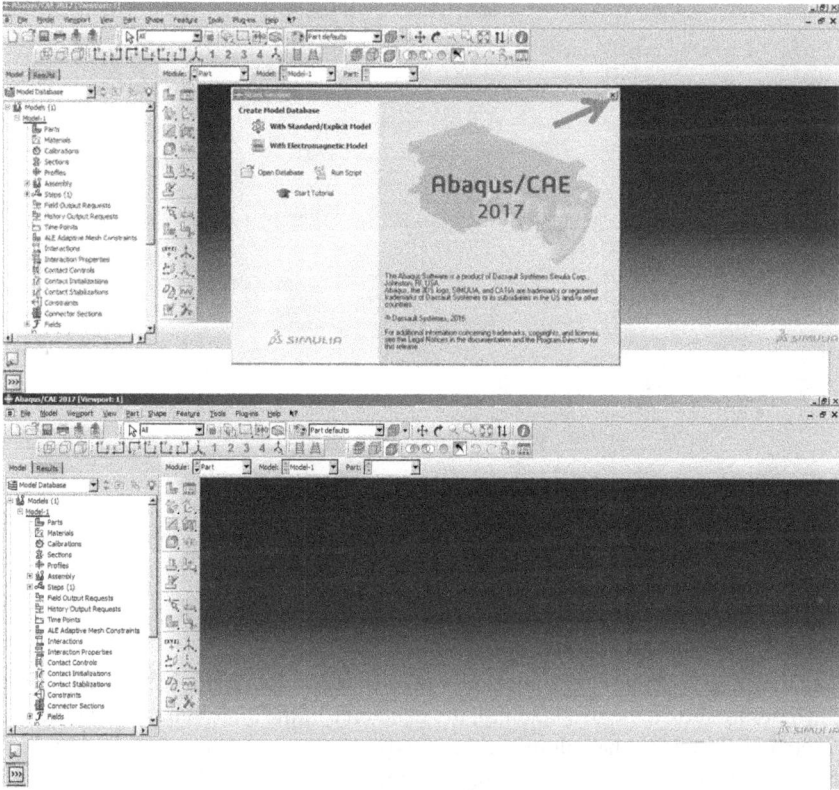

FIGURE 2.1 Create a new model database.

FIGURE 2.2 Create a new part.

FIGURE 2.3 Create Beam as 3D solid part.

2.4.1.3 Define an I Shape with Dimensions

Use the "Create Lines", "Connected tool" located in the upper-left corner of the Sketcher toolbox to begin sketching the geometry of the plate. The user can select a starting corner for the column at the viewport or enter the X and Y coordinates. Next, create a line with the following coordinates: (0, 0), (10, 0) (10, 70), (140, 70), (140, 0), (150, 0), (150, 150), (140, 150), (140, 80), (10, 80), (10, 150), (0, 150), and (0, 0). Alternatively, the user can define the dimension of the geometry by clicking on the add dimension tool. Once finished sketching the section for the dimension, right click and click on "Cancel Procedure" to exit the sketcher. Click on "Done" in the prompt area, and it will be displayed, as shown in Figure 2.4.

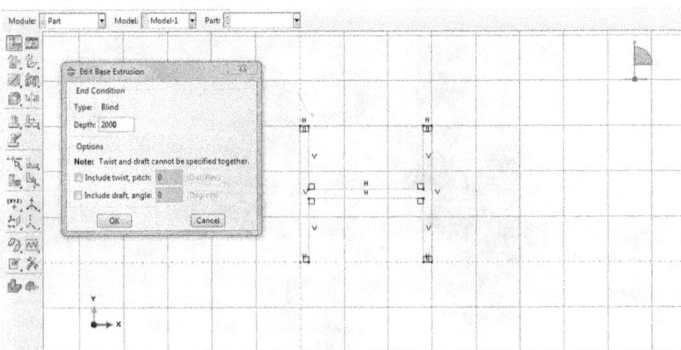

FIGURE 2.4 Define I shape section and dimensions.

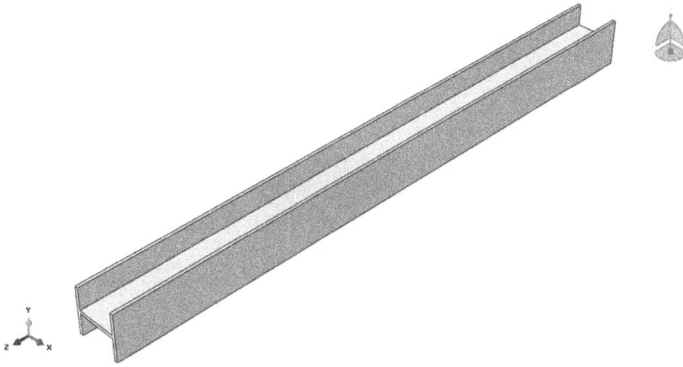

FIGURE 2.5   Modeling part.

The user then needs to define the depth for the extrusion of the beam. Once the depth of the beam has been defined, it will be displayed, as shown in Figure 2.5.

Repeat the same procedure for all parts, including the column, box, plates, jack plate, and stiffener.

2.4.1.4   Create Partition

In order to mesh each part, it is better to create a partition for each part to obtain a better result.

In the "Part" section, from the toolbar, choose "Create Partition", and then choose the type as a "Face" and use the "Sketch Method" (see Figure 2.6).

Select the front faces to partition and click "Done".

Select an edge or axis that appears in the required orientation.

Sketch the partition geometry by drawing the line in each section (drawn by the user), as shown in Figure 2.6, and click on "Done" to complete the process of creating a partition.

Then change the type to "Cell" and choose the method as "Extrude/Sweep edges" (see Figure 2.7).

Select the cells to partition and click on "Done". Select the line which needs to sweep and click on "Done".

Choose the sweep method as "Extrude Along Direction". Select an extrude direction in the z-direction, as shown in Figure 2.7, and press "Ok".

Then click on "Create Partition" to complete the partition definition, and it will be displayed as shown in Figure 2.8.

2.4.2   Property Module

In this module, the material properties for the analysis should be defined and assign those properties to the available parts.

Note: If the "Done" button in the prompt area does not appear, right click on the "viewport" until it appears.

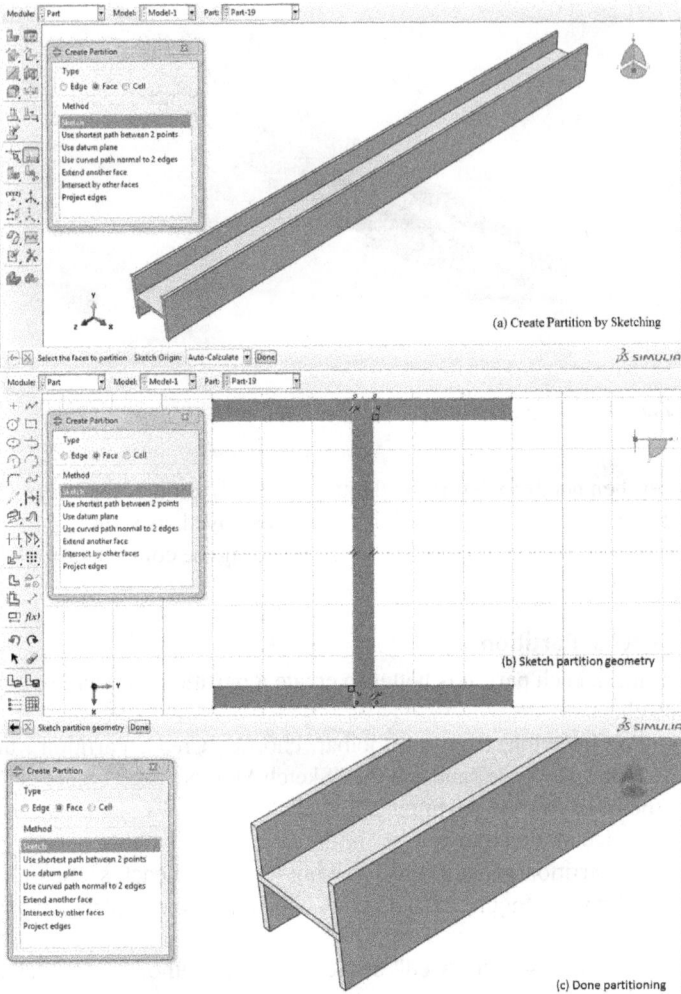

FIGURE 2.6 Create partition by using the face sketching method.

2.4.2.1 Material Properties

The property module is used to add materials and define related properties. In this example, all the members of the frame are made of steel and assumed to be linear elastic with a density of $7850\,kg/m^3$, Young's modulus of $21{,}000$ N/mm^2, and Poisson's ratio of 0.3. Thus, a single linear elastic material is created with these properties to define steel material.

In the module list located under the toolbar, select "Property" to open the Property module. The cursor changes to an hourglass while the property module is loading.

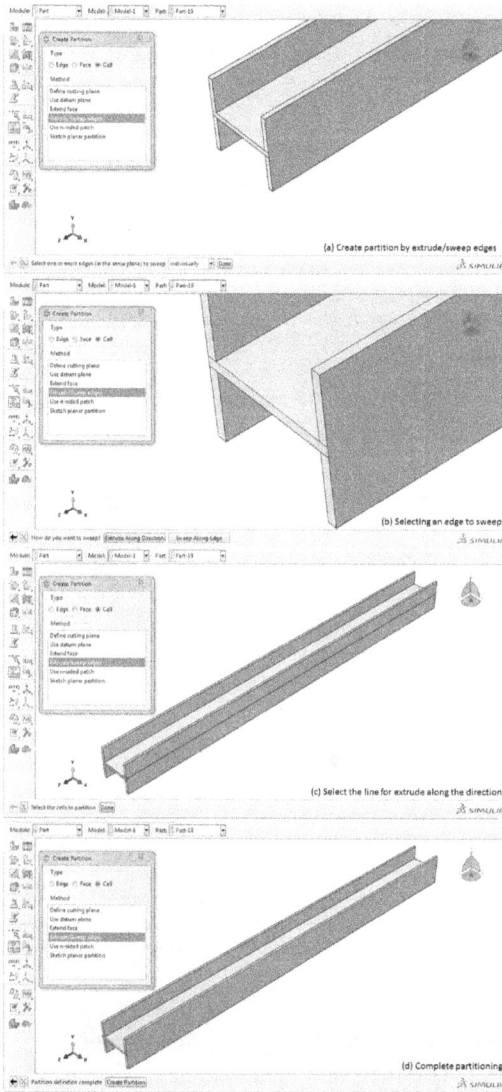

FIGURE 2.7 Create partition by using the extrusion method.

From the main menu bar, select "Material → Create", in order to create a new material.

The "Edit Material" dialog box appears.

Name the material *Steel*.

From the general menu, select density and type the value as 7850×10^{-9} (see Figure 2.9).

From the material editor's menu bar, select "Mechanical → Elasticity → Elastic". The software displays the Elastic data form (see Figure 2.10).

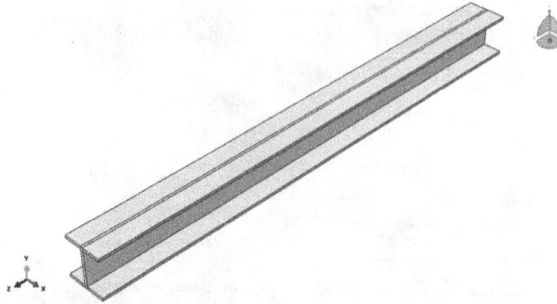

FIGURE 2.8   Partition completed.

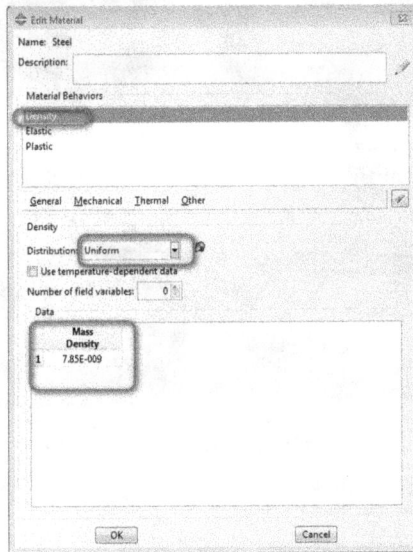

FIGURE 2.9   Identify the material density for steel.

Type the value of 210,000 for Young's modulus and 0.3 for Poisson's ratio in the respective fields. Use the [Tab] button or move the cursor to a new cell and click to move between cells (see Figure 2.11). Click "OK" to exit the material editor. Save the changes by clicking on the [Save] button.

2.4.2.2   Section Properties (Type, Thickness, and Material Assignment)

The section properties of a model are defined by creating sections in the Property module. After the section is created, one of the following methods can be used to assign the section to the part.

i. Select the region from the part and assign the section to the selected region,
ii. Use the Set toolset to create a homogeneous set containing the region and assign the section to the set.

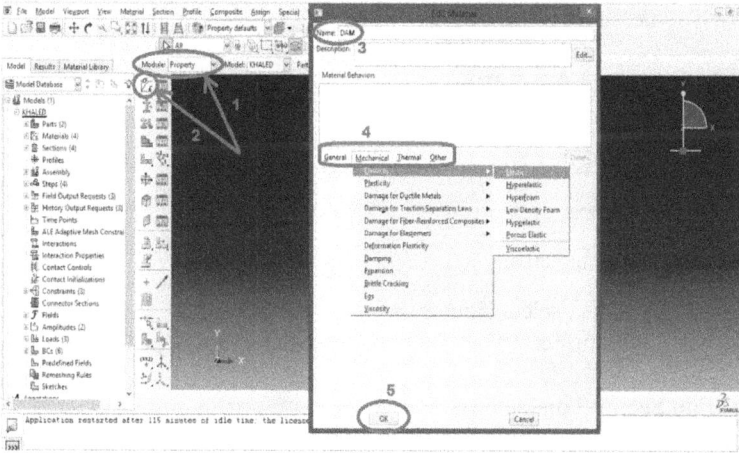

FIGURE 2.10 Define elasticity for steel

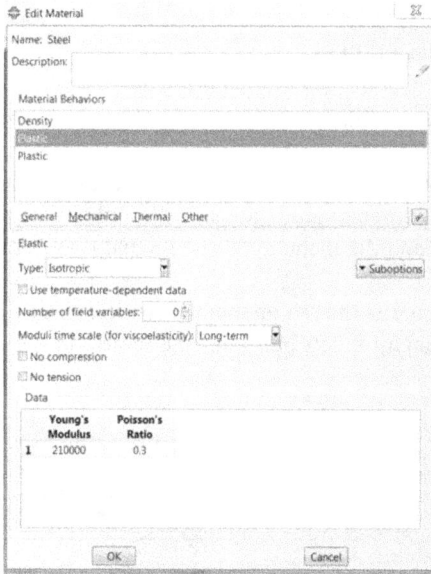

FIGURE 2.11 Identify the material elasticity for steel.

To define a beam section, follow the steps as shown in Figure 2.12.

From the main menu bar, select "Section → Create". The "Create Section" dialog box appears.

In the "Create Section" dialog box:

a. Name the section *Steel*.

b. In the Category list, select Solid.

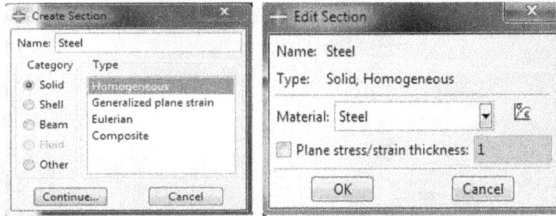

FIGURE 2.12   Define section.

c. In the Type list, select Homogeneous.

d. Click on "Continue". The "Edit Section" dialog box appears.

In the "Edit Section" dialog box,

Accept the default selection of Steel for the Material associated with the section. If other materials have been previously defined, click on the arrow next to the "Material" text box to see a list of available materials and scroll through the "Materials" to see a list of available materials and select the required material.

Click on "OK".

2.4.2.3   Section Assignment

Next, assign the defined section to the corresponding part. The Assign menu is used in the Property module to assign the Steel section to the beam. To assign the section to the beam, as depicted in Figure 2.13, perform the following steps:

From the main menu bar, select "Assign → Section". The software displays prompts in the prompt area to guide the user throughout the procedure.

Alternatively, expand the menu under Beam and double click on the "Section Assignments".

Select the entire part as the region that the section will be applied for it.

Click and hold down the left button of the mouse in the upper-left corner of the viewport.

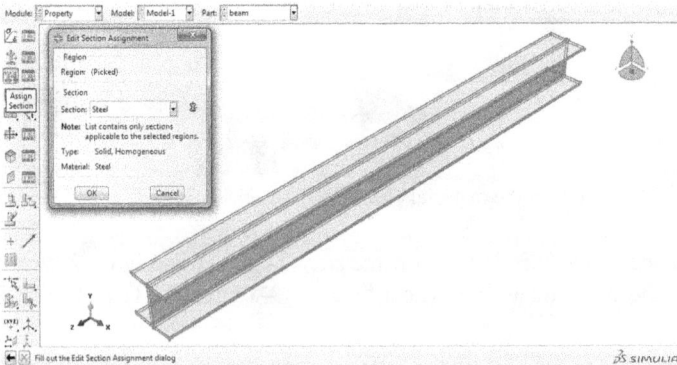

FIGURE 2.13   Assign the beam section.

Drag the mouse pointer to create a box around the plate.

Release the left mouse button. The software highlights the entire plate.

Right click on the viewport or click on "Done" in the prompt area to accept the selected geometry. The "Assign Section" dialog box appears containing a list of existing sections.

In the "Section", scroll to "Steel" and click on "OK".

The color of the part changes to green once the section is assigned.

2.4.3 ASSEMBLY MODULE

Each part is oriented in its own coordinate system and it is independent of the other parts in the model. Although a model may contain many parts, it can only have one assembly.

Define the geometry of the assembly by creating instances of a part and then positioning the instances relative to each other in a global coordinate system. An instance may be independent or dependent. Independent part instances are meshed individually, while the mesh of a dependent part instance is associated with the mesh of the original part.

2.4.3.1 Assemble Part Instances into the Model

In the module list located under the toolbar, click "Assembly" to open the Assembly module. The cursor changes to an hourglass while the Assembly module loads as shown. From the main menu bar, select "Instance → Create". The "Create Instance" dialog box appears.

In the opened window, under the "Instance Type" box, chose "Dependent" (mesh on the part).

In the dialog box, select "Plate" and click on "Apply".

In the dialog box, select "Beam" and click on "Apply".

In the dialog box, select "Column" and click on "Apply".

In the dialog box, select "Box" and click on "Apply".

In the dialog box, select "Jack Plate" and click on Apply.

In the dialog box, select "Stiffener" and click on "Apply".

Click on the "Linear Pattern" icon to copy the part to several elements.

Click on "Translate Instance" to move the elements to the required position.

Click on "Rotate Instance" to rotate any part about any axis to suit the right direction. The model will be assembled, as shown in Figure 2.14.

2.4.4 STEP MODULE

After the assembly is completed, the configuration of the analysis should be defined. In this simulation, it is required to identify the static response of the plate under applied 500 kN load at the adjacent sides. The software generates the initial step automatically; however, the rest of the analysis steps should be defined by the user as well as for any requested analysis output.

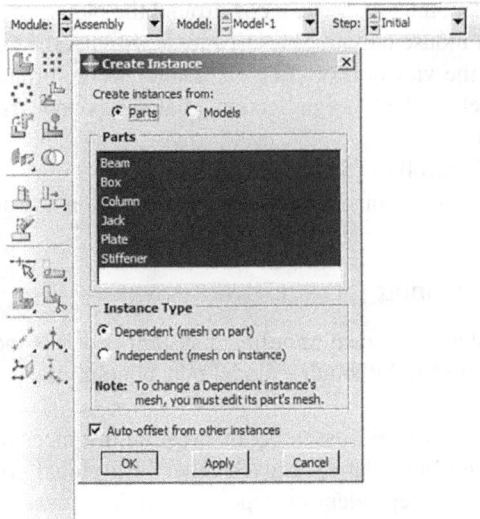

FIGURE 2.14   Assembly of the entire individual parts into a single model.

There are two kinds of analysis steps in the ABAQUS software:

i. General analysis steps, which can be used to analyze the linear or non-linear response.
ii. Linear perturbation steps, which can be used only to analyze linear problems. However, only general analysis steps are available in ABAQUS/Explicit. In this step, create a static, general step following the initial step of the analysis.

2.4.4.1   Create an Analysis Step: Step 1

As the assembly of the model has been done, then move to the Step module to configure the required analysis. In this simulation, the main aim is to evaluate the static response of the steel shear wall. Therefore, this is a single event, so only a single analysis step is needed for simulation. Thus, the analysis will consist of two steps:

The initial step, which boundary conditions that constrain the end of the plate, is undertaken.

The analysis step, which distributed load at the other end of the plate, is applied.

The ABAQUS software generates the initial step automatically, but the Step module needs to be operated by the user in order to create the analysis step. The Step module also allows the user to request output (field output and history output) for any steps in the analysis.

In the Module list located under the toolbar, click on "Step" to open the Step module.

From the menu bar, then select "Step → Create" to create a step. The "Create Step" dialog box appears with a list of all general procedures and a default step name for Step 1 (see Figure 2.15).

Select "General" as the procedure type.

From the list of available procedures, select "Static, General", and then click on "Continue".

Then, the "Edit Step" dialog box appears.

The "Basic" tab is selected by default. Check the "Nlgeom" as "On" and set the "Time Period" as 100 seconds. Enter 10,000 as the maximum number of increments and thereafter consider 0.1, 1e-5, and 0.1 as initial, minimum, and maximum increment sizes, respectively

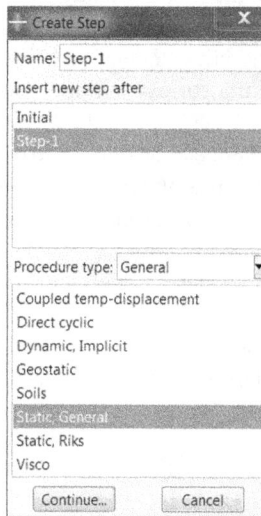

FIGURE 2.15   Define analysis step.

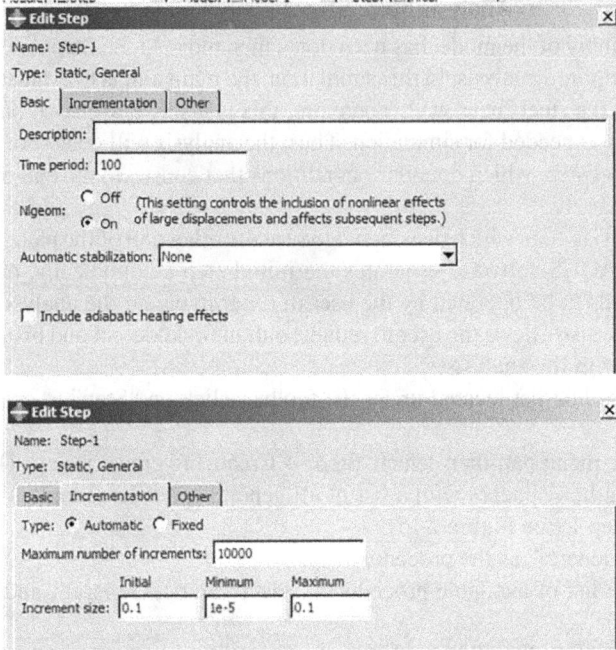

FIGURE 2.16   Time period, nonlinear geometry, and incrementation.

Click on "OK" to create the step and to exit the "Edit Step" dialog box (see Figure 2.16).

To obtain the required outputs at the visualization module, it is necessary to request the history outputs in the "Step Module". Select "Tools → Set" to define two sets of regions. For the first one, name it "Displacement" and choose the upper vertex of the beam, and click "Done" to define the set. Perform the same for a set named "Force" and the corresponding region to all faces at the bottom of the model.

Click on "History Output Requests Manager" to create history output requests.

The "Create History" dialog box appears.

Name the history output as "Displacement" and click on "Continue" (see Figure 2.17).

The "Edit History Output Request" dialog box appears.

At the domain, select "Set".

Click on the field next to the domain and scroll through the "Set" to view a list of available sets and select "Displacement" as set.

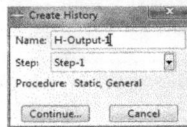

FIGURE 2.17   Create History output.

Expand the "Displacement/Velocity/Acceleration" and toggle on "U3" under "U, Translations, and Rotation". Do the same for a new history output request named "Force" and choose "Force" as set. Consider "RF3" as the requested variable. Click on "OK" to exit from the dialog box. Figure 2.18 shows the procedure.

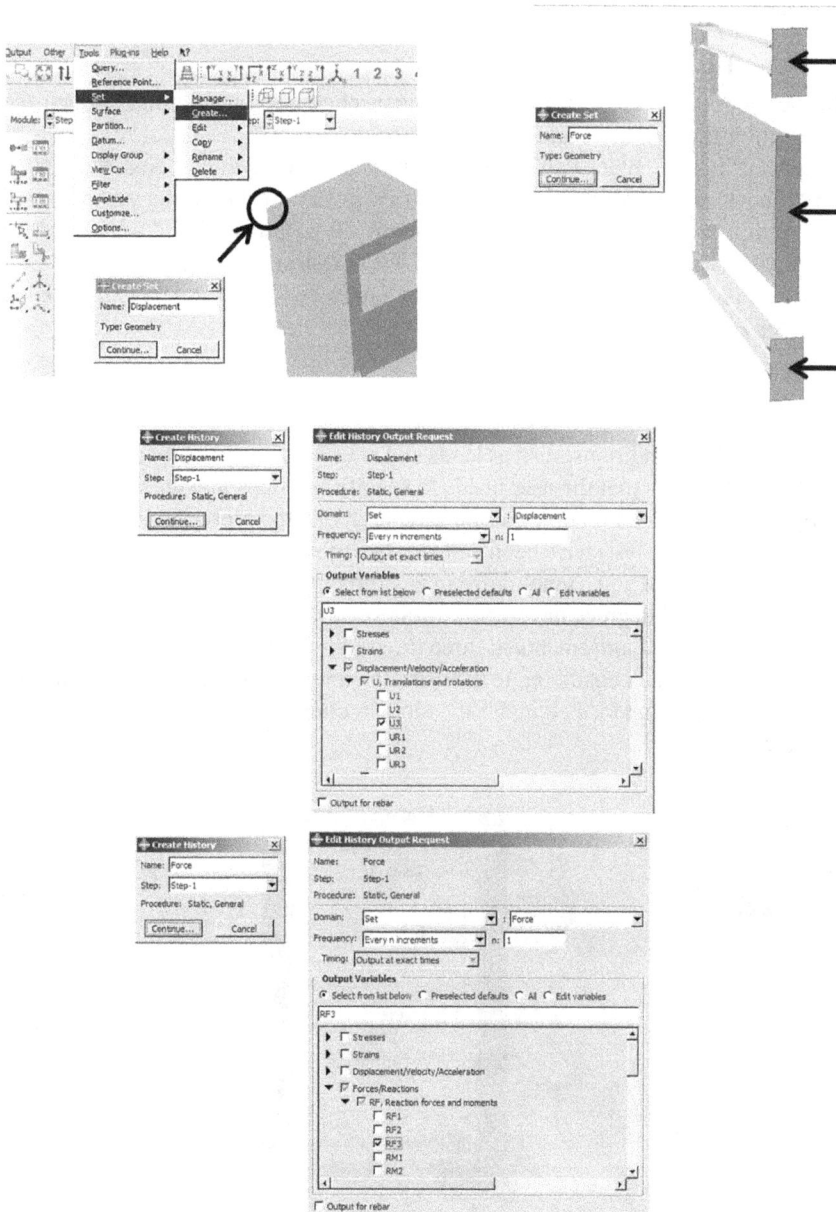

FIGURE 2.18   Create Set and History output.

2.4.5 INTERACTION MODULE

In the ABAQUS software, there are many interactions having a specific definition. This command can be used to identify the type of connection and constraints between two similar and different materials such as the tie and rigid connection, embedded region, shell to solid coupling, etc. In this example, a tie connection is considered to use.

2.4.5.1 Tie Constraint

In the module list located under the toolbar, click on "Interaction" to open the Interaction module (see Figure 2.19).

Select "Find Contact Pair" at the toolbox to open its dialog box. Then click on "Find Contact Pairs". The software tries to find all faces that are near and list them in the dialog box. Click on column "Type", then "Edit" to open "Edit Multiple Cells" in the dialog box. Select "Tie Constraint" as type and click "Ok" to accept and close it. Finally, click on "Ok" to close "Find Contact Pairs" dialog box (see Figure 2.20).

2.4.6 LOAD MODULE

The prescribed conditions, such as loads and boundary conditions, are step dependent, which means that the user needs to specify the steps in which they become active. Then, the load module can be used to define the prescribed conditions as the steps in the analysis have been defined. In this model, the bottom of the frame is fully constrained and cannot move in any direction. Also, the type of boundary conditions for all degrees of freedom is chosen as a pinned boundary condition. This means that all movements in three directions are restricted.

To apply boundary conditions to the structure, follow the steps as described below:

In the module list located under the toolbar, click on "Load" to open the Load module.

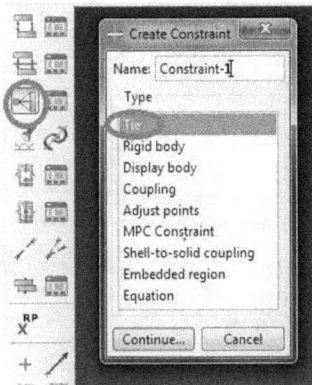

FIGURE 2.19 Interaction module.

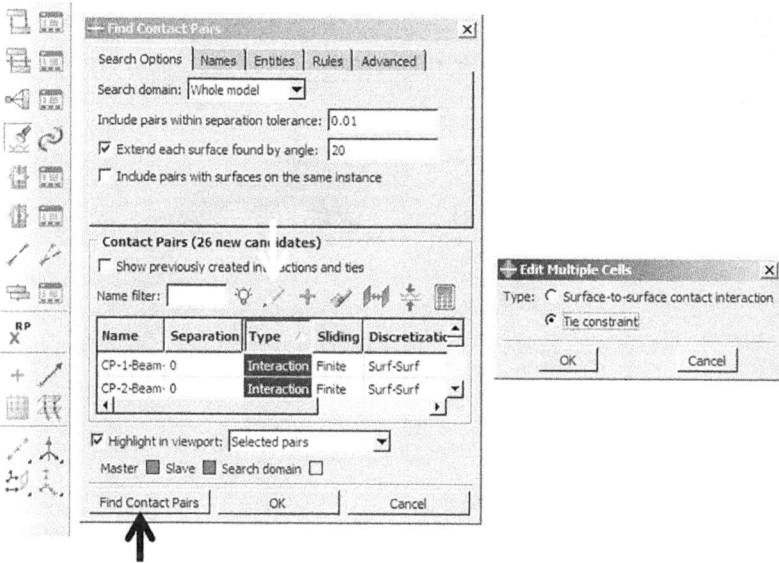

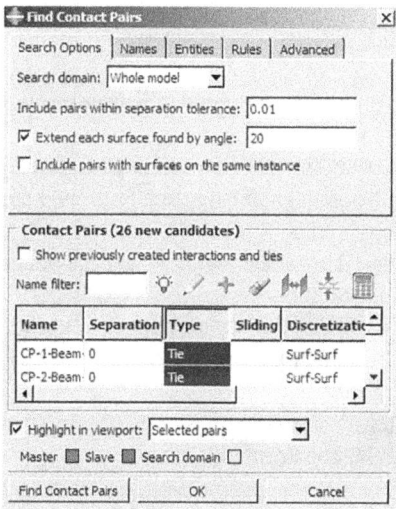

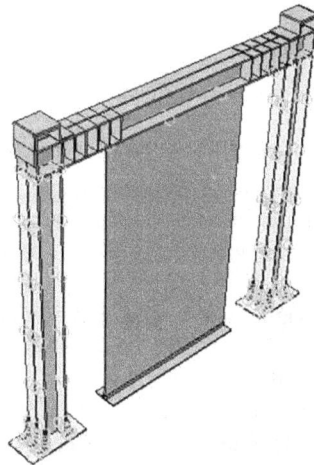

FIGURE 2.20 Create Tie connection between all parts by Find Contact Pair option.

From the main menu bar, select "BC → Create". The "Create Boundary Condition" dialog box appears.

In the "Create Boundary Condition" dialog box:

From the list of steps, select "Step-1" as the step in which the boundary condition will be activated. All the mechanical boundary conditions specified in the initial step must have zero values. This condition is enforced automatically by the software.

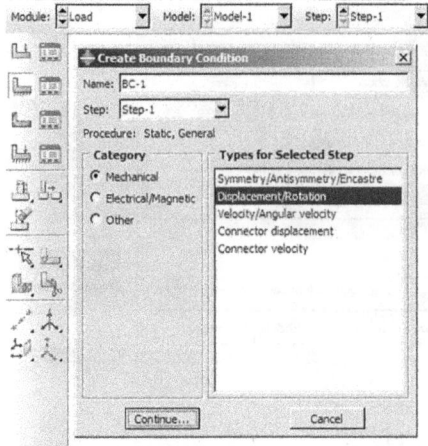

FIGURE 2.21   Create boundary condition.

In the Category list, accept "Mechanical" as the default category selection (see Figure 2.21).

For "Types" for the selected step list, select "Displacement/Rotation", and click "Continue".

The software displays prompts in the prompt area to guide the user throughout the procedure.

To apply a prescribed condition to a region, the user can either select the region directly in the viewport or apply the condition to an existing set. Sets are a convenient tool that can be used to manage large and complicated models.

In the viewport, select the base surfaces. This is the region where the boundary condition will be applied, as shown in Figure 2.22.

Right click on the viewport or click on "Done" in the prompt area to indicate the end of selecting regions. The "Edit Boundary Condition" dialog box appears.

In the dialog box:

Toggle on "U1, U2, and U3" and enter a zero value in each text field since all translational degrees of freedom need to be constrained.

FIGURE 2.22   Select the base surface.

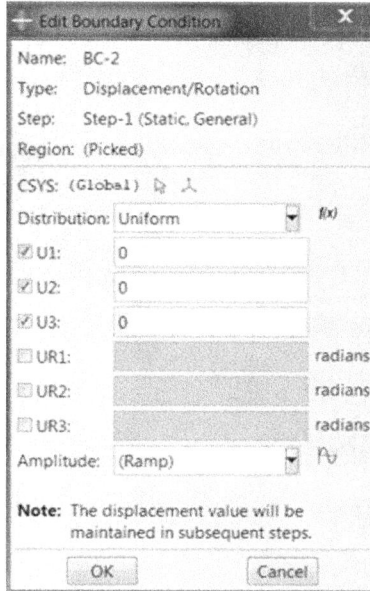

FIGURE 2.23 Define the boundary condition for the base surface.

Click on "OK" to create the boundary condition and to close the dialog box.

The software displays arrowheads at the vertex to indicate the constrained degrees of freedom (see Figure 2.23).

As the frame is constrained, then loading can be applied to the plate at the top of the frame. In this simulation, a distributed force of 80 N/mm² is applied in the negative direction of the z axis. To apply the distributed force to the plate, perform the following steps:

Double click on "Loads" in the Model tree to open its dialog box.

In the "Create Load" dialog box:

From the list of steps, select "Step 1" from the step in which the load will be exerted.

In the Category list, choose "Mechanical" from the default category selection.

In the "Types for the Selected Step" list, select "Pressure".

Click on "Continue" (see Figure 2.24).

The software displays prompts in the prompt area to guide the user throughout the procedure. It asked the user to select a region or a point that the loads are needed to apply. As a boundary condition, the region to which the load is applied can be selected either directly in the viewport or from a list of existing sets. Select the region directly in the viewport.

In the viewport, select the left surface of the plate as the region where the load needs to apply.

Click on the "viewport" or click "Done" in the prompt area to finish selecting the regions. The "Edit Load" dialog box appears.

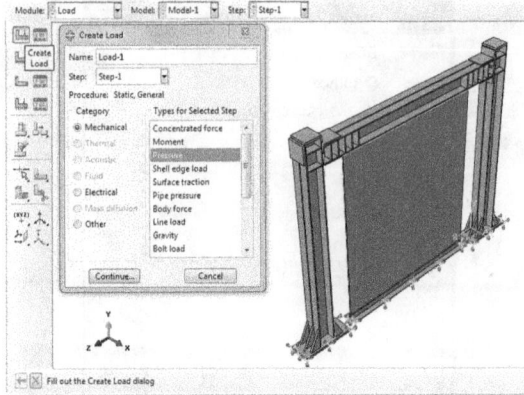

FIGURE 2.24   Create pressure loading.

In the dialog box, enter a value of 80. Click "Create Amplitude" to define the cyclic amplitude of loading.

Create amplitude:

Select the "Tabular type" of amplitude and click "Continue". Input the increment of the amplitude data. Click "OK" to exit from the dialog box (see Figure 2.25).

2.4.7   MESH MODULE

The Mesh Module is one of the most important modules in the modeling process since the accuracy of the results depends on the number of elements and element size. The finite element mesh should be generated in this module and this module can be used to generate meshes as well as to verify them. The larger number of mesh elements is resultant of the smaller mesh element size which leads to more accurate results. ABAQUS/CAE uses a number of different meshing techniques. The default meshing technique assigned to the model is indicated by the color of the model that is displayed when the Mesh module is entered. If the model displays in orange, it cannot be meshed without the assistance of the user. This command is used to mesh the whole structure into small and equal parts and elements.

2.4.7.1   Mesh: Seed the Part (50 mm Elements)

As mentioned before, the mesh module generates the finite element mesh. Various meshing techniques are used to create the mesh, and choose element shape, and element type.

2.4.7.2   Assign an ABAQUS Element Type

In the module list located under the toolbar, click "Mesh" to open the Mesh module.

At the context bar, click "Part", to unclick the assembly.

From the main menu bar, select "Mesh → Element Type".

In the viewport, select the entire frame as the region to be assigned to an element type.

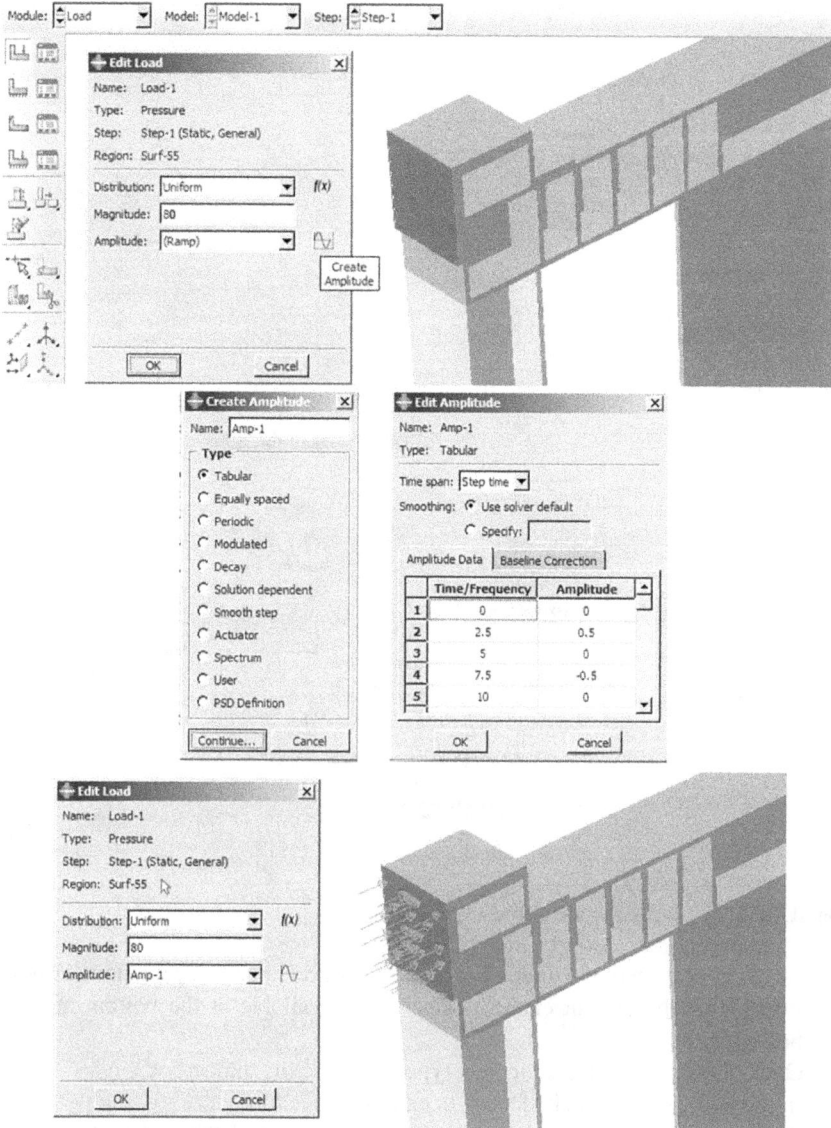

FIGURE 2.25 Input the load value and amplitude.

In the prompt area, click "Done". The "Element Type" dialog box appears, as shown in Figure 2.26.

In the dialog box, select the following:

- Standard as the Element Library selection (the default).
- Linear as the Geometric Order (the default).
- 3D stress as the Family of elements.

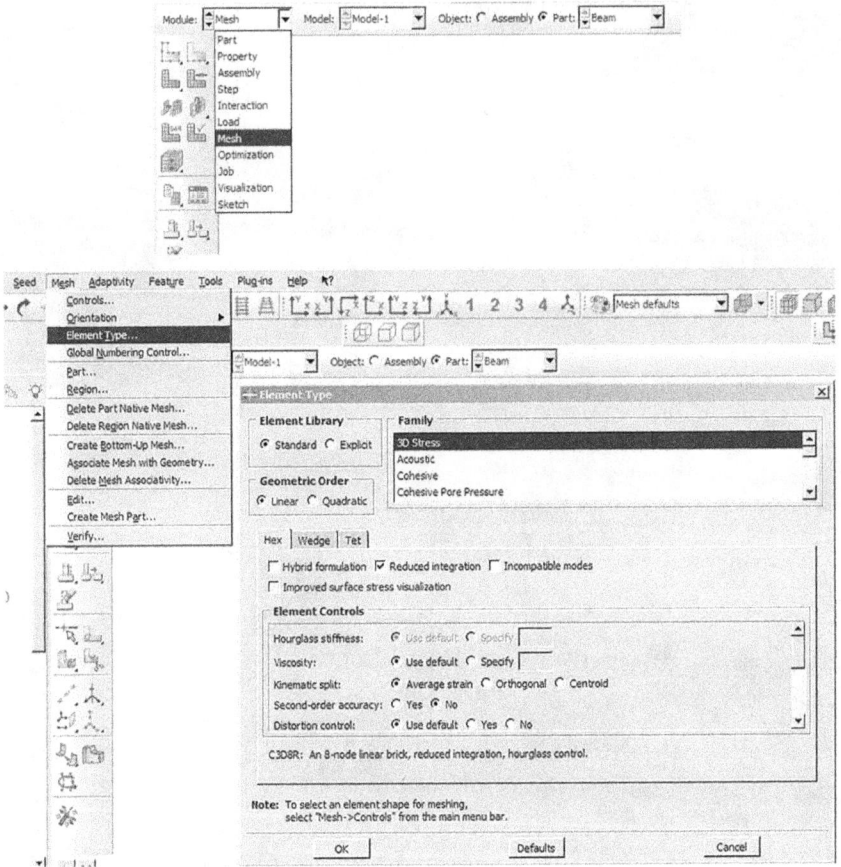

FIGURE 2.26   Selecting the element type.

In the lower portion of the dialog box, examine the element shape options. A brief description of the default element selection is available at the bottom of each tabbed page.

Click "OK" to assign the element type and close the dialog box.

In the prompt area, click "Done" to end the procedure.

In the next step, the mesh can be created. First seed the edges of the part instance, followed by mesh in the part instance. Select the number of seeds based on the desired element size or the number of elements that are required along an edge, and the software places the nodes of the mesh at the seeds whenever possible.

2.4.7.3   Seed and Mesh the Model

From the main menu bar, select "Seed → Part" to seed the part instance.

Alternatively, select the "Seed Part" in the upper-left corner of the meshing toolbox. The "Global Seeds" dialog box will appear.

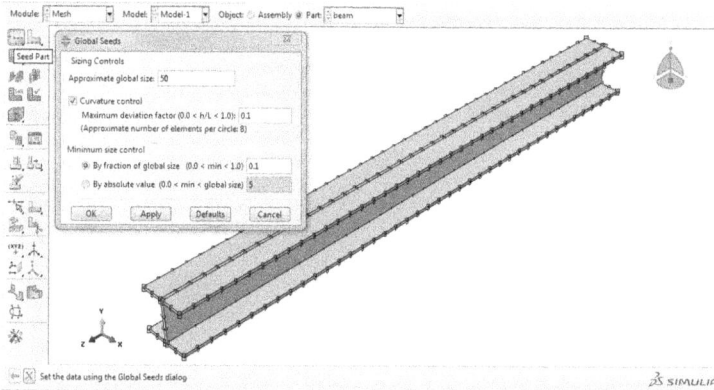

FIGURE 2.27   Assign the approximate global size.

Type "50" as the appropriate value for the approximate global size of the mesh elements.

Click "OK" to accept the seeding, as shown in Figure 2.27.

Note: more control for the resulting mesh can be gained by seeding each edge of the part instance individually, but it is not necessary for this example. The prompt area displays the default element size that the software uses to seed the part instance. This default element size is based on the size of the part instance. Therefore, relatively large seed values will be used so that only one element will be created per region.

The ABAQUS software offers a variety of meshing techniques to mesh the models with different geometries. Various meshing techniques provide different levels of automation and user control. There are three (3) types of mesh generation techniques available in the ABAQUS software as follows:

i. Structured meshing applies pre-established mesh patterns to particular model geometry. It is applicable to complex models; however, it must generally be partitioned into simpler regions to use this technique.

ii. Swept meshing extrudes an internally generated mesh along a sweep path or revolves it around an axis of revolution. Like structured meshing, swept meshing is limited to models with specific topologies and geometries.

iii. Free meshing is the most flexible meshing technique and uses no pre-established mesh patterns. It can be applied to almost any model shape.

Select the "Assign Mesh Controls" tool in the meshing toolbox, and the mesh controls dialog box appears. Choose Hex from the element shape. Choose Structured meshing technique. To mesh the part, choose "Mesh → Part" and click "Yes" in the prompt area to perform the mesh as shown in Figure 2.28.

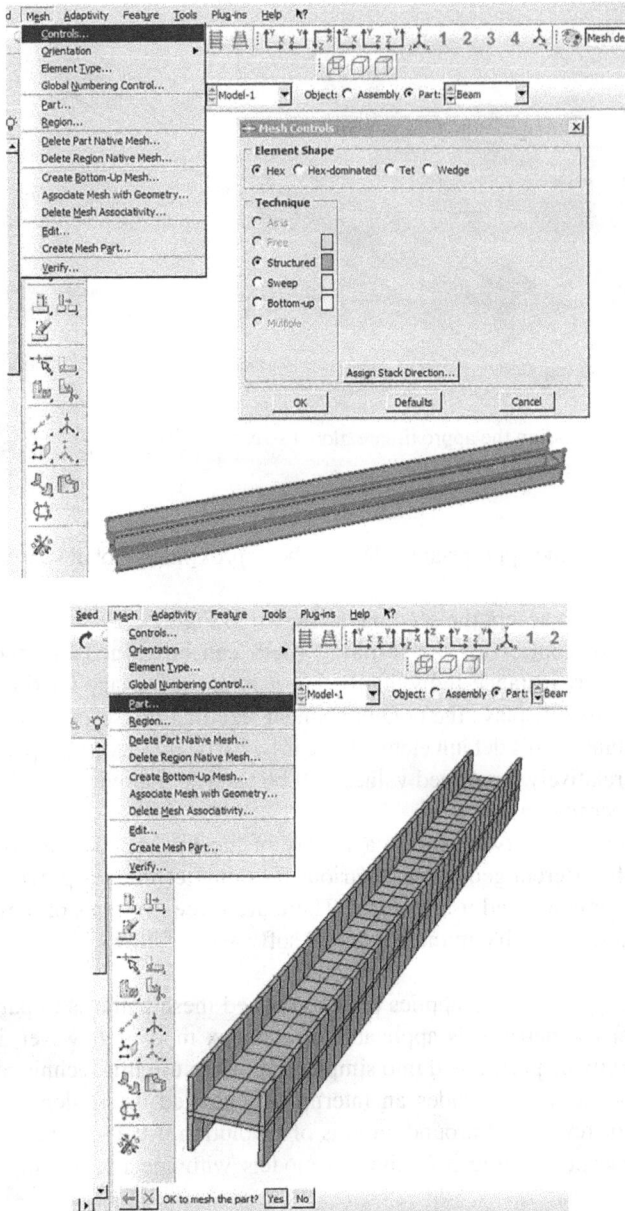

FIGURE 2.28 Choose element shape and meshing technique and mesh part.

2.4.7.4 Verify Mesh

Meshing quality should be checked. Select "Mesh → Verify" as shown in Figure 2.29. Select the part and then click on the middle mouse button to open "Verify Mesh" dialog box (see Figure 2.30).

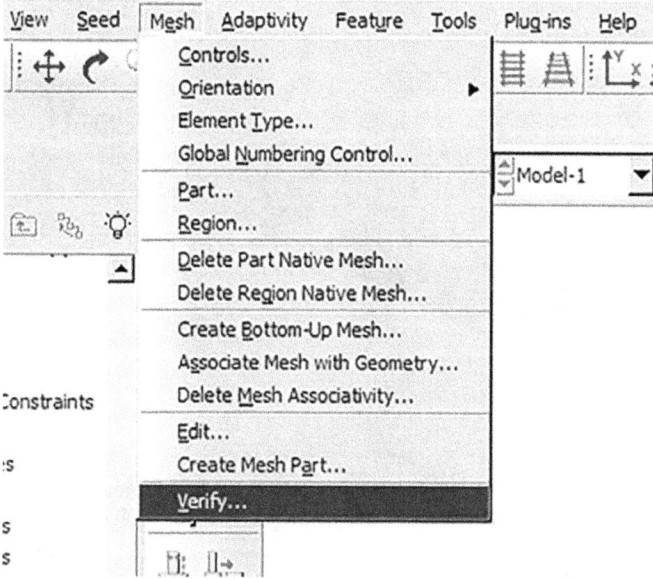

FIGURE 2.29   Verify mesh.

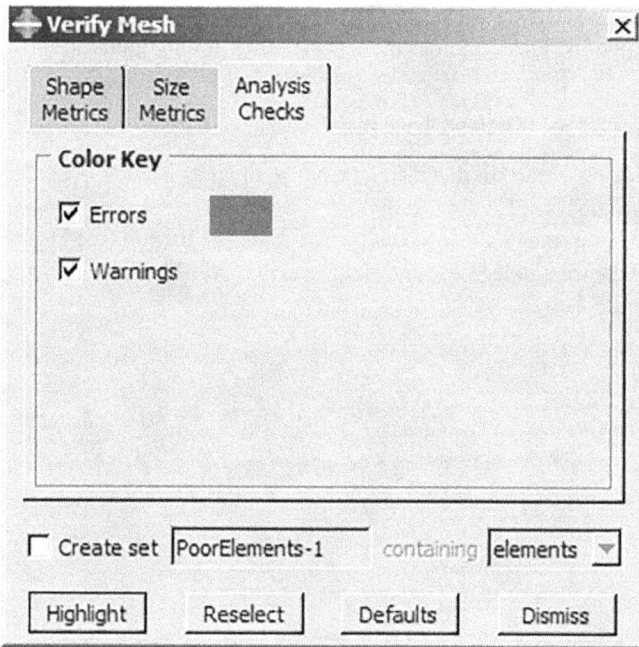

FIGURE 2.30   Select part for mesh verification.

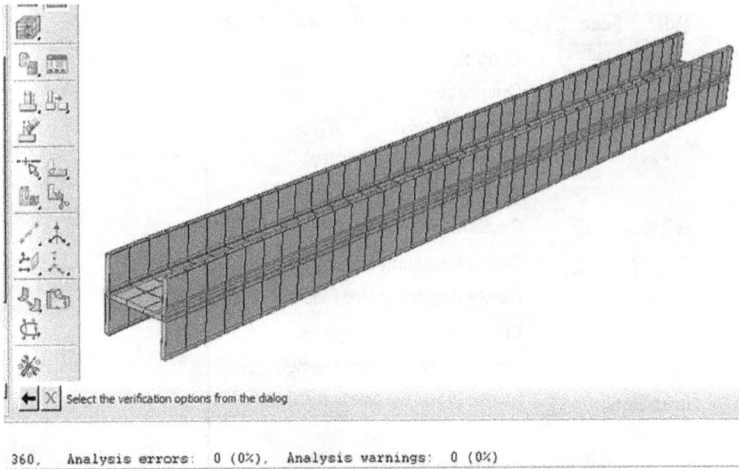

```
360,    Analysis errors:  0 (0%),   Analysis warnings:  0 (0%)
```

FIGURE 2.31 Checking mesh part for analysis warnings and errors.

Click "Highlight" to show all elements included analysis warnings and analysis errors in the message area. In this example, there is no element contains warning and error as shown in Figure 2.31.

2.5 ANALYSIS: JOB MODULE

In the Module list located under the toolbar, click Job to open the Job module.

2.5.1 CREATE AN ANALYSIS JOB: JOB-1

Double click on "Jobs" in the "Model tree" to open its box. Accept the name and click "Continue" (see Figure 2.32).

Click "Ok" in the "Edit Job" dialog box to define the job (see Figure 2.33).

Expand the tree under jobs, and right click on "Job-1". Then, click on "Submit" (see Figure 2.34).

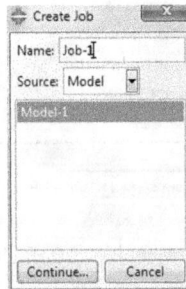

FIGURE 2.32 Define a job.

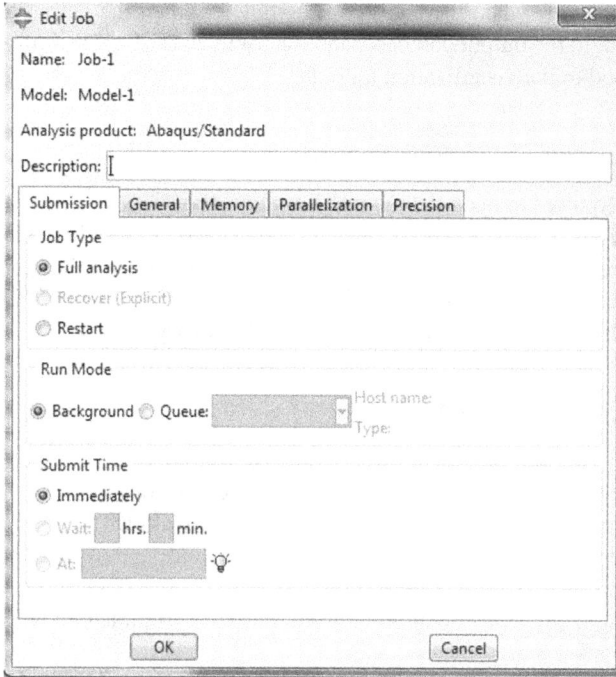

FIGURE 2.33   Create job by default.

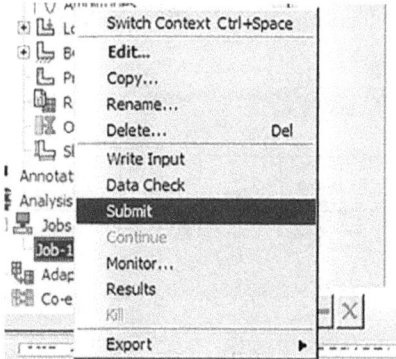

FIGURE 2.34   Job submission.

The user can indicate the job's status after job submission. The Status column for the overhead hoist problem shows one of the following:

- "None" while the analysis input file is being generated.
- "Submitted" while the job is being submitted for analysis.
- "Running" while the software is analyzing the model.

- "Completed" when the analysis is completed, then the output has been written to the output database, and the user can click on the" Results" to proceed to the visualization module.

2.5.2 MONITOR SOLUTION IN PROGRESS

Click on "Monitor" to open the job monitor dialog box once the job is submitted.

The top half of the dialog box displays the information available in the status (*.sta) file that the software creates for the analysis (see Figure 2.35).

If the status on top of the "Job Monitor" window is shown as "Completed", then this job is free of errors and was executed properly.

2.6   VISUALIZATION MODULE

Graphical postprocessing is showing the vast volume of data created during a simulation. For any realistic model, it is impractical to interpret results in the tabular form of the data file. The ABAQUS software allows the user to view the results graphically using a variety of methods, including deformed shape plots, contour plots, vector plots, animations, and X–Y plots. Therefore, using this model, most of the results can be viewed, and various plots can be generated.

2.6.1   VIEW THE RESULTS OF THE ANALYSIS

When the job is completed, the user can view the results of the analysis with the Visualization module. Right click on "Job-1" underneath of Jobs, then click "Results". The software loads the Visualization module, opens the output database

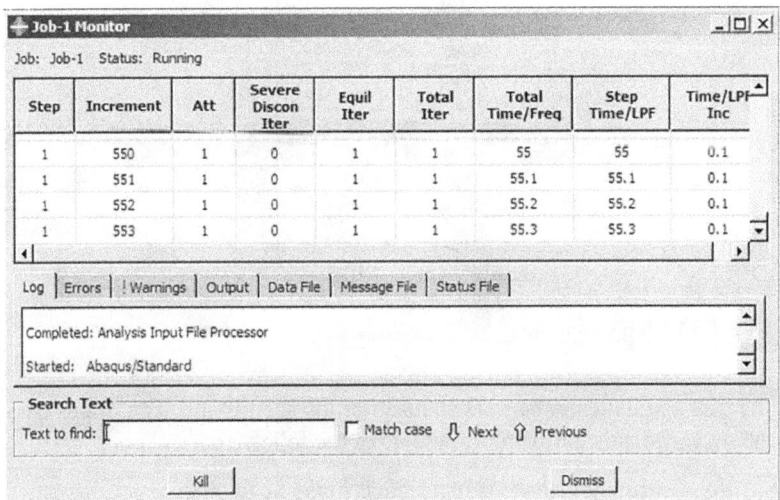

FIGURE 2.35   Monitoring.

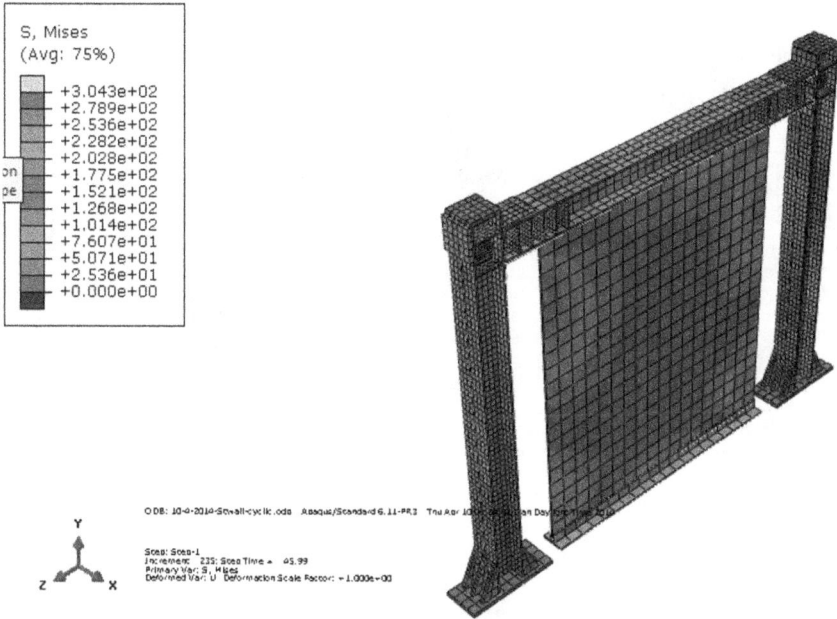

FIGURE 2.36 Stress contour.

created by the job, and displays a fast plot of the model. A fast plot is a basic representation of the undeformed shape of the model. Alternatively, the visualization option can be clicked in the module list located under the toolbar. The stress contour can also be observed by clicking the "Plot Contours" on the "Deformed Shape" from the visualization toolset, as shown in Figure 2.36.

2.6.2 VISUALIZATION/RESULTS MODULE

In the visualization module, the user can see the plot contours on the deformed shape and also the animation of the structure after applying the load. The ABAQUS software allows the user to plot the graph of each output. In order to draw the hysteresis graph, perform the following process:

From the menu bar, select "Result → History Output".

In the "History Output" opened window, select all "Reaction Forces" and click on "Save As" (see Figure 2.37).

Name the output "Reaction Force" and save the operation of SUM (XY, XY, ...) and click "OK" (Figure 2.38).

Then, select "Spatial Displacement" in "History Output" and click "Save As".

Name the output "Displacement" and select "as is" in the "Save Operation" and click "OK" (see Figure 2.39).

Double click on "XY Data" and Check the "Operate on XY data" in the list of sources.

Click on "Continue" (see Figure 2.40).

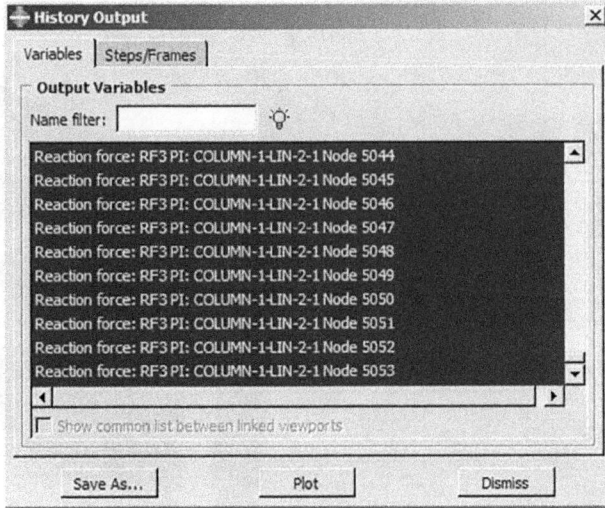

FIGURE 2.37   Select the reaction force history outputs.

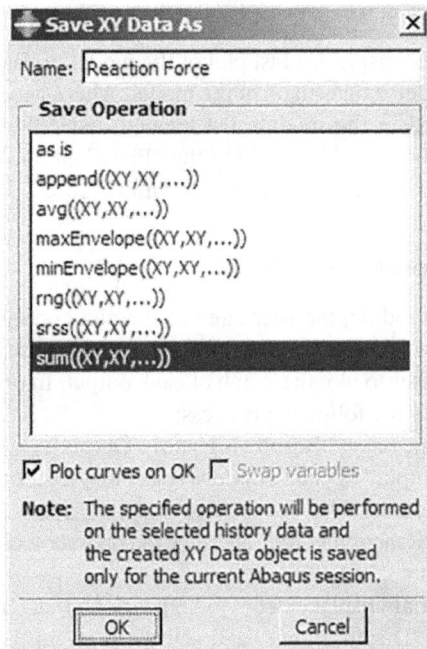

FIGURE 2.38   Save the reaction force with SUM operation.

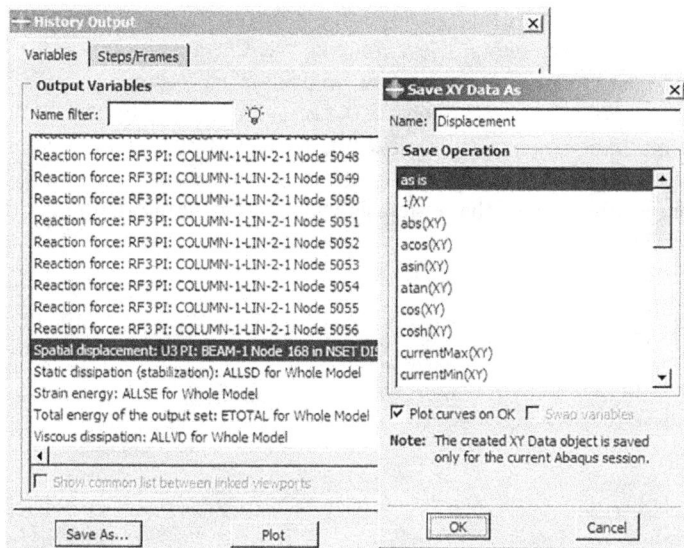

FIGURE 2.39 Save the displacement records with "as is" operation.

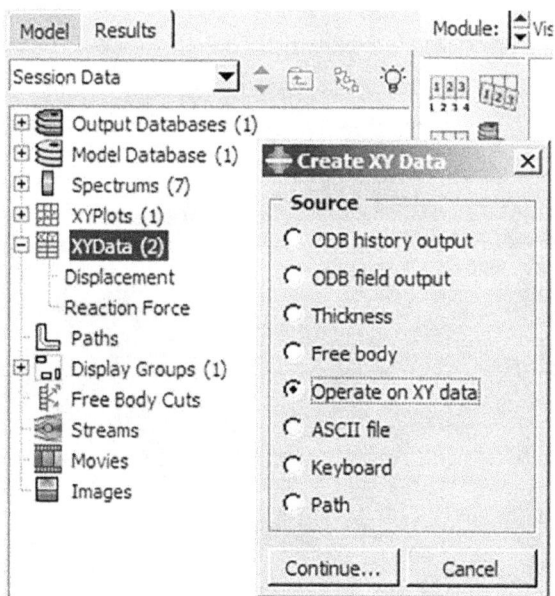

FIGURE 2.40 Create XY data.

The "Operate on XY Data" dialog box appears.

From the list of operators, select combine "X,X".

First, double click on "Displacement" and put minus (–) then only double click on "Reaction Force" in order to obtain the actual graph (see Figure 2.41).

Click on "Plot Expression" to draw the hysteresis plot, and the graph is displayed in the viewport, as shown in Figure 2.42.

The larger the area of the hysteresis graph is indicating the higher dissipate energy of the system.

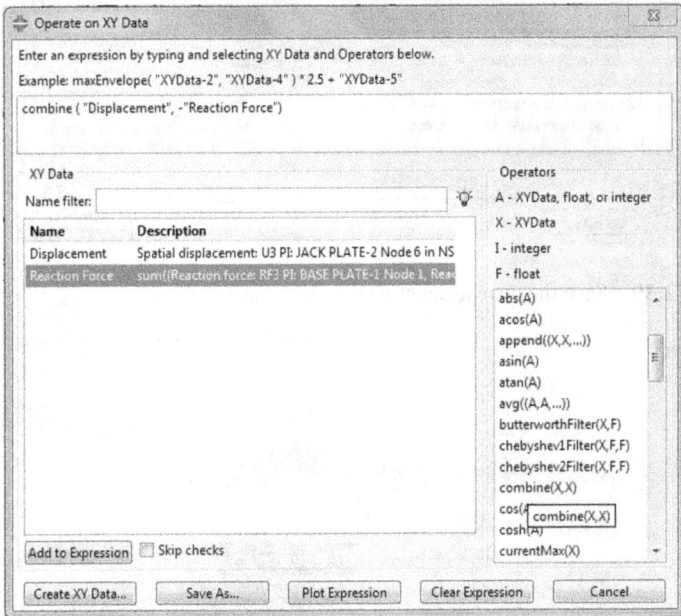

FIGURE 2.41   Operate on XY data.

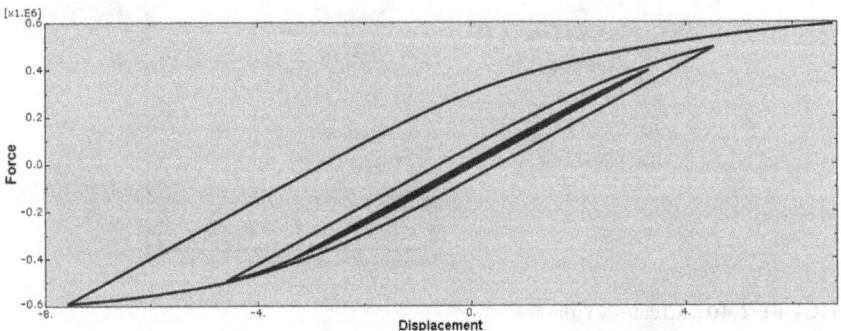

FIGURE 2.42   Hysteresis graph of the reaction force vs. displacement.

3 Performance of Reinforced-Concrete Frame with Embedded CFRP Rod under Cyclic Load

3.1 INTRODUCTION

The application of fiber-reinforced polymers (FRPs) for retrofitting and strengthening of concrete sections has steadily increased and consequently become a valid option. The FRP has shown to be an excellent strengthening material for repairing purposes and achieving higher strength for RC structures compared to other conventional strengthening materials and techniques. There are various forms of FRPs such as sheets, plates, cables, and more recently, in bar form. Recently, the Carbon-Fiber-Reinforced Polymer (CFRP) and Glass-Fiber-Reinforced Polymer (GFRP) rods have been frequently implemented in various concrete structures to avoid unpredictable and brittle failures due to the lack of strength in RC sections.

Researcher: Ahmed Abdulrahman Fatikhan (ahmedalkhateeb861@gmail.com)

3.2 PROBLEM DESCRIPTION

This chapter presents the modeling and analysis of reinforced concrete (RC) frame strengthened by embedded CFRP using the ABAQUS finite element package. In order to evaluate the effect of implementing CFRP instead of steel rebars on the performance of the RC frame, the frame is subjected to lateral cyclic load and the response of the frame is investigated.

For this purpose, a bare frame was modeled (Figure 3.1).

The size and dimension details of the multistory frame (beam, column, slab, and foundation) are shown in Figure 3.2.

The position of the CFRP rods is shown in Figure 3.3.

3.2.1 PROBLEM STATEMENT

The ordinary RC frame exhibits low resistance to seismic loading as this kind of building lacks lateral loading resistance. Previous techniques developed for strengthening the RC frame have been inadequate for preventing the seismic effect.

DOI: 10.1201/9781003219491-3

FIGURE 3.1   Frame structure formed by Lu et al. (2008).

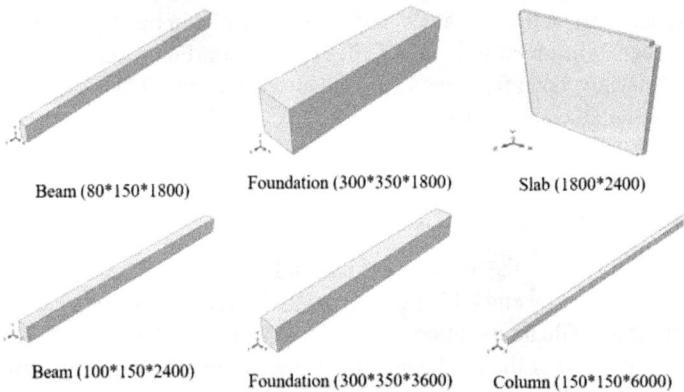

Beam (80*150*1800)   Foundation (300*350*1800)   Slab (1800*2400)

Beam (100*150*2400)   Foundation (300*350*3600)   Column (150*150*6000)

FIGURE 3.2   The dimensions for the beam and column of the frame model.

The application of using FRP rods as the main reinforcement is rarely used in many areas, especially in strengthening the RC frame. Accordingly, this study is important in terms of investigating the behavior of the RC frame by replacing the FRP rod instead of steel bars.

3.2.1.1   Material Properties

The elastic and plastic behavior of the concrete material was modeled based on concrete damage plasticity (CDP). The mechanical properties of the reinforcing

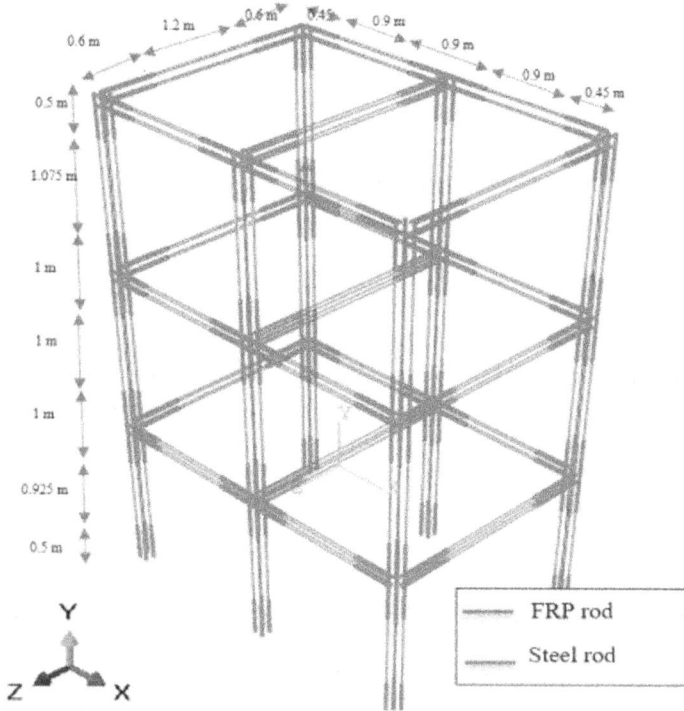

FIGURE 3.3 The position of the steel and FRP rod.

bars and shear links of steel are defined the same. The mass density of steel is 7850 kg/m^3, Young's modulus is 210,000 N/mm^2, Poisson's ratio is 0.3, and the maximum yield stress is 470 MPa. The properties of the FRP rod are shown in Table 3.1.

3.3 OBJECTIVES

- To develop the finite element model of the RC frame with embedded CFRP and rod subjected to cyclic load using the software.
- To evaluate the effect of embedded CFRP and GFRP rods on the capacity of frame building under lateral load.

TABLE 3.1
Properties of the FRP Rod

FRP Type	Diameter (mm)	Density (kg/m³)	Ultimate Tensile Strength (MPa)	Young's Modulus (GPa)	Poisson's Ratio
CFRP	13	1650	2079	138	0.26
GFRP	13	2170	716	44.3	0.2

3.4 MODELING

3.4.1 Part Module

The part module is used to create various parts of the model. In this case, the model is divided into six parts: beam1, beam2, column, foundation, slabs, and steel rebars. This module allows for the creation of the geometry required for the considered problem. To create a 3D geometry, first a 2D section sketch should be created and then manipulating it to obtain the solid geometry. The following sections demonstrate the step-by-step procedure for creating each of these parts.

3.4.1.1 Create a New Model Database

From the "Start menu", open the ABAQUS software and close the "Start Session" dialog box (Figure 3.4).

When the "Part" module is loaded, it displays the Part module toolbox on the left side of the software main window. Each module displays its own set of tools in the module toolbox.

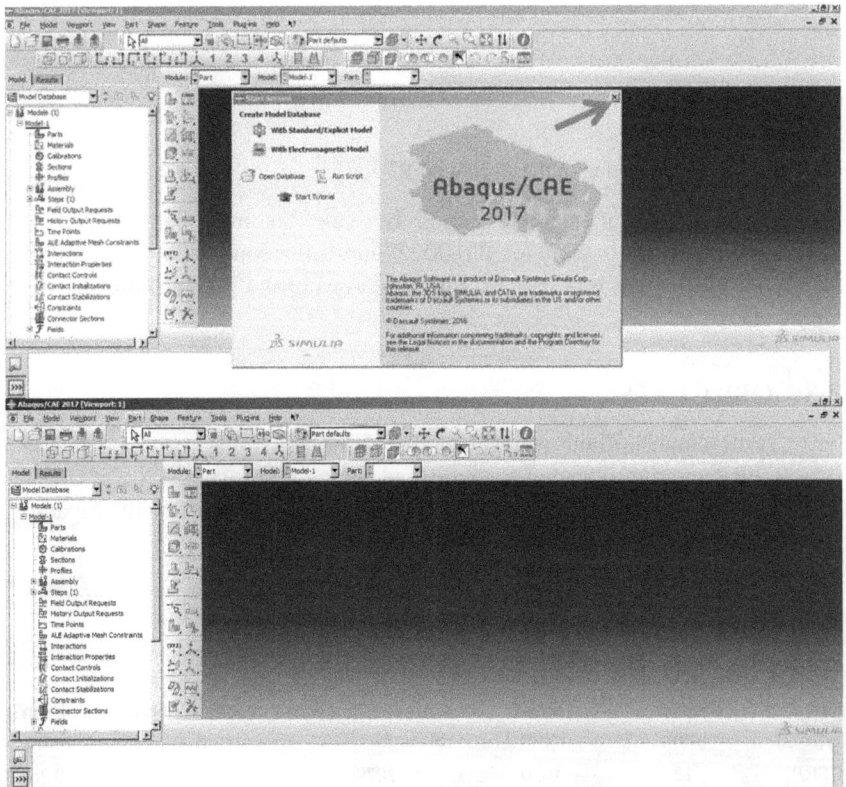

FIGURE 3.4 Getting started.

3.4.1.2 Create a New Model Database and a New Part

From the main menu bar, select "Part → Create", to create a new part. Then, the "Create Part" dialog box appears.

Use the "Create Part" dialog box to name the part; to choose its modeling space, type, and base feature; and to set the approximate size. The name of the part can be edited once it has been created, but the modeling space, type, or base feature cannot be changed.

Name the part "Beam" and choose "3D", "Deformable" type, and "Solid" as the "base feature" (see Figure 3.5).

Enter 3000 for "Approximate Size". The value entered in the approximate size text field at the bottom of the dialog box sets the approximate size for the new part.

Click "Continue" to exit the "Create Part" dialog box.

Repeat the above procedure for the column and foundation while steel reinforcements and the FRP rod should be created as a "Wire" base feature (see Figure 3.5).

3.4.1.3 Define a Rectangle with Dimensions

Use the "Create Lines: Rectangle (4 lines)" tool located in the upper-left corner of the "Sketcher" toolbox to begin sketching the geometry of the rectangle. The user can choose a starting corner for the rectangle at the viewport or enter the x- and y-coordinates. Create a rectangle in the middle of the sketcher without any dimension. Define the dimension of the geometry for the rectangle, as shown in Figure 3.6 by using the add dimension tool located in the lower corner of the sketcher.

Once sketching the section for the dimension is finished, right click and click on "Cancel Procedure" to exit the sketcher. Click on "Done" in the prompt area to exit the sketcher, and it will be displayed as shown in Figure 3.7, in which the user needs to define the depth for the extrusion of the beam.

Once the depth of the beam has been defined, it will be displayed as shown in Figure 3.8.

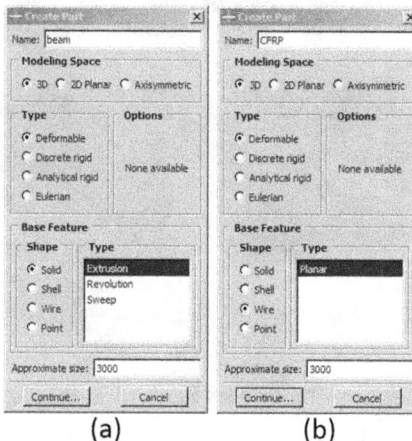

(a)   (b)

FIGURE 3.5   Create new parts: solid base feature and wire base feature.

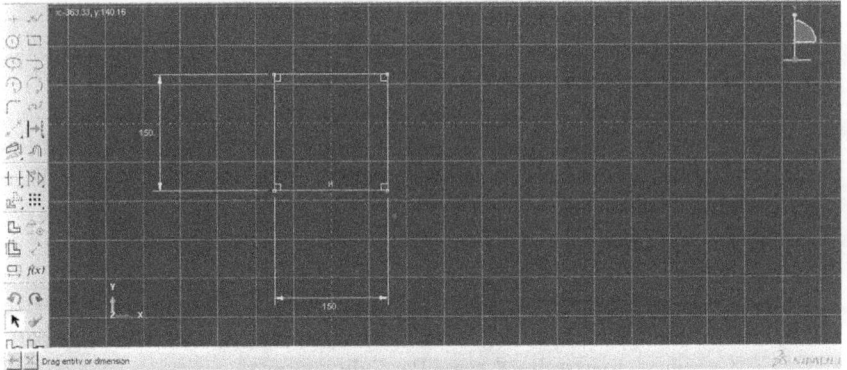

FIGURE 3.6 Define a rectangle and dimensions of the geometry to the rectangle.

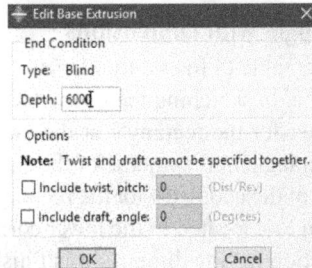

FIGURE 3.7 Depth of the beam.

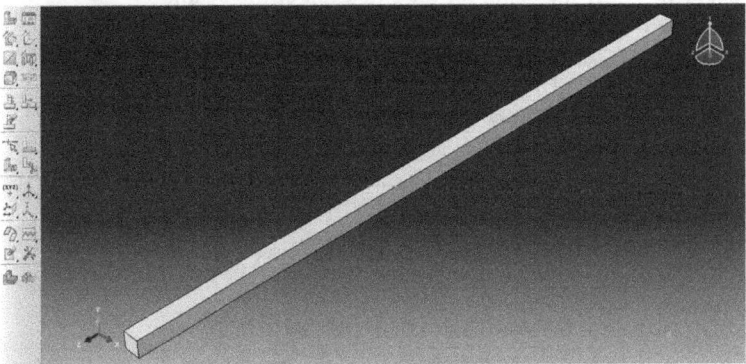

FIGURE 3.8 Beam 3D model.

3.4.1.4 Define Section of the Rebars Section with Dimensions

Use the "Create Lines: Connected" tool located in the upper-right corner of the Sketcher toolbox to begin drawing the geometry of rebars (steel and FRP). The user can pick a starting corner for the rebar (steel and FRP) at the viewport or

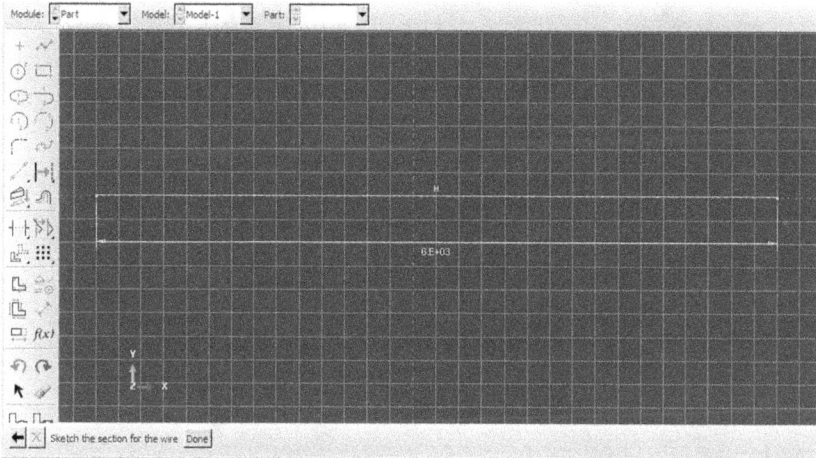

FIGURE 3.9 Define bar dimension.

enter the x- and y-coordinates. Create a line with known coordinates, or the user can alternatively define the dimension of the geometry, as shown in Figure 3.9 by clicking on the add dimension tool. Once sketching the section for the dimension is done, then right click and click on "Cancel Procedure" to exit the sketcher. Click "Done" to complete creating the rebars (steel and FRP) part.

3.4.2 PROPERTY MODULE

In this module, the material properties for the analysis should be defined and those properties need to be assigned to the available parts.

3.4.2.1 Material Properties

The property module is used to create a material and to define its properties. In this problem, there are three types of materials that include concrete, steel, and the FRP rod. The FRP rod is assumed to be linear elastic, but concrete and steel are defined as elastic and plastic materials. To define a material, the following steps should be followed:

In the module list located under the toolbar, select "Property" to open the property module. The cursor changes to an hourglass while the property module loads.

From the main menu bar, select "Material → Create" to create a new material. Then, the "Edit Material" dialog box appears.

Name the material "CFRP."

From the material editor's menu bar, select "Mechanical → Elasticity → Elastic". The software displays the Elastic data form.

Enter the value of 138,000 for "Young's Modulus" and 0.28 for "Poisson's Ratio" in the respective cells. Use the [Tab] button or move the cursor to a new cell and click to move between cells (see Figure 3.10).

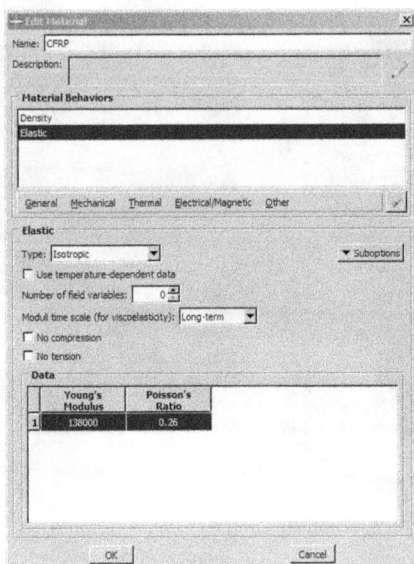

FIGURE 3.10   Material properties for CFRP.

To define Concrete properties, double click on "Materials" in the Model tree and define a new material named "Concrete" and define its elasticity as shown in Figure 3.11.

To demonstrate the nonlinear material behavior of concrete in software, it is necessary to include the concrete damaged plasticity.

In the next step, click on "Mechanical → Plasticity → Concrete Damaged Plasticity".

Enter the values for each text field (Dilation angle, Eccentricity, f_{b0}/f_c, K, and Viscosity Parameter) as stated in Figure 3.11.

Click on the "Compressive Behavior" tab, enter or copy and paste the yield stress and inelastic strain values into the cells.

Then, click on the "Sub Options → Compressive Damage". The Sub Option Editor for the "Concrete Compression Damage" dialog box appears.

Enter or copy and paste the compressive damage parameters and inelastic strain values into the cells. Click on "OK" to exit the dialog box.

Next, click on the "Tension Behavior" tab, enter or copy and paste the yield stress and cracking strain values into the cells.

Then, click on the "Sub Options → Tension Damage" to open the "Sub Option Editor" for the "Concrete Tension Damage" dialog box.

Enter or copy and paste the tension damage parameters and cracking strain values into the cells. Click on "OK" to exit the dialog box.

Click "OK" to exit the material editor (see Figure 3.11).

The properties of steel reinforcement are created with similar steps as mentioned above. However, instead of concrete damaged paucity, normal plasticity

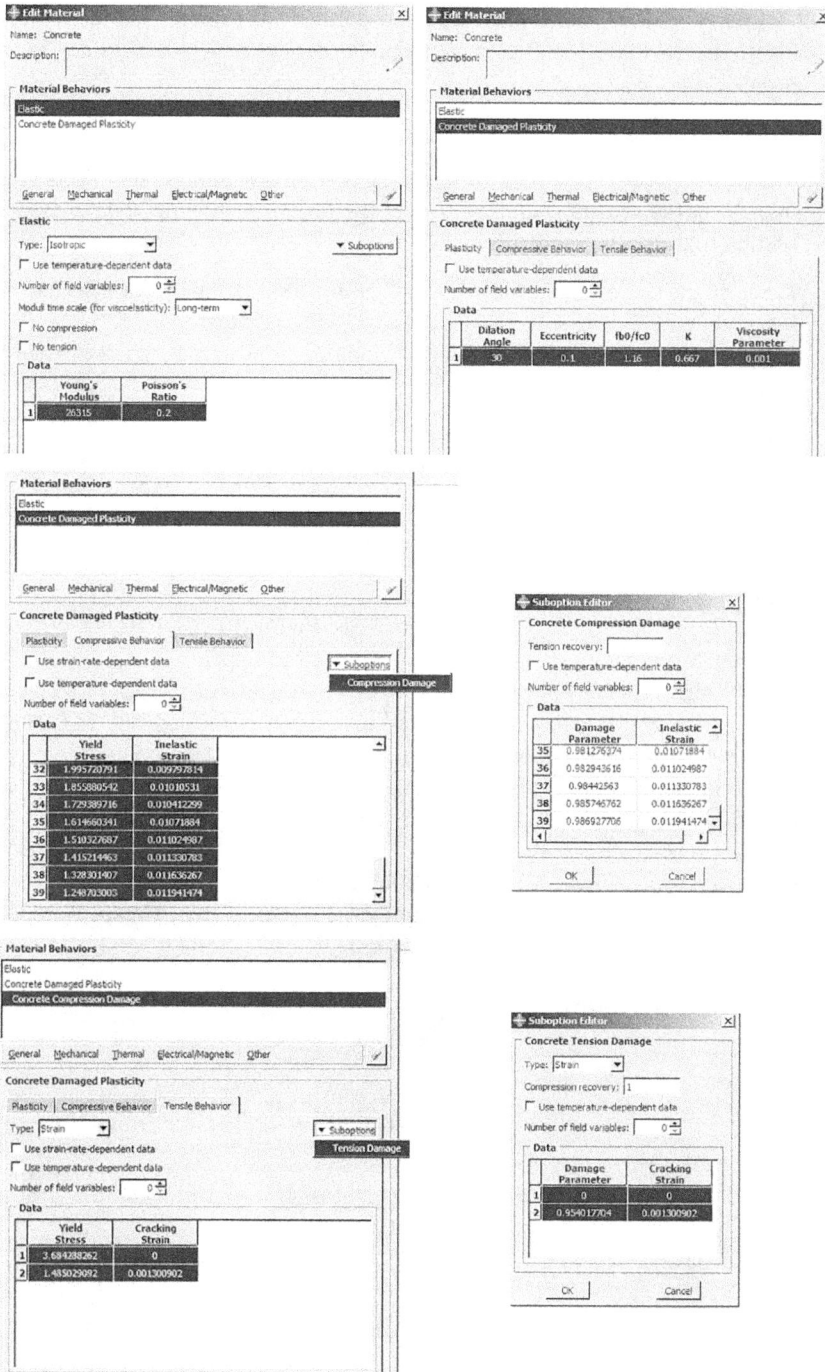

FIGURE 3.11 Material properties for concrete.

and property of steel material should be defined by selecting "Mechanical → Plasticity → Plastic". Then, the software displays the plastic data form (Figure 3.11).

3.4.2.2 Section Properties

The section properties of a model are defined by creating sections in the "Property" module. After the section is created, one of the following two methods to assign the section to the parts can be used:

  i Select the region from the part and assign the section to the selected region, or
  ii Use the "Set" toolset to create a "Homogeneous" set containing the region and assign the section to the set.

Define a beam section for concrete:

From the main menu bar, select "Section → Create". The "Create Section" dialog box appears.

In the "Create Section" dialog box:

Name the section "Concrete".

In the "Category" list, select "Solid".

In the "Type" list, select "Homogeneous" (see Figure 3.12).

Click on "Continue". The "Edit Section" dialog box appears.

In the "Edit Section" dialog box:

Click the arrow next to the "Material" text box and scroll through the "Material" to view a list of available materials and to select the required materials. To define each material, it is necessary to do the same process for "Create Section" and then in the "Plane stress/strain thickness" field, accept default.

Click on "OK" (see Figure 3.13).

Define a truss section for steel and FRP rod:

From the main menu bar, select "Section → Create". The "Create Section" dialog box appears.

In the "Create Section" dialog box:

Name the section "FRP".

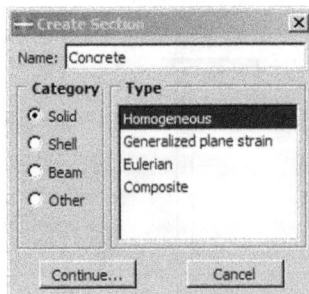

FIGURE 3.12   Define concrete section properties.

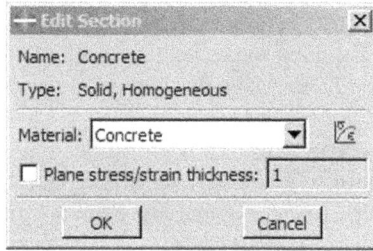

FIGURE 3.13 Edit concrete section properties.

FIGURE 3.14 Define steel and FRP section type.

In the "Category" list, select "Beam".

In the "Type" list, select "Truss" (see Figure 3.14).

Click "Continue". The "Edit Section" dialog box appears.

In the "Edit Section" dialog box:

Click the arrow next to the "Material" text box and scroll through the "Material" to see a list of available materials and to select the required material. To define each material, it is necessary to perform the same process as per "Create Section" and then determine the material by clicking on the "Continue". Insert the value of the "Cross-Sectional Area" of the "FRP" rod and Click OK (see Figure 3.15).

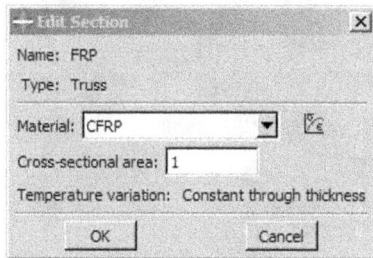

FIGURE 3.15 Edit steel and FRP section properties.

3.4.2.3 Section Assignment

Next, assign the defined section to the corresponding part. The Assign menu is used in the property module to assign the section to the parts. To assign the section to the beam, perform the following steps:

From the main menu bar, select "Assign → Section", the software displays prompts in the prompt area to guide the user throughout the procedure.

Alternatively, expand the menu under the "Beam" and double click on the "Section Assignments".

Select the entire part as the region that the section will be applied.

Click and hold down the left button of the mouse in the upper-left corner of the "viewport".

Drag the mouse pointer to create a box around the beam.

Release the left mouse button. The software highlights the entire beam.

Right Click on the viewport or click "Done" in the prompt area to accept the selected geometry. The "Assign Section" dialog box appears.

In "Section", scroll to "Concrete" and click "OK" as shown in Figure 3.16. The part changes color to green once the section is assigned, as shown in Figure 3.16.

3.4.3 Mesh Module

The finite element mesh should be generated in this module. When the default meshing technique is assigned to the model, the color of the model appears. If the software displays the model in orange color, it means that the user is needed to define the mesh.

3.4.3.1 Mesh

The "Mesh" module is used to generate the finite element mesh.

Assign an ABAQUS element type:

In the Module list located under the toolbar, click on "Mesh" to open the "Mesh" module.

At the context bar, click "Part", to unclick the assembly.

From the main menu bar, select "Mesh → Element Type".

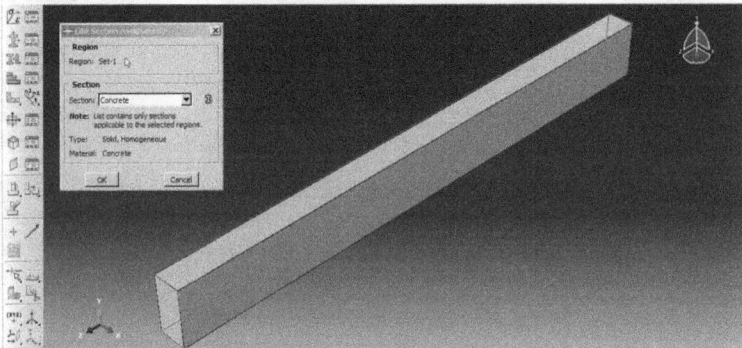

FIGURE 3.16 Section assignment to the beam.

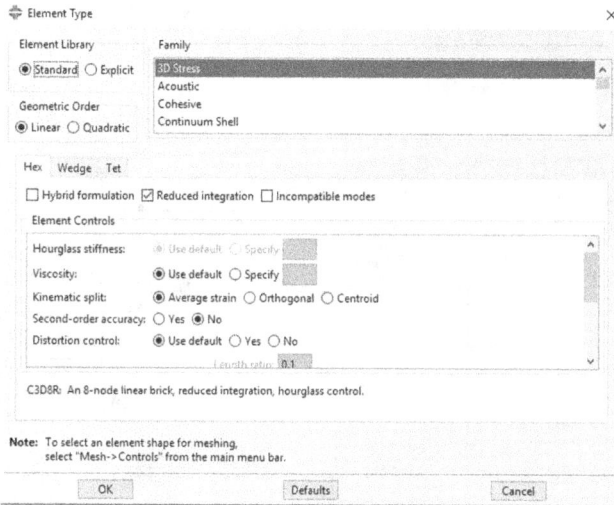

FIGURE 3.17   Select the element type.

In the viewport, select the entire frame as the region to be assigned an element type.

In the prompt area, click "Done". The "Element Type" dialog box appears, as shown in Figure 3.17.

In the dialog box, select the following:

- "Standard" as the "Element Library" selection (the default).
- "Linear" as the "Geometric Order" (the default).
- "3-D stress" as the "Family" of elements.

In the lower portion of the dialog box, examine the element shape options. A brief description of the default element selection is available at the bottom of each tabbed page.

Then, the mesh can be created. Meshing is basically a two-stage operation. First, seeding the part is required, and second, meshing the part.

Seed and mesh the model:

From the main menu bar, select "Seed → Part" to seed the part instance.

Alternatively, select the "Seed Part" in the upper-left corner of the meshing toolbox. The "Global Seeds" dialog box will appear, as shown in Figure 3.18.

Type the appropriate value for the approximate global size of the mesh elements. In this case, the user specifies the value of 100.

Click "OK" to accept the seeding.

The software offers a variety of meshing techniques to mesh models of different geometries, which includes structure meshing, sweep meshing, and free meshing as demonstrated in previous chapters.

Select the "Assign Mesh Controls" tool in the meshing toolbox, and the "Mesh Controls" dialog box appears, as shown in Figure 3.19.

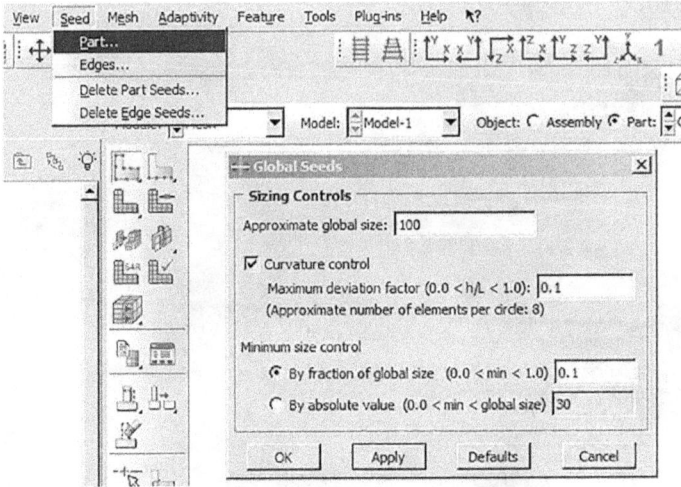

FIGURE 3.18 Assign the approximate global size for the mesh.

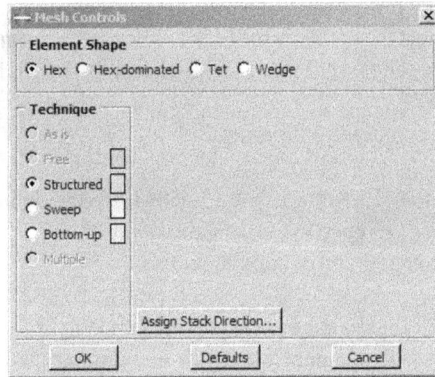

FIGURE 3.19 Choose the element shape and meshing technique option.

Choose "Hex" from the element shape and choose "Structured" meshing technique, as shown in Figure 3.19.

From the main menu bar, select "Mesh → Part" to mesh the part.

Once selection is done, click on "Yes" in the prompt area to confirm the mesh of the part instance. Once meshing is done, the plate color changes to blue and the meshed geometry is displayed as shown in Figure 3.20.

3.4.4 ASSEMBLY MODULE

In this module, all the parts that were created earlier are combined (assembly) to obtain the required model. The constraints and loads are also applied to the model once all the individual part instances are assembled.

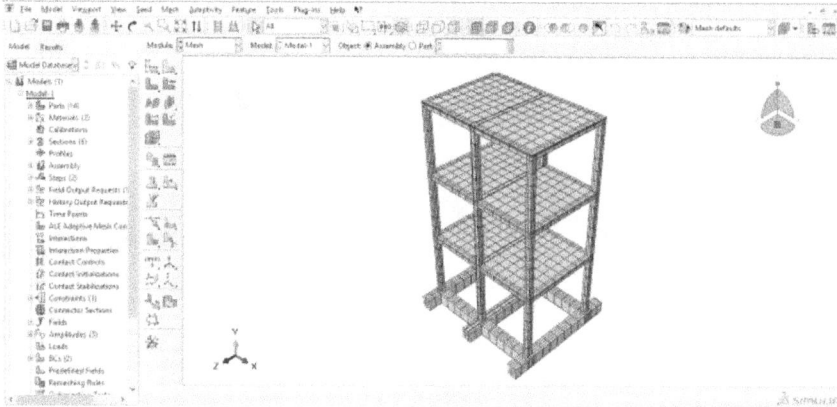

FIGURE 3.20 Meshed frame.

3.4.4.1 Assemble Part Instances into the Model

In the module list located under the toolbar, click "Assembly" to open the "Assembly module". The cursor changes to an hourglass while the "Assembly module" loads. From the main menu bar, select "Instance → Create". The "Create Instance" dialog box appears.

In the opened window, under the "Instance Type" box, choose "Dependent (mesh on the part)".

In the dialog box, select all the part instances and click "OK" (see Figure 3.21).

Use "Linear Pattern", "Translate", and "Rotate" tools to make the final assembly as shown in Figure 3.22.

3.4.5 INTERACTION MODULE

This module is used to define various interactions within the model or interactions between the regions of the model and its surroundings. The interaction used in this study is the Tie connection between all concrete parts. Another type of constraint utilized in this study is the embedded region, which has been used to represent the connection between the concrete and steel and CFRP bars.

3.4.5.1 Tie Constraint

Structural welding is a heating/fusing process, which enables the parts to be connected with the supplementary molten metal at the joint. A relatively small depth of material is molten, and upon cooling, the structural steel and weld metal acts as one continuous part where it is joined.

In ABAQUS software, fully constrained contact behavior can be defined using tie constraints. A tie constraint provides a simple way to bond surfaces together permanently and ties two separate surfaces together so that there is no relative motion between them. Moreover, it is a surface-based constraint using a master-slave formulation. The constraint prevents slave nodes from separation or

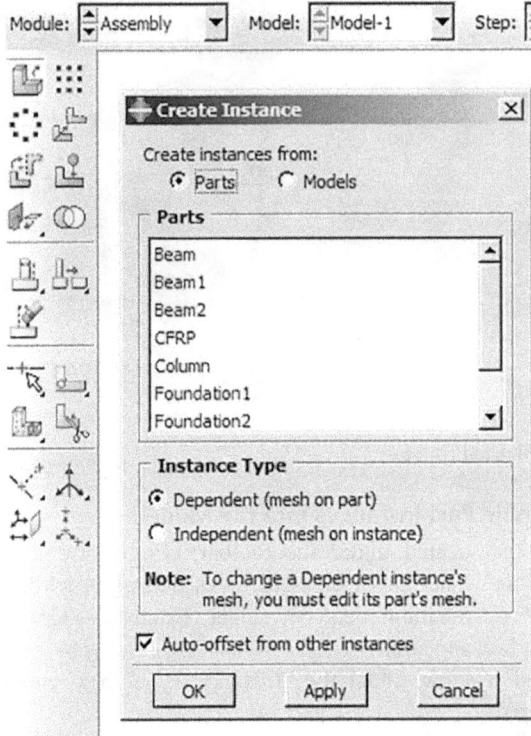

FIGURE 3.21   Assemble part instances into the model.

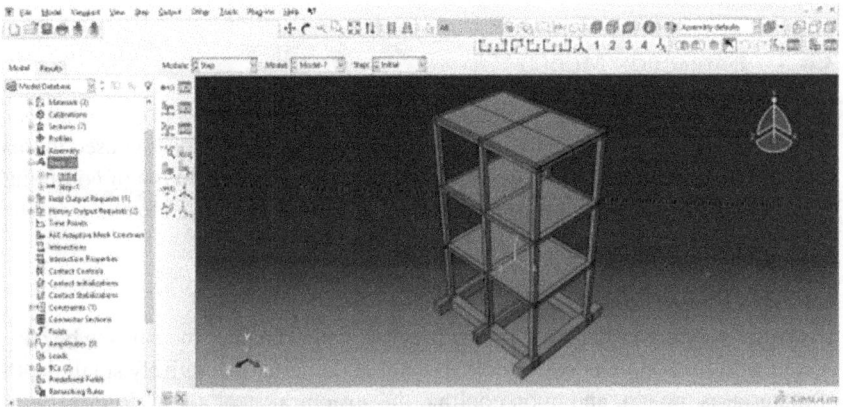

FIGURE 3.22   Complete assembly of the model.

relative sliding to the master surface. This type of constraint allows two regions to fuse together even though the meshes created on the surfaces of the regions may dissimilar. In this study, the tie connection has been used to demonstrate the contact between all concrete parts. It is required to make sure that in the dialog

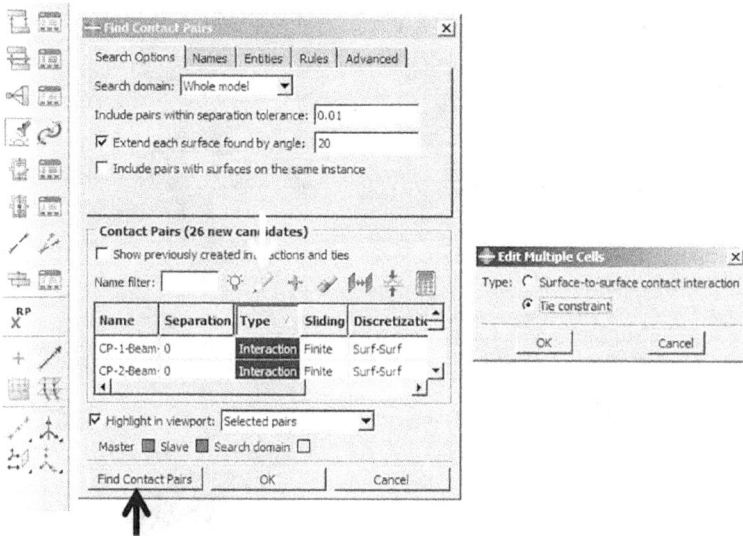

FIGURE 3.23 Find contact pairs.

box, the "Tie rotational degree of freedoms are used (if applicable)" is checked to allow the connection to have a possibility of rotation in the default value.

Tie surfaces:

In the module list located under the toolbar, click "Interaction" to open the "Interaction" module.

Select "Find Contact Pair" in the toolbox to open its dialog box. Then, click "Find Contact Pairs". The software tries to find all near faces and lists them in the dialog box. Click on the column "Type", then "Edit" to open the "Edit Multiple Cells" dialog box. In the box, select "Tie" constraint from the type and click on "Ok" to accept and close it (see Figure 3.23).

Finally, click "Ok" to close the "Find Contact Pairs" dialog box (see Figure 3.24).

3.4.5.2 Embedded Region Constraint

An embedded region constraint can be used to embed a region within a host region or within the whole model. An embedded region constraint can be created by specifying the embedded region, the host region, a weight factor round-off tolerance, and an absolute exterior tolerance or fractional exterior tolerance.

3.4.5.3 Create an Embedded Region Constraint

From the main menu bar, select "Constraint → Create" to create a constraint. The "Create Constraint" dialog box appears with a list of all types of constraints and a default constraint named "FRP".

Select "Embedded region" as the type of constraint and click on "Continue" (see Figure 3.25).

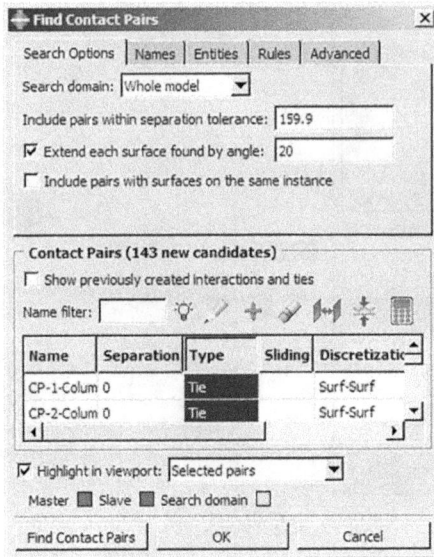

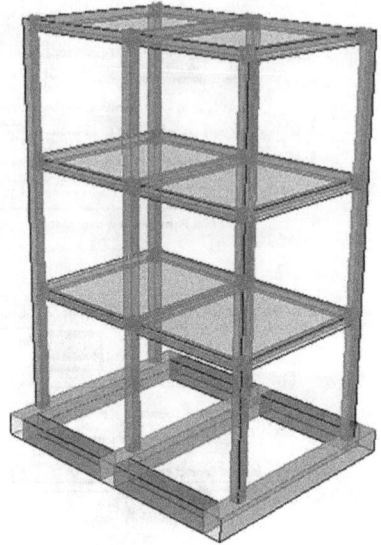

FIGURE 3.24   Tie constraint by find contact pairs definition.

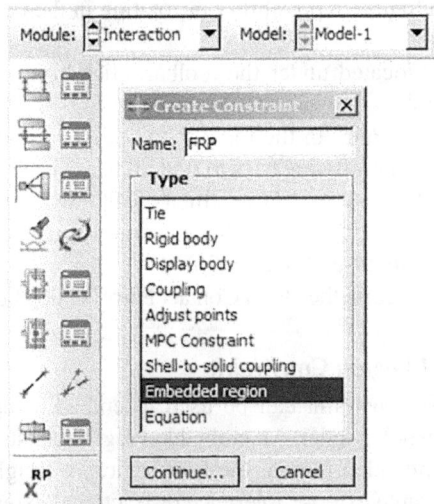

FIGURE 3.25   Create embedded region constraint.

Hide all concrete parts and then select all rods as the embedded parts. Then click on "Done" in the prompt area. Click on "Select Region" as a selection method for the host region, and choose all concrete parts. Finally, click on the middle mouse button to open the "Edit Constraint" dialog box.

Then, the "Edit Constraint" dialog box appears.

In the constraint editor, some options could be checked:

Weight factor round off tolerance: The user can specify a small value below the weighting factors which will be zeroed out. The default value is 10^{-6}.

Absolute exterior tolerance: The user can specify the absolute value to determine which nodes on the embedded region may lay outside the host region. If this parameter is omitted or has a value of 0.0, the fractional exterior tolerance will apply.

Fractional exterior tolerance: The user can specify the fractional value to define which node on the embedded region may lie outside the host region. The fractional value is based on the average element size within the host region. The default value is 0.05.

Leave the options unchanged and click on "OK" to define the constraint definition and to close the editor (see Figure 3.26).

3.4.6 STEP MODULE

After finishing the assembly and interaction definitions, the configuration of the analysis should then be defined.

In this simulation, the static response of the frame is considered as mentioned before, the software generates the initial step automatically, and the rest of the analysis step should be carried out by the user.

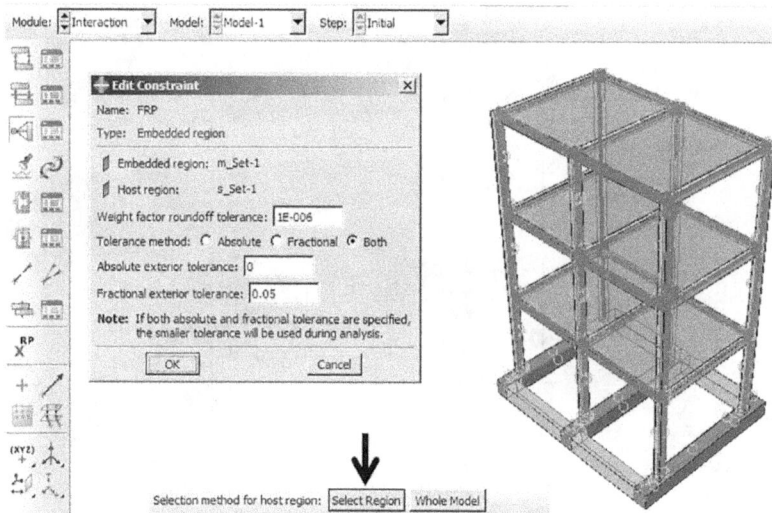

FIGURE 3.26 Concrete to bars embedded region constraint.

Create a static, general step that follows the initial step of the analysis. This module is used to perform many tasks, mainly to create analysis steps, and specify output requests.

3.4.6.1 Create an Analysis Step: Apply Load

In this simulation, the main aim is to evaluate the static response of the frame by the applied cyclic load at the top of the frame. Thus, the analysis consists of the following two steps:

- **The initial step:** The boundary conditions that constrain the end of the plate are applied as shown in Figure 3.27.
- **The analysis step:** A distributed load at the other end of the plate is applied.

The software generates the initial step automatically, but the "Step" module needs to be operated by the user in order to create the analysis step. The "Step" module also allows the user to request output for any steps in the analysis.

In the module list located under the toolbar, click on "Step" to open the "Step" module.

From the main menu bar, select "Step → Create" to create a step. The "Create Step" dialog box appears with a list of all general procedures and a default step named Step-1.

Change the step name to "Apply Load."

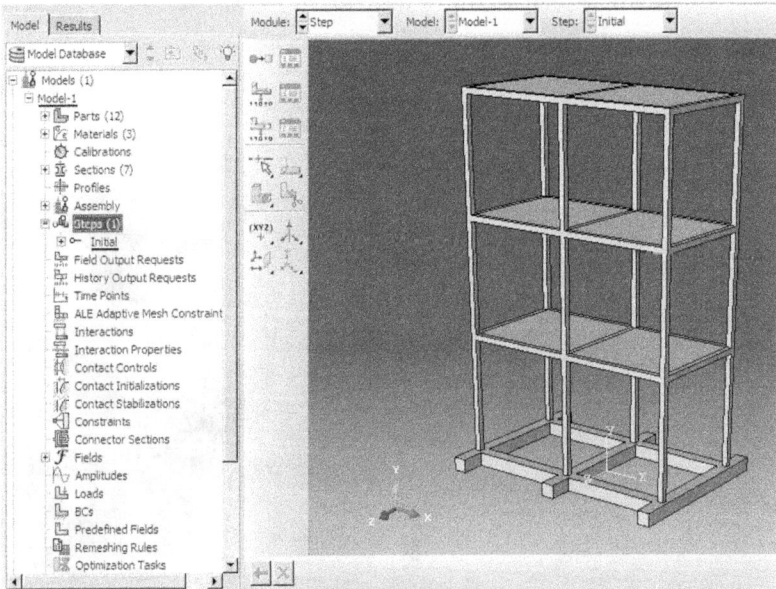

FIGURE 3.27   Initial step.

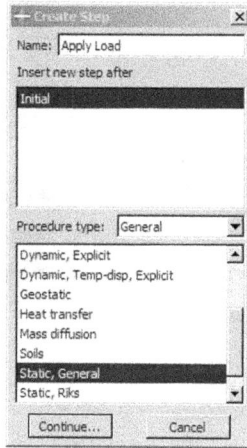

FIGURE 3.28   Analysis step.

Select "General" as the "Procedure type".

Scroll through the available list, select "Static, General" and click on "Continue" (see Figure 3.28).

Then, the "Edit Step" dialog box appears.

The "Basic" tab is selected by default.

In the "Description" field, type "This is a load step to apply cyclic load", then set the "Time period" to 8 and toggle "Nlgeom" On.

Click on the "Incrementation" tab and change the increment size, as shown in Figure 3.29. Click on "OK" to create the step and to exit the "Edit Step" dialog box.

At the amplitude field, a tabular form of the amplitude-time step is defined as shown in Figure 3.30. This option allows arbitrary time variations of load, displacement, and other prescribed variable values to be given throughout a step.

Create amplitude:

Under the "Model tree", double click on the "Amplitude".

The "Create Amplitude" dialog box appears.

Name the amplitude as "my data".

Select the "Tabular" type of amplitude and click on "Continue".

Input the incremental of the amplitude data, as shown in Figure 3.30.

Click on "OK" to exit from the dialog box.

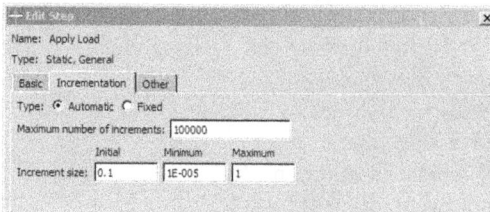

FIGURE 3.29   Step configurations and time incrementation.

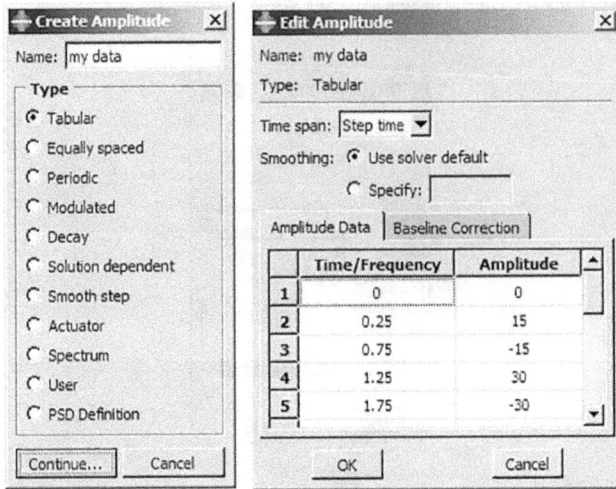

FIGURE 3.30 Create amplitude.

To obtain the required outputs at the visualization module, it is necessary to request the field outputs and history outputs in the "Step Module".

First two sets of regions should be defined. Select "Tools →Set→ Create" and name it "Displacement", then choose a top vertex as the region and click on "Ok". Repeat the same for the surfaces at the bottom of the foundations and name the set as "Reaction Force" (see Figure 3.31).

Double click on "History Output Requests" in the "Model tree". Name it "Displacement" for the current step and click on "Continue". Select "Set" as "Domain" and "Displacement" as the set name. Then, choose "U1" as the output variable. Do the same for the new history output name "Reaction Force" and "RF1" as the output variable (see Figure 3.32).

3.4.7 LOAD CONDITION MODULE

The prescribed conditions, such as loads and boundary conditions, are step dependent. Then, the load module can be used to define the prescribed conditions. In this model, the foundation is considered fully constrained, and the cyclic load is applied in the x-direction.

To apply boundary conditions to the frame, follow these steps:

In the module list located under the toolbar, click on "Load" to open the "Load" module.

From the main menu bar, select "BC → Create". The "Create Boundary Condition" dialog box appears.

In the "Create Boundary Condition" dialog box:

Name the boundary condition as "BC-1".

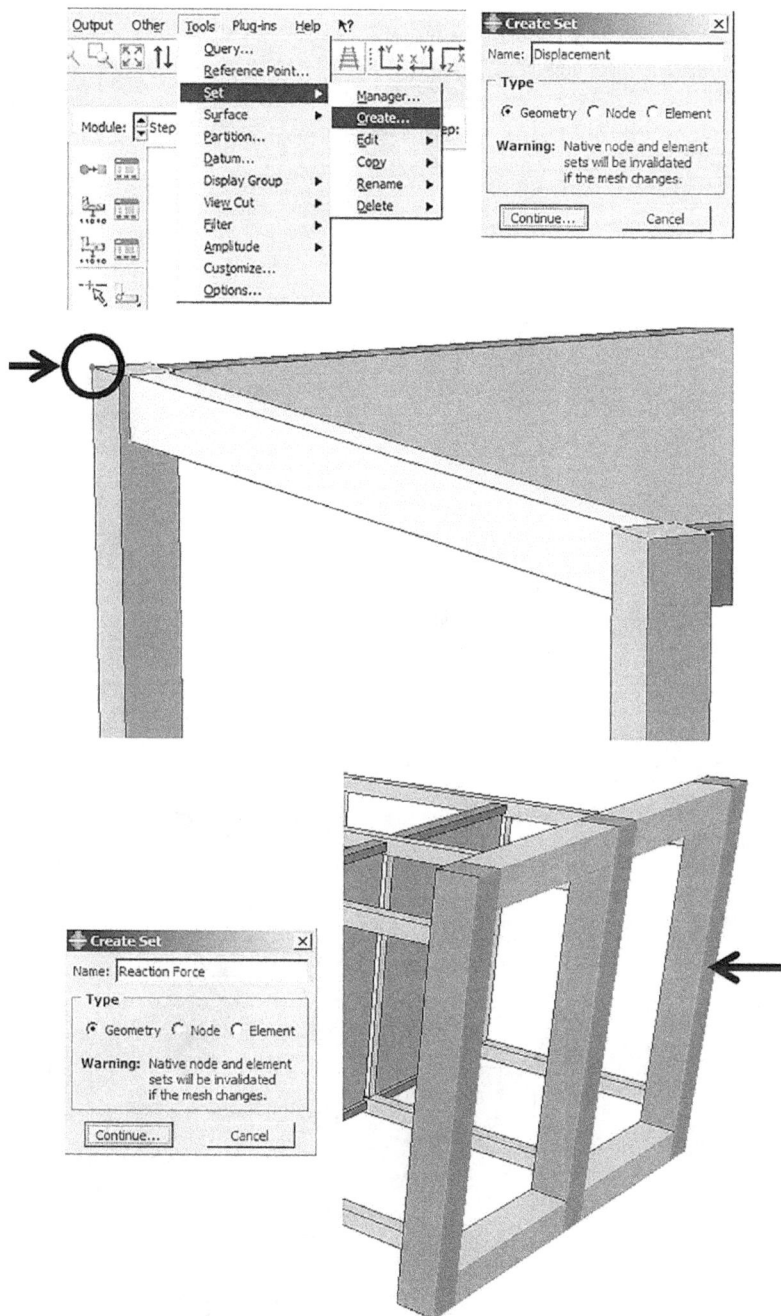

FIGURE 3.31 Create sets.

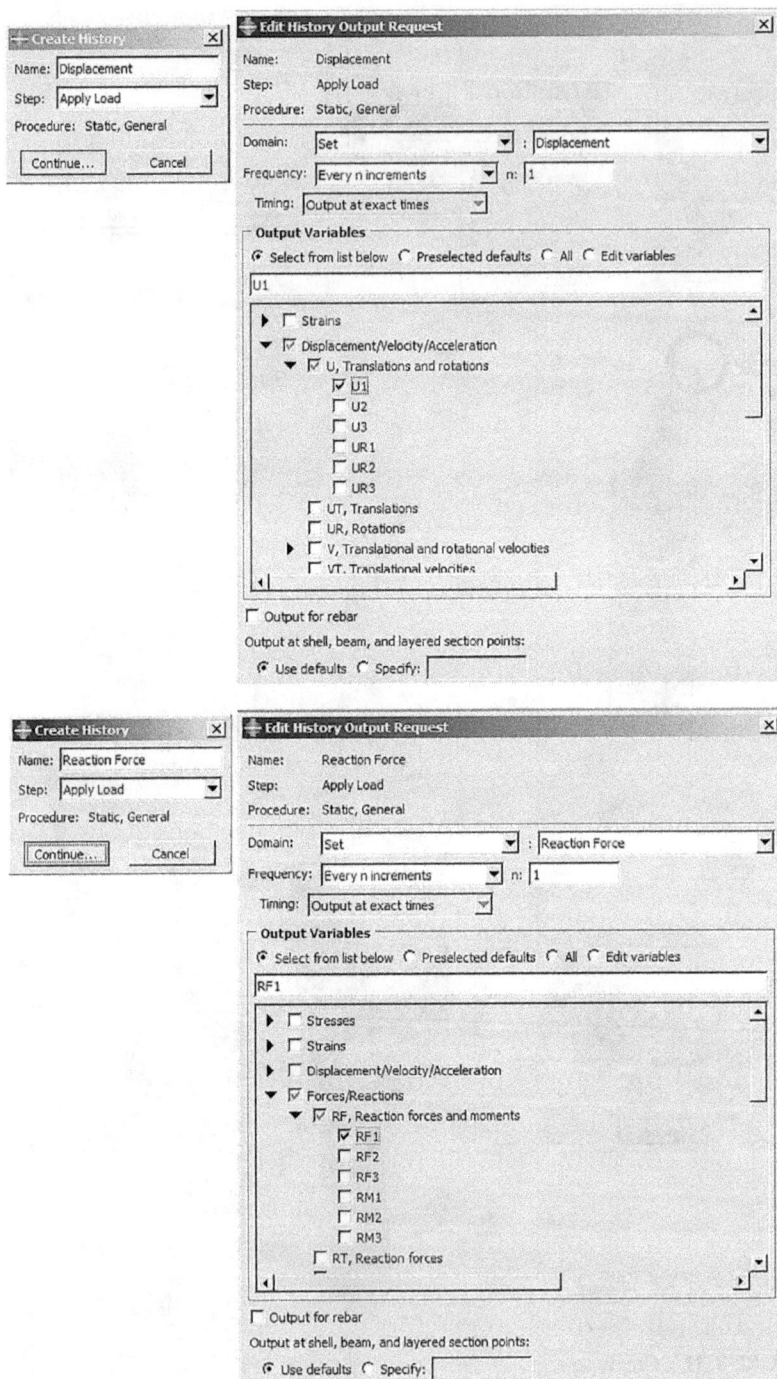

FIGURE 3.32   Edit history output request.

From the list of steps, select "Initial" to activate the boundary condition. All the mechanical boundary conditions specified in the "Initial" step must have zero values. This condition is enforced automatically by the software.

In the "Category" list, accept "Mechanical" as the default category selection.

In the "Types" for the selected step, select "Symmetry/Axisymmetry/Encastre" and click on "Continue". The software displays prompts in the prompt area to guide the user throughout the procedure (see Figure 3.33).

In the viewport, select the bottom of the Foundations. This is the region that the boundary condition needs to apply, as shown in Figure 3.34.

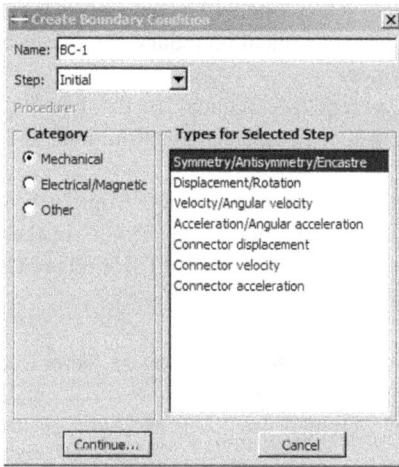

FIGURE 3.33   Create boundary condition.

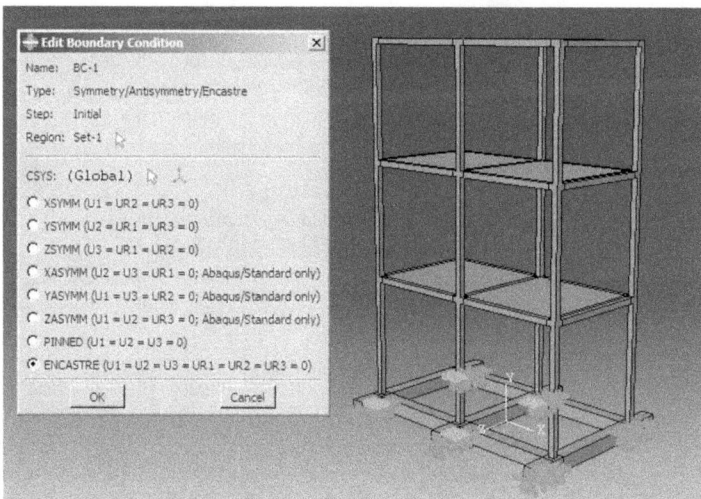

FIGURE 3.34   Edit boundary condition.

Select "ENCASTRE" since all the horizontal and vertical degrees of freedom need to be constrained as fixed support.

Click on "OK" to create the boundary condition and to close the dialog box.

The boundary conditions at the column base are shown in Figure 3.34.

3.4.7.1 Apply Cyclic Loading to the Frame

Again, select "BC → Create". The "Create Boundary Condition" dialog box appears.

In the "Create Boundary Condition" dialog box:

Name the boundary condition as "BC-2".

From the list of steps, select "Step-1" as the step that the boundary condition will be activated. All the mechanical boundary conditions specified in the Initial step must have zero values.

In the category list, accept "Mechanical" as the default category selection.

In the "Types" list, select "Displacement/Rotation" and click on "Continue" (see Figure 3.35).

In the viewport, select the node in the top floor to apply the load in the x-direction.

Right click on the viewport or click on "Done" in the prompt area to indicate the end of selecting regions. The "Edit Boundary Condition" dialog box appears.

In the dialog box:

Check "U1" and enter "1" as the multiplication factor to the amplitude created in "Step Module", as shown in Figure 3.36.

Choose "my data" (created in Step Module) as the "Amplitude" (see Figure 3.36).

Click on "OK" to save the load and to close the dialog box (see Figure 3.36).

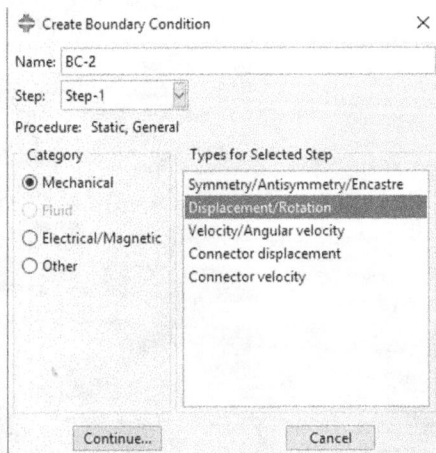

FIGURE 3.35 Create boundary condition.

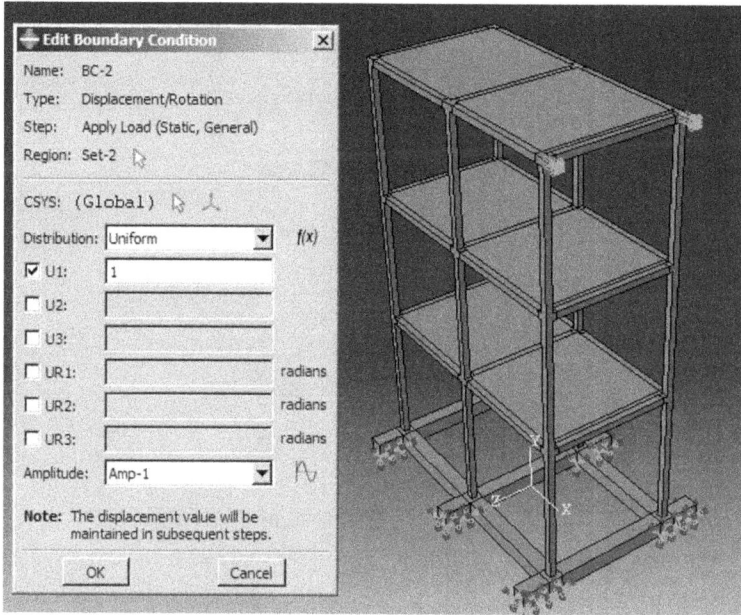

FIGURE 3.36   Edit the boundary condition to apply the cyclic load.

3.5   ANALYSIS: JOB MODULE

The "Job" module can be used to create and manage analysis jobs and submit them for analysis.

3.5.1   CREATE AN ANALYSIS JOB: JOB-1

Double click on "Jobs" in the "Model tree". The "Create Job" dialog box appears with a list of models in the model database.

Name the job as "Job-1" and click on "Continue" (see Figure 3.37).

The "Edit Job" dialog box appears. In the "Submission" tabbed page, select "Full Analysis" as the "Job Type" (see Figure 3.38).

Click on "OK" to accept all other default job settings in the job editor and to close the dialog box.

Click on "Submit" to start checking the input file and running the analysis (see Figure 3.39).

After job submission, the information in the Status column is updated to indicate the job's status. The Status column for the overhead hoist problem shows one of the following points (note that the analysis input file is being generated):

- **Submitted:** while the job is being submitted for analysis.
- **Running:** while the software analyzes the model.

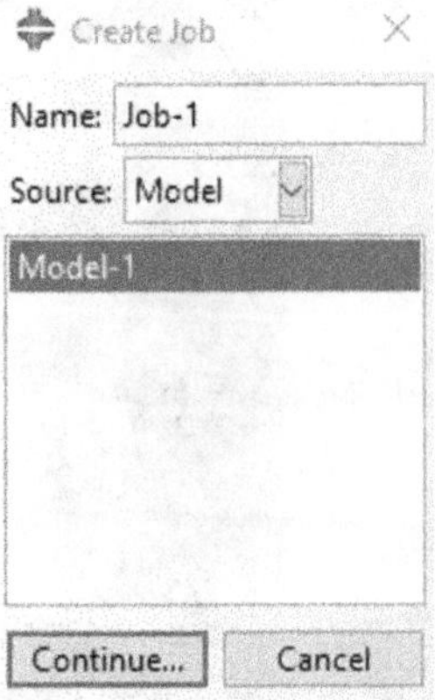

FIGURE 3.37   Define a job.

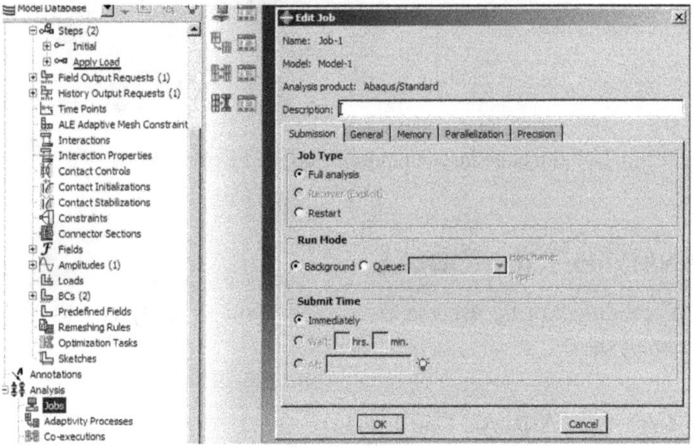

FIGURE 3.38   Create job.

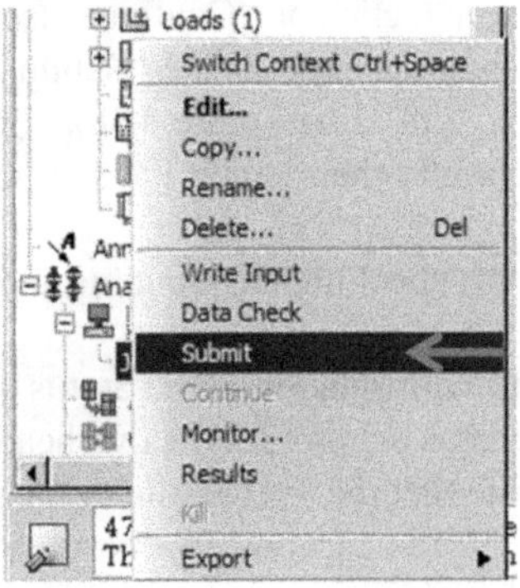

FIGURE 3.39   Submit Job-1 for analysis.

- **Completed:** when the analysis is done, and the output has been written to the output database, and the user can click on "Results" to proceed to the visualization module.
- **Aborted:** if the software finds a problem with the input file or the analysis and aborts the analysis. In addition, the software reports the problem in the message area.

3.5.2 Monitor the Solution in Progress

Click on "Monitor" to observe errors/warnings by right click on the job and select "Monitor".

The top half of the dialog box displays the information available in the status (*.sta) file that the software creates for the analysis.

If the status on top of the "Job Monitor" window is shown as "Completed", then this job is free from any errors and it was executed properly.

3.6 VISUALIZATION MODULE

Graphical postprocessing is one of the main features of the software to show the vast volume of data that created during simulation. For any complex problems, it is impractical to attempt to interpret the results in the tabular form of the data file and it is required to plot some graphs for the results. The ABAQUS software allows the user to view the results graphically using a variety of methods, including deformed shape plots, contour plots, vector plots, animations, and X–Y plots.

3.6.1 View the Results of the Analysis

When the job is completed, the user can view the results of the analysis through the "Visualization" module. The software loads the "Visualization" module, opens the output database created by the job, and displays a fast plot of the model. To open the "Visualization" module, right click on completed job and select "Results" as shown in Figure 3.40.

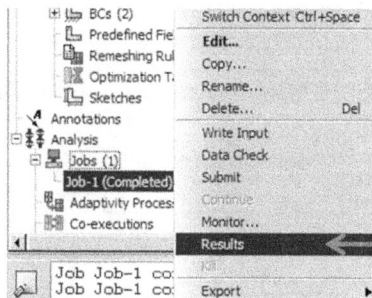

FIGURE 3.40   View the results of the analysis.

3.7   ANALYSIS RESULT

3.7.1   Hysteresis Results

From the main menu bar, select "Result → History output" or double click on
the "XY Data" from the "Result tree". Select "ODB history output" to open the
history output dialog box, as shown in Figure 3.41.

The "History Output" dialog box appears.

Select all the displacement results, as shown in Figure 3.42, and the graphs are
plotted in the viewport and click "Save As", the "Save XY Data As" dialog box
appears.

Name the data as "Displacement" as shown in Figure 3.43 and choose "Save
Operation" as "as is" and click "OK" to save the graph.

Again open "History Output" box and select all "Reaction Force: RF1"
variables (see Figure 3.44).

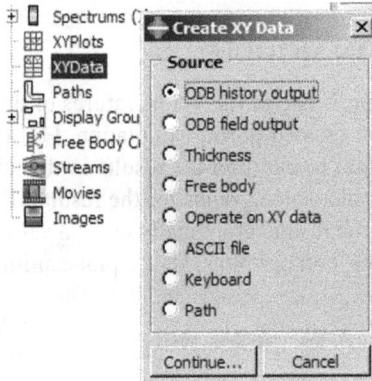

FIGURE 3.41   Create XY Data.

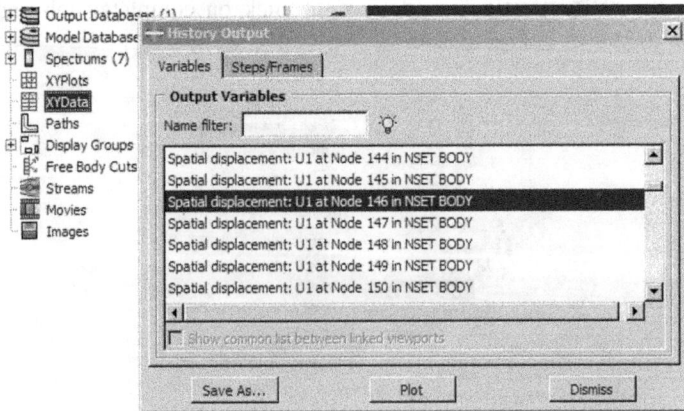

FIGURE 3.42   Select the displacement results.

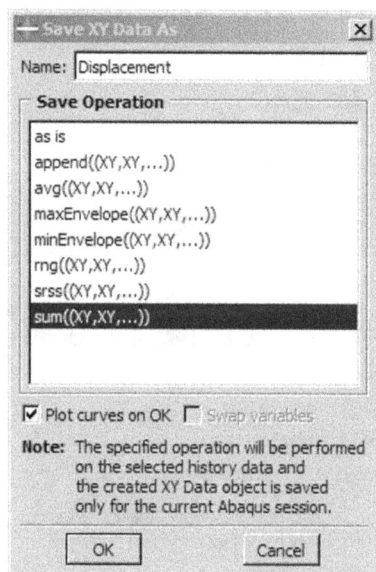

FIGURE 3.43 Save displacement diagram.

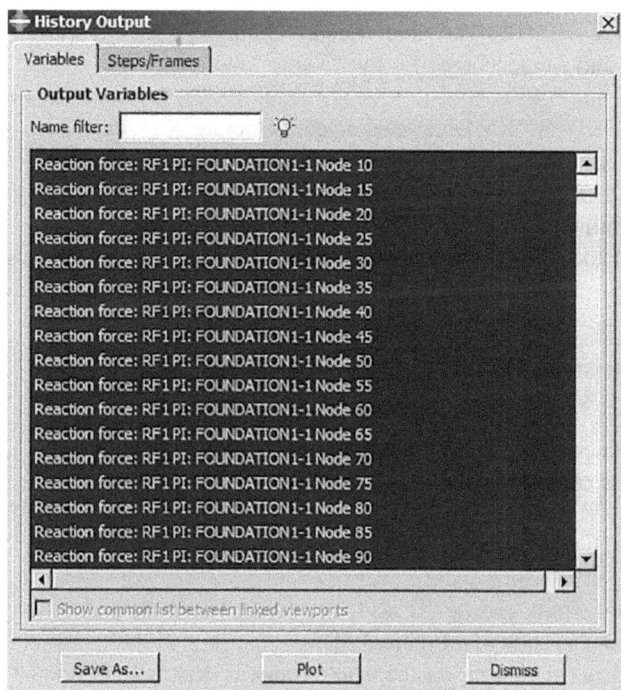

FIGURE 3.44 Select reaction forces results.

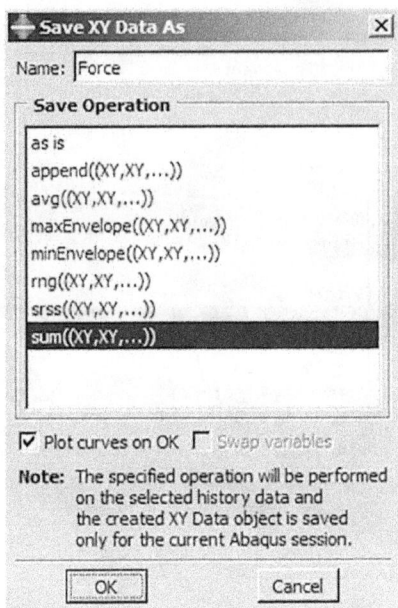

FIGURE 3.45   Save reaction forces as "Force".

Then click on "Save As" and choose "Sum ((XY,XY,...))" from the "Save Operation" and enter "Force" for the "Name" and click "Ok" to save the plot as shown in Figure 3.45.

Double click on "XY-Data" and select "Operate XY" data from source and click on "Continue" in the dialog box, and then choose "Combine" from "operation" and thereafter select "Displacement" and "Force" plots, respectively. Consider a minus before "Force" as depicted in Figure 3.46.

Click on the "Plot Expression" to plot the hysteresis diagram as shown in Figure 3.47.

3.7.2   STRESSES IN FRAME

The stress contour for the whole frame is shown in Figure 3.48.

The stress contour for concrete is shown in Figure 3.49.

The stress contour for the rods is presented in Figure 3.50.

3.7.3   TOTAL STRAIN IN FRAME

To show total strain, choose "E" as the "primary" variable and "Max. Principal" as its invariant as shown in Figure 3.51.

The total strain contour for the whole frame is depicted in Figure 3.52.

The total strain contour for concrete parts is shown in Figure 3.53.

The total strain contour for bars is depicted in Figure 3.54.

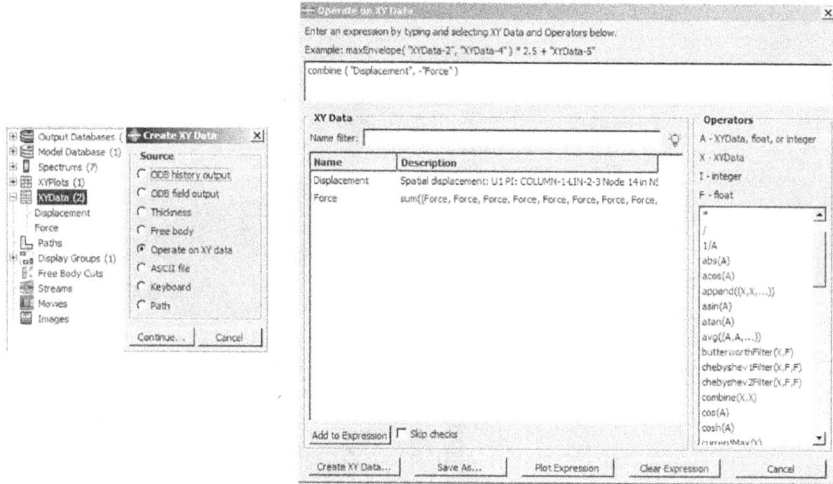

FIGURE 3.46   Create the Force–Displacement plot.

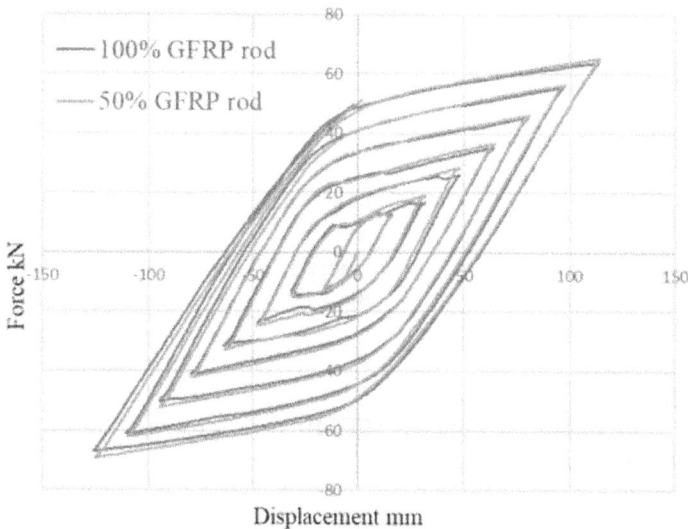

FIGURE 3.47   Hysteresis loops in the Force–Displacement plot.

3.7.4   Plasticity Contour Plots in Frame

Choose "PEEQ" as the "primary" variable as shown in Figure 3.55.

The plastic strain contour for the concrete parts is shown in Figure 3.56.

The plastic strain in rods is shown in Figure 3.57.

Interpretive Solutions for Dynamic Structures

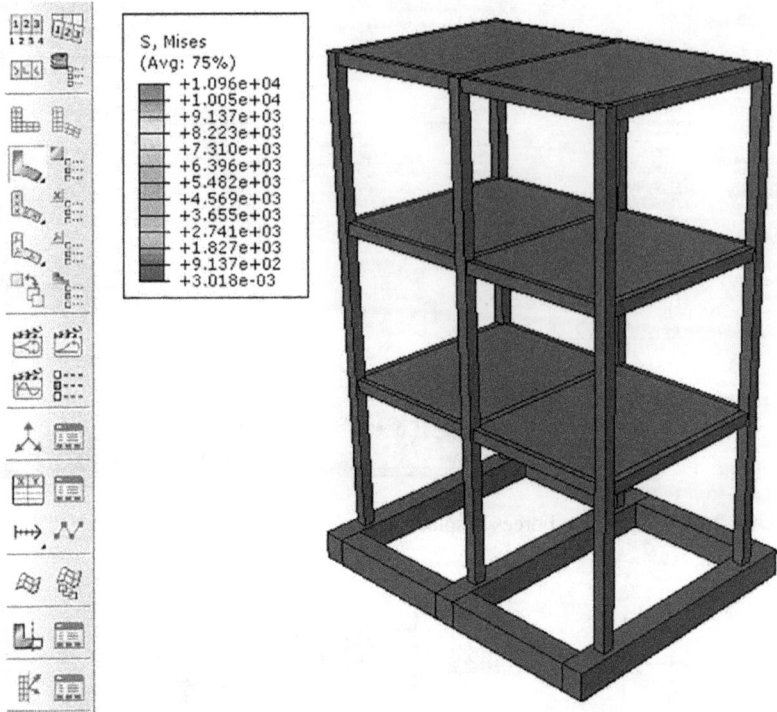

FIGURE 3.48 Von Mises stress contour plot in the whole model.

3.8 DISCUSSIONS

The numerical results have shown the superior performance of the CFRP rod in terms of ultimate loads, energy dissipation, ductility index, and plastic hinge formation, which is around 34% improvement for ultimate loads, 6.61% increase in energy dissipation, and 1.12 for ductility. Besides, frames reinforced with CFRP bars showed different behaviors compared to the frames reinforced with GFRP bars due to the low elastic modulus of GFRP bars, which lead to an increase in deflection dramatically.

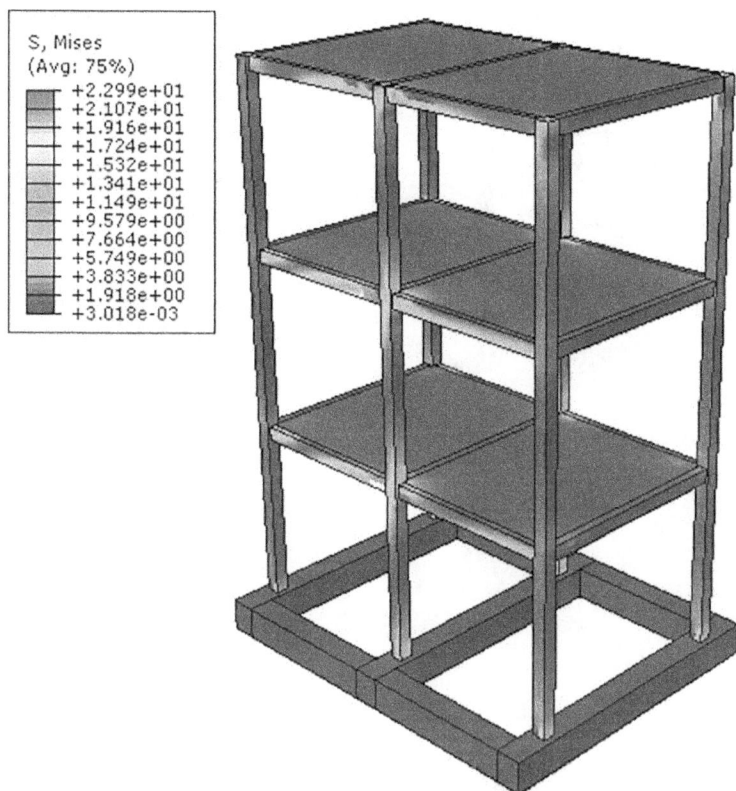

FIGURE 3.49 Von Mises stress contour plot in concrete parts.

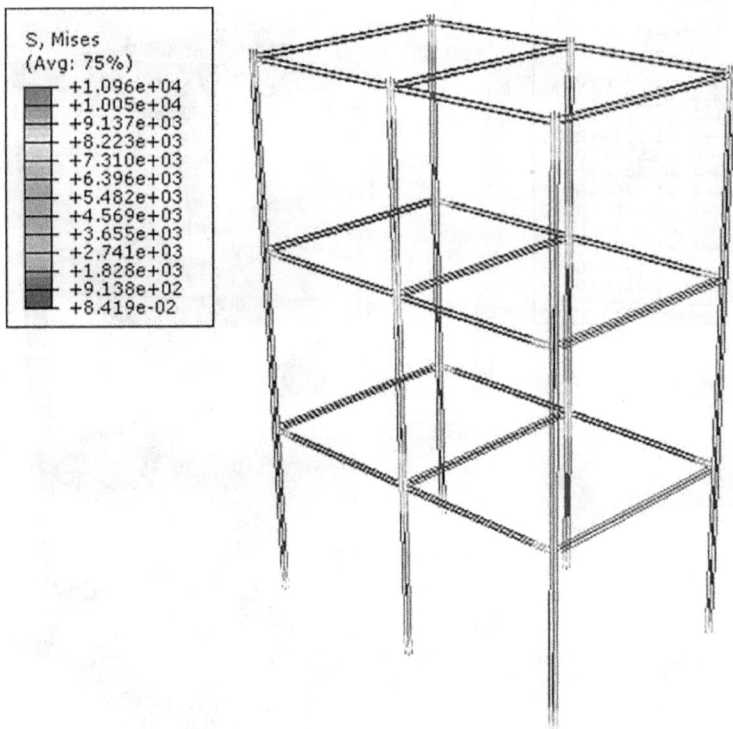

FIGURE 3.50   Von Mises stress contour plot in rods.

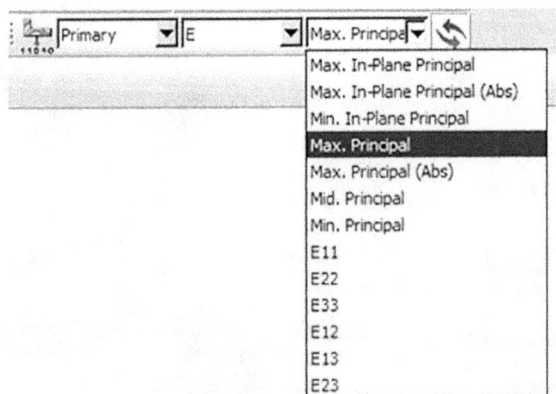

FIGURE 3.51   Set Maximum principal strain as the primary variable in contour plots.

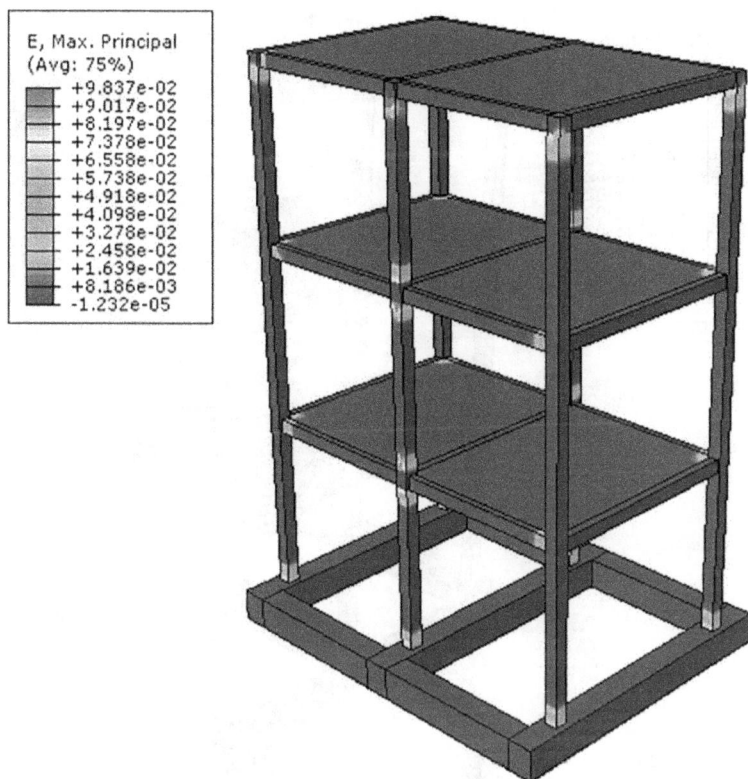

E, Max. Principal
(Avg: 75%)
+9.837e-02
+9.017e-02
+8.197e-02
+7.378e-02
+6.558e-02
+5.738e-02
+4.918e-02
+4.098e-02
+3.278e-02
+2.458e-02
+1.639e-02
+8.186e-03
-1.232e-05

FIGURE 3.52   Maximum principal strain in the whole model.

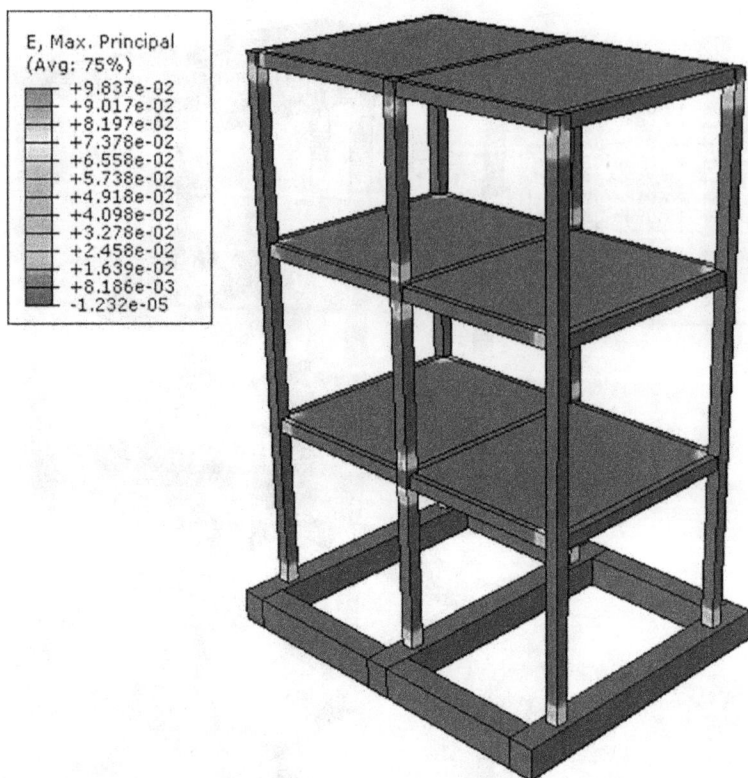

FIGURE 3.53   Maximum principal strain in concrete parts.

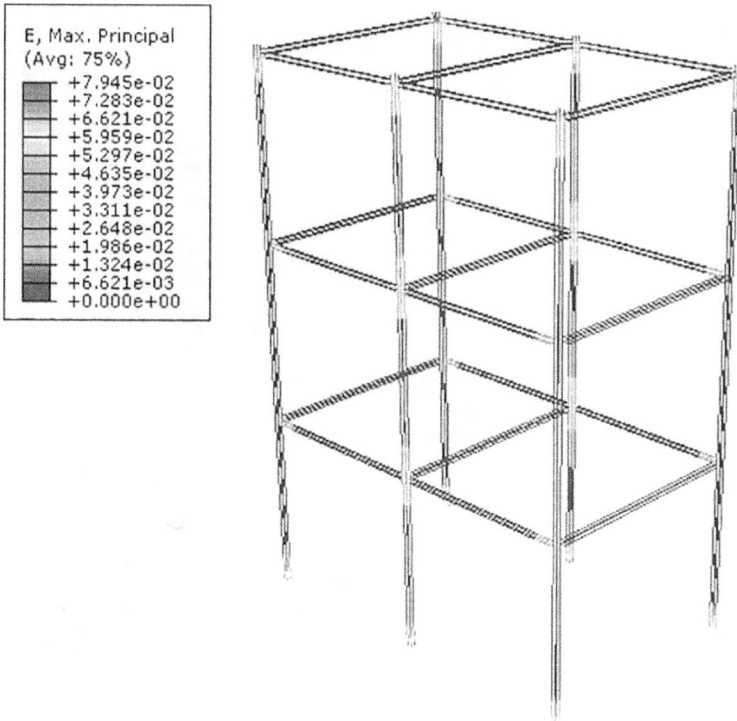

E, Max. Principal
(Avg: 75%)
+7.945e-02
+7.283e-02
+6.621e-02
+5.959e-02
+5.297e-02
+4.635e-02
+3.973e-02
+3.311e-02
+2.648e-02
+1.986e-02
+1.324e-02
+6.621e-03
+0.000e+00

FIGURE 3.54 Maximum principal strain in rods.

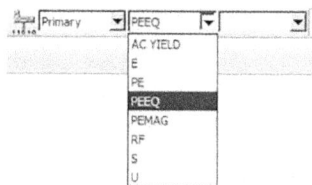

Primary   PEEQ
AC YIELD
E
PE
PEEQ
PEMAG
RF
S
U

FIGURE 3.55 Set Plastic strain contour plot as the primary variable in contour plots.

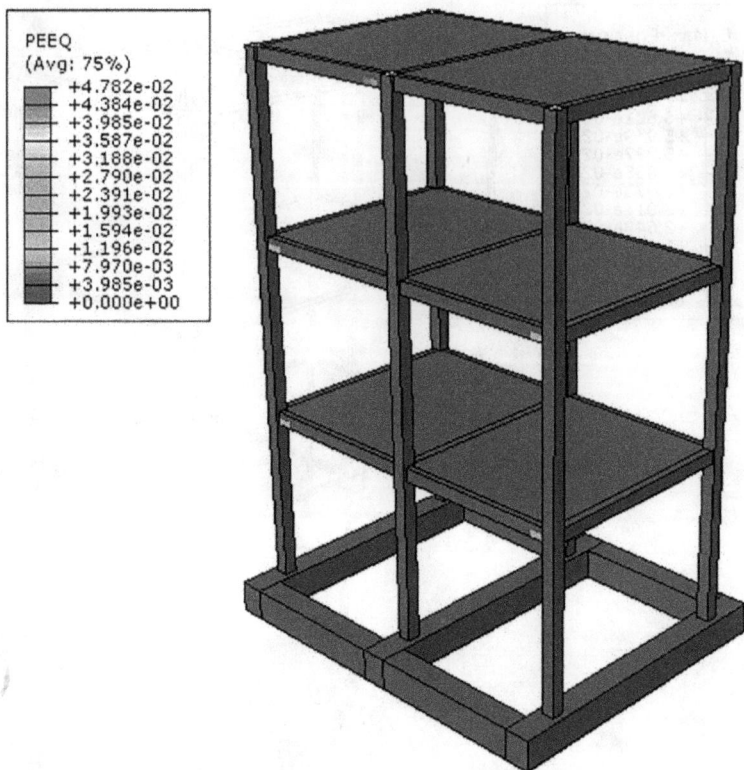

FIGURE 3.56 Plastic strain contour plot in concrete parts.

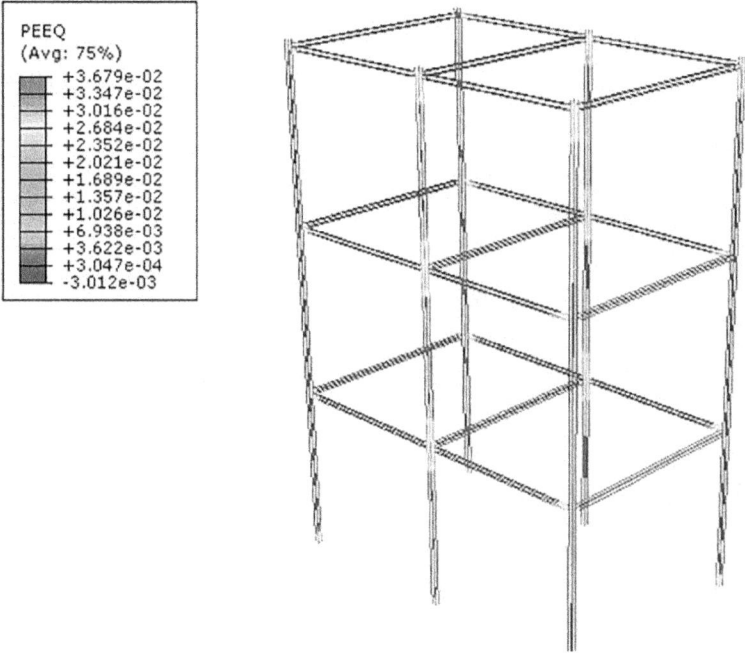

FIGURE 3.57 Plastic strain contour plot in rods.

4 Behavior of Precast Beam–Column Dowel Connection under Cyclic Loads

4.1 INTRODUCTION

This chapter presents an example for simulation of a precast beam–column connection and performing nonlinear cyclic load analysis using the ABAQUS/ Standard finite element program to study the force–displacement capacity of a dowel-connected precast frame under the action of cyclic loading. Both ends of the precast beam are connected to the precast column by a single dowel that protruded from the top of the column. Then, the force–displacement results for the frame obtained from numerical analysis are plotted.

Researcher: Woon Kai Siong (woonks@utar.edu.my)

4.2 PROBLEM DESCRIPTION

In this study, a precast frame with a special joint is considered. The precast beam and column reinforced concrete are connected by a single dowel bar at the corbel. Figure 4.1 illustrates the three-dimensional finite element model of the precast frame using the ABAQUS software.

4.2.1 GEOMETRIC PROPERTIES

The dimension and reinforcement details of the precast column, beam, rubber bearing pad, and dowel bar are presented in Table 4.1 and Figure 4.2. A 25 mm concrete cover is adopted in this model.

4.2.2 MATERIAL PROPERTIES

A "Concrete Damage Plasticity (CDP)" model and a "Classical Plastic" model are adopted to simulate the plasticity behavior of concrete "C25/30" and steel reinforcement. Hence, the material properties of concrete, rubber bearing pad, steel reinforcement, and dowel bar are listed in Tables 4.2–4.4, respectively.

DOI: 10.1201/9781003219491-4

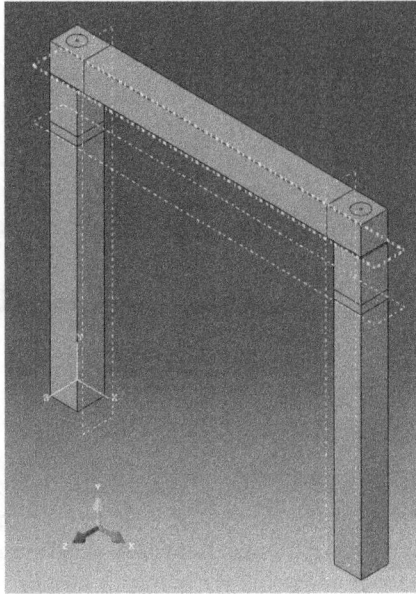

FIGURE 4.1   Three-dimensional model of the precast frame.

TABLE 4.1
Geometric Properties of the Components

Component	Dimension (mm)	Main Reinforcement	Shear Link
Precast column	2000 (height)×200 (width)×200 (thick)	4H20	R8–80
Precast beam	200 (height)×2300 (length)×200 (thick)	Top: 2H12 Bottom: 2H12	R8–100
Rubber bearing pad	200 (width)×200 (length)×10 (thick)	-	-
Dowel bar	Ø16 with 500 length	-	-

4.3   OBJECTIVES

- To develop the finite element model of the precast beam–column connection with a single dowel bar.
- To conduct the nonlinear analysis and study the deformation behavior of the precast beam–column connection with a single dowel bar.
- To identify the force–displacement hysteretic curve of the precast beam–column connection with a single dowel bar.

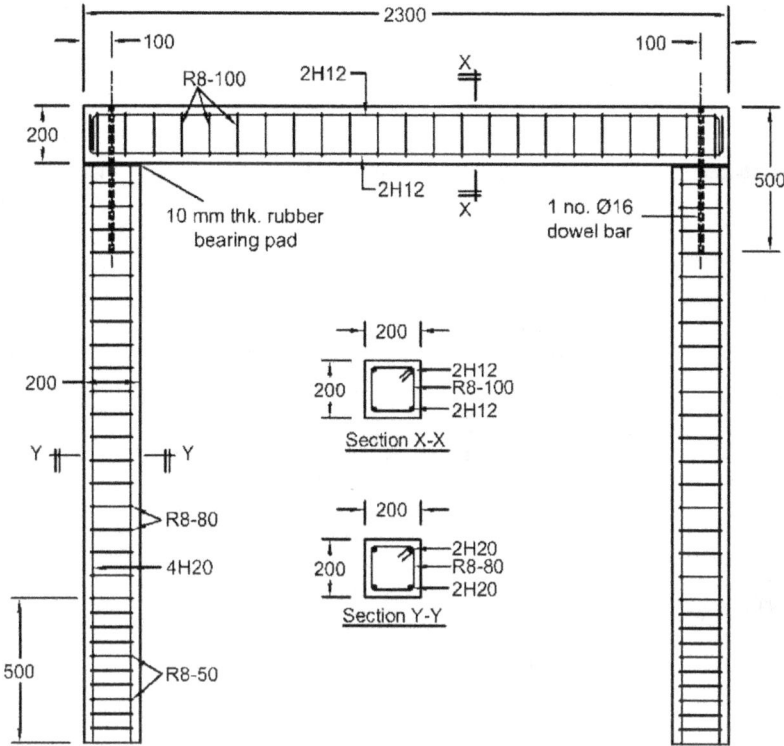

FIGURE 4.2 Details of the precast reinforced concrete frame.

TABLE 4.2
Properties of Concrete C25/30 for the CDP Model

Parameters of the Material C25/30		Parameters of the CDP Model	
Density [ton/mm³]	2.4×10^{-9}	Dilation angle	36°
Elasticity of concrete		Eccentricity	0.1
E [N/mm²]	26,600	f_{b0}/f_c	1.16
V	0.2	K	0.667
		Viscosity parameter	0.001

Concrete compression hardening		Concrete compression damage	
Stress [N/mm²]	Inelastic strain	Compression damage parameter	Inelastic strain
15	0	0	0
19.64	0.000306	0.22335	0.000306
24.67	0.000681	0.337494	0.000681
28.44	0.001181	0.433877	0.001181

(Continued)

TABLE 4.2 (*Continued*)
Properties of Concrete C25/30 for the CDP Model

Concrete compression hardening		Concrete compression damage	
Stress [N/mm²]	Inelastic strain	Compression damage parameter	Inelastic strain
29.86	0.001681	0.509579	0.001681
30	0.001932	0.543121	0.001932
29.89	0.002181	0.573888	0.002181
28.66	0.002931	0.653647	0.002931
26.25	0.003931	0.734294	0.003931
23.83	0.004931	0.792499	0.004931
21.67	0.005931	0.834764	0.005931
19.81	0.006931	0.865923	0.006931
18.22	0.007931	0.889323	0.007931
16.85	0.008931	0.907233	0.008931
15.68	0.009931	0.921189	0.009931
14.66	0.010931	0.932247	0.010931
13.88	0.011931	0.941142	0.011931
12.99	0.012931	0.948396	0.012931
12.29	0.013931	0.954384	0.013931
11.67	0.014931	0.959382	0.014931
11.11	0.015931	0.963595	0.015931
10.6	0.016931	0.96718	0.016931
10.15	0.017931	0.970254	0.017931
9.73	0.018931	0.972911	0.018931
9.35	0.019931	0.975222	0.019931
9	0.020916	0.977216	0.020916

Concrete tension stiffening		Concrete tension damage	
Stress [N/mm²]	Cracking strain	Tension damage parameter	Inelastic strain
2.9	0	0	0
2.23	2.75×10^{-5}	0.23	2.75×10^{-5}
1.3	0.00033	0.55	0.00033
0.29	0.000846	0.9	0.000846

TABLE 4.3
Properties of the Rubber Bearing Pad

Property	Density (ton/m³)	Young's Modulus (N/mm²)	Poisson Ratio
Value	1.23×10^{-5}	300	0.49

TABLE 4.4
Properties of Steel Reinforcement and Dowel Bar

Property	Density (ton/m³)	Young's Modulus (N/mm²)	Poisson Ratio	Yield Stress (N/mm²)	Failure Stress (N/mm²)	Plastic Strain at Yield	Plastic Strain at Failure
High tensile steel and dowel bar	7.85×10^{-9}	200,000	0.3	400	460	0	0.156
Mild steel	7.85×10^{-9}	200,000	0.3	200	250	0	0.156

4.4 MODELING

4.4.1 PART MODULE

In this study, the modeling of the precast reinforced concrete frame comprises parts, representing a "Precast Beam", "Precast Column", "Steel Reinforcement", "Dowel", and "Rubber Pad".

4.4.1.1 Create a New Model Database

Start ABAQUS/CAE software from programs in the "Start menu". Select "With Standard/Explicit Model" from the "Start Session" dialog box that appears. This step allows the user to start modeling whereby the user can create a new file and save it under any name in a new folder.

When the "Part" module has finished loading, it displays the "Part" module toolbox on the left side of the software main window. Each module displays its own set of tools in the module toolbox.

Precast beam:
Figure 4.3 shows the dimension of the precast beam in the software.

4.4.1.2 Create a New Model Database and a New Part

From the menu bar, select "Part → Create" in order to create a new part.

The "Create Part" dialog box appears. Use the "Create Part" dialog box to name the part; to choose its modeling space, type, and base feature; and to set the approximate size. The name of the part may be edited once it has been created, but the modeling space, type, or base feature cannot be changed.

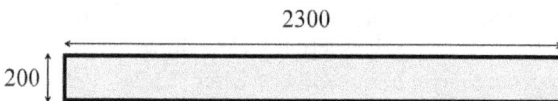

FIGURE 4.3 Dimension of the precast beam.

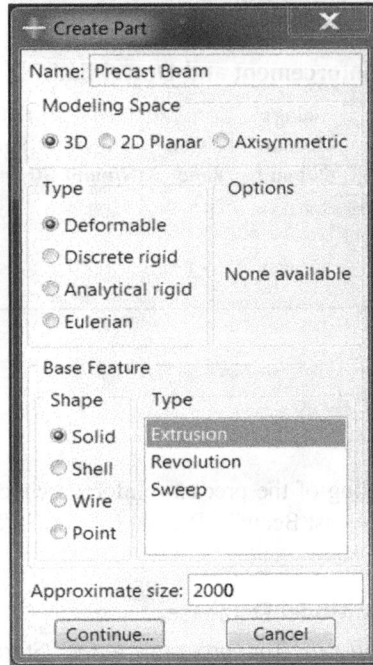

FIGURE 4.4 Create the precast beam.

Name the part "Precast Beam". Choose "3D" from "Modeling space", "Deformable" from "Type", "Solid" from "Shape", and "Extrusion" from Type as the base feature. Enter an "Approximate size" of "2000" (see Figure 4.4). The value entered in the approximate size text field at the bottom of the dialog box sets the approximate size of the new part. Click "Continue" to enter sketcher.

4.4.1.3　Define Section of the Precast Beam with Dimension

Use the "Create Lines: Rectangle (4 lines)" tool which is located at the upper-right corner of the sketcher toolbox to begin drawing the geometry of the precast beam, as shown in Figure 4.5.

The user can choose a starting corner for the rectangular at the viewport or enter the x- and y-coordinates. Then, select the opposite corner for the rectangle or enter the x- and y-coordinates. Instead of using known coordinates, the user can also define the dimension of the geometry by clicking on the add dimension tool. Once finished sketching the section for the dimension, right click and click on "Cancel Procedure" to exit the sketcher. Click "Done" in the prompt area and it will be displayed, as shown in Figure 4.6.

The user needs to define "200" as the depth for the extrusion of the precast beam. Once the depth of the precast beam has been defined, it will be displayed, as shown in Figure 4.7.

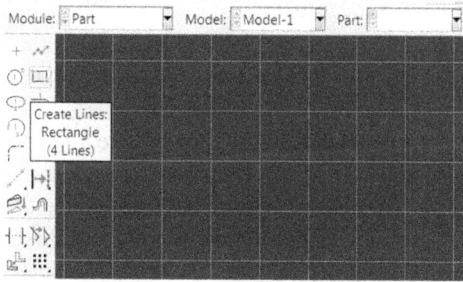

FIGURE 4.5 ABAQUS software sketcher to draw precast beam section.

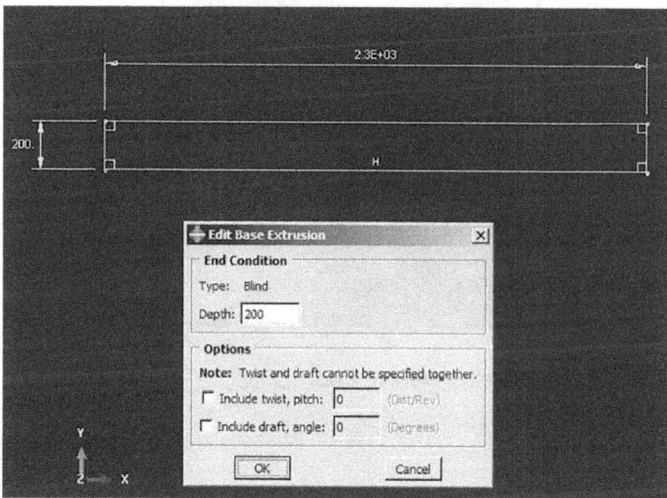

FIGURE 4.6 Edit base extrusion.

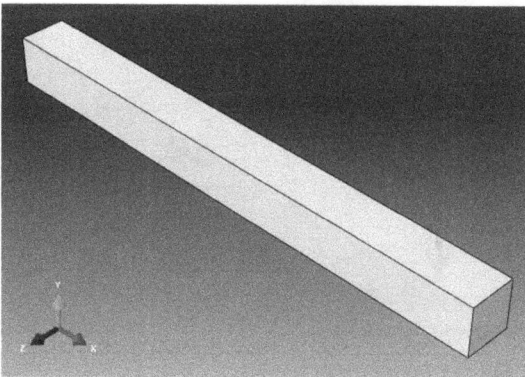

FIGURE 4.7 Precast beam.

4.4.1.4   Create a Circular Hole

A circular hole needs to be created in order to allow the dowel bar to slot into the precast beam. Select "Cut → Circular Hole" from the shape menu, as shown in Figure 4.8.

Click on "Through All" to choose the type of hole and then select the top surface of the precast beam as the plane of the hole, as shown in Figure 4.9.

Click "OK" to agree upon the arrow shows the direction of the hole. Next, select the top plane of the left edge as the first edge to locate the hole with a distance of "100 mm". Then, select the top plane of the bottom edge as the second edge to locate the hole with a distance of "100 mm". The selected edges are highlighted in pink color. Insert "16 mm" as the diameter of the hole. Subsequently, a through-hole is created in the precast beam, as shown in Figure 4.10. Repeat the same procedure to create another circular hole at the other end of the precast beam.

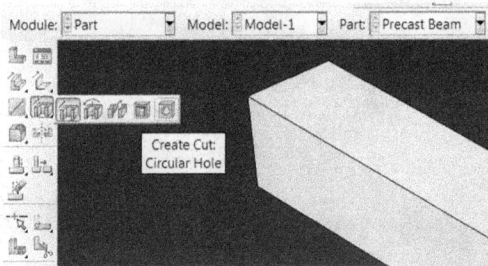

FIGURE 4.8   Create a circular cut hole.

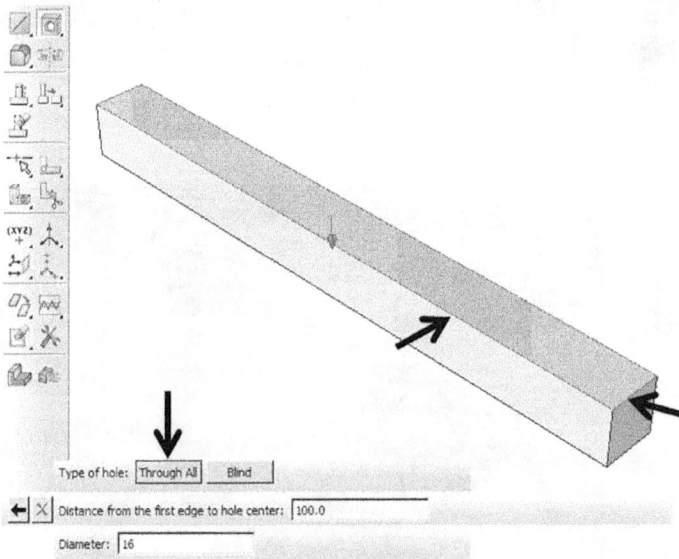

FIGURE 4.9   Select the top surface as the plane of the hole.

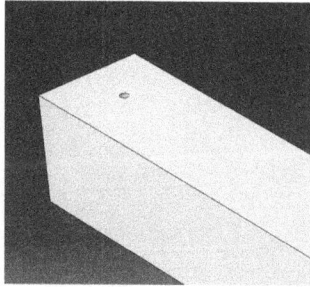

FIGURE 4.10 A circular hole is created on the precast beam.

4.4.1.5 Create Partition

In this example, the presence of prefabricated holes on the concrete beam and concrete column creates irregularities in the geometry, which causes complexity in meshing. In order to simplify meshing, it is suggested to create the partition lines at a distance of 250 and 2050 mm from the left side of the beam, respectively, so that the generated mesh elements are uniform in shape and smooth and also avoid meshing errors. The advantage of partitioning the part's cell/surface is that it allows the user to select an appropriate area of interaction surfaces easily and to manage changes of mesh sizes throughout the entire model. A line needs to be created in order to partition the cell part. The procedures to create a partition using the datum plane approach include the following steps:

Click on "Tools → Datum → Plane → Offset from Plane" from the toolbox, as shown in Figure 4.11.

Click on the left side of the beam to offset the principal plane, as shown in Figure 4.12, with an offset value of "250 mm".

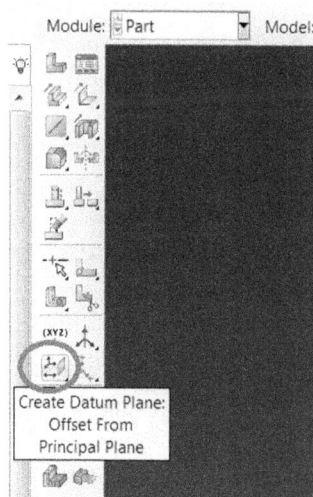

FIGURE 4.11 Create a datum plane.

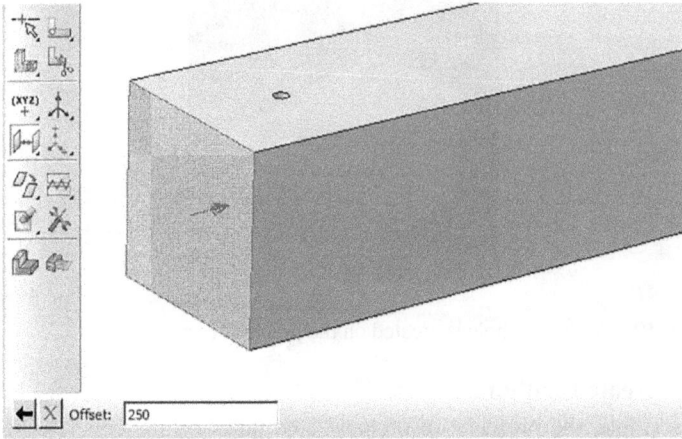

FIGURE 4.12   Select the left side plane to be offset.

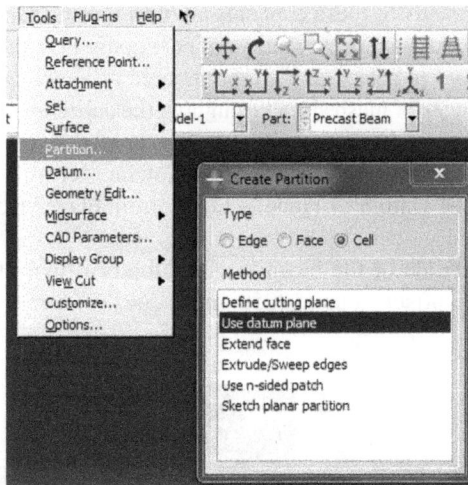

FIGURE 4.13   Create a partition using a defined datum plane.

From the main menu bar, select "Tools → Partition". The "Create Partition" dialog box appears.

Select "Cell" from "Type" to create a partition and apply the "Use datum plane" from method, as shown in Figure 4.13.

Select the datum plane that needs to create the partition along the plane, as shown in Figure 4.14. Click "Create Partition" to complete the definition of the partition.

Repeat the same steps to create a partition at the other end of the precast beam by selecting the cell that needs to be partitioned with an offset value of "2050 mm" from the "YZ plane".

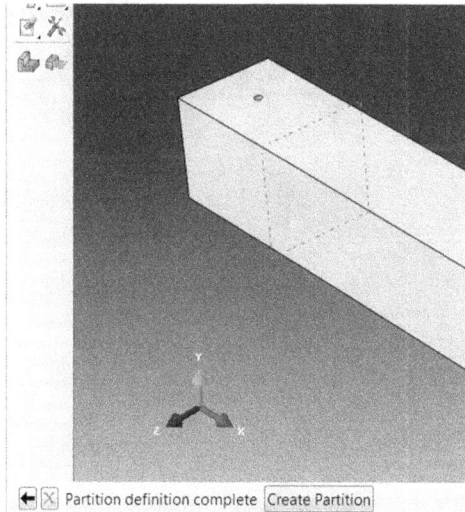

FIGURE 4.14 Complete definition of partition.

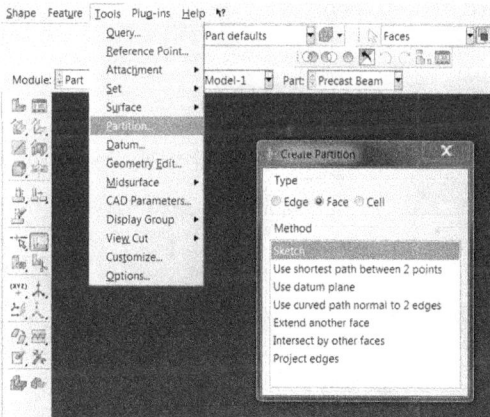

FIGURE 4.15 Define face sketch partition.

The procedures to create a partition using the sketch surface approach include the following steps:

From the main menu bar, select "Tools → Partition". The "Create Partition" dialog box appears.

Select "Face" from "Type" to create a partition and apply the "Sketch" from "Method", as shown in Figure 4.15.

Select the faces of the cell to draw the partition line. The "Sketch Origin" can be defined as "Auto-Calculate", as shown in Figure 4.16.

Then, select the proper edge or axis to be defined as vertical on the right to open the sketcher.

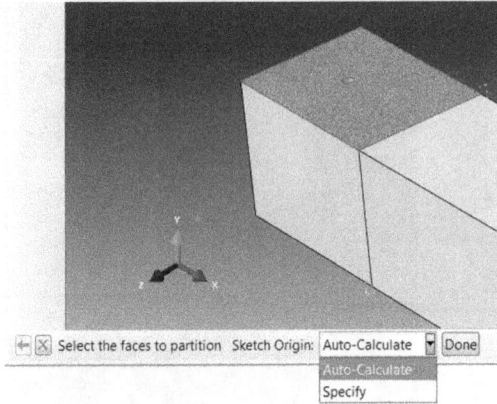

FIGURE 4.16 Select the face to be partitioned.

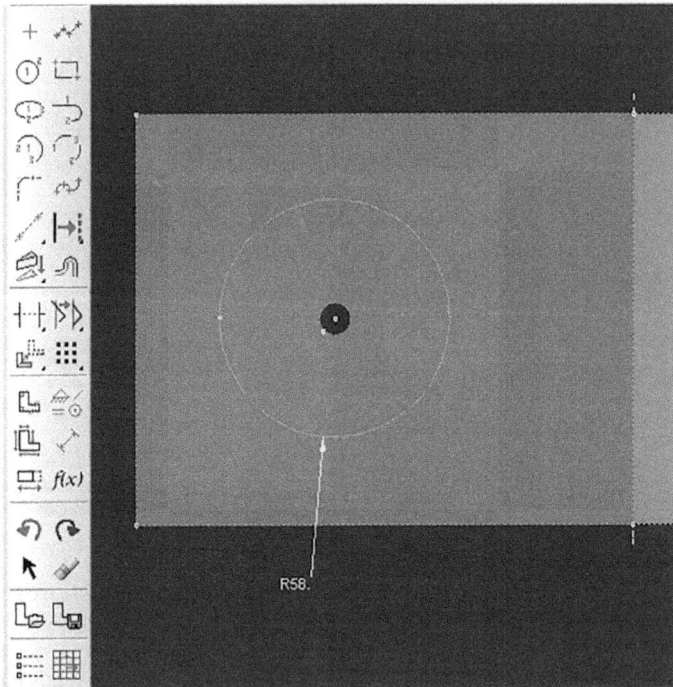

FIGURE 4.17 Draw a circle in the sketcher.

Click "Create Circle: Center and Perimeter" from the toolbox to draw a circle on the selected surface, as shown in Figure 4.17.

In order to have good meshing around the circular hole, it is recommended that the distance between the inner and outer ring is approximately "1–2× desirable

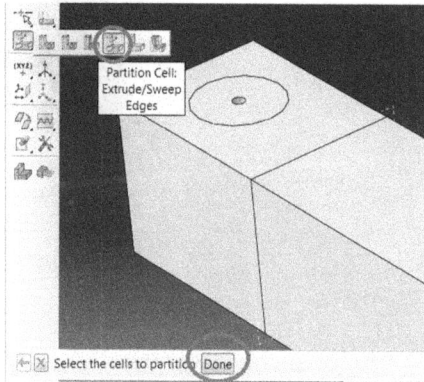

FIGURE 4.18 Define extrude edge partition.

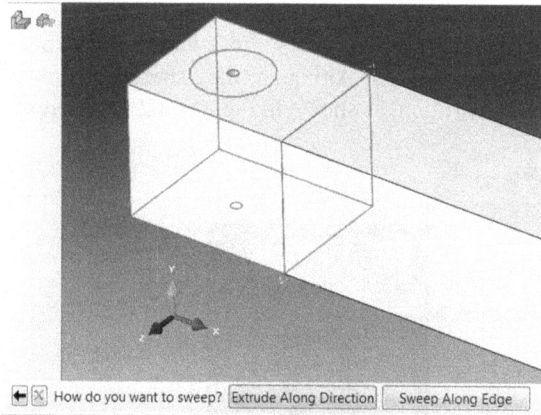

FIGURE 4.19 Use extrude along direction.

mesh size". Hence, the proposed radius of the circle is set to "58 mm". Click "Done' once sketching the partition geometry is completed.

Select "Partition Cell: Extrude/Sweep Edges" from the toolbox, as shown in Figure 4.18. Select the cell to be partitioned and click "Done". Select the drawn circular line as the edge to sweep individually, then click "Done".

Click "Extrude Along Direction", as shown in Figure 4.19.

Then, select any edge of the cell that is parallel to the required extrudes direction, which is along the y-axis in this case. Once the arrow shows the selected extrude direction, click "OK", and then click "Create Partition" to complete the definition of the partition. Figure 4.20 shows the partitioned precast beam under the Render Model: Wireframe view.

Repeat the same steps to create an extrude edge partition at the other end of the precast beam by selecting the cell that wants to be partitioned.

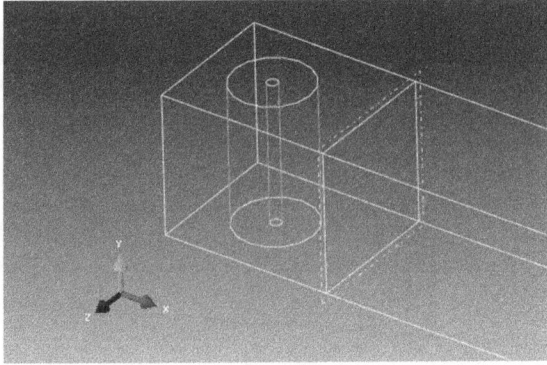

FIGURE 4.20   Wireframe view of the partitioned precast beam.

Precast column, rubber bearing pad, and dowel bar:

The steps for creating the precast column, rubber bearing pad, and dowel bar are similar to the preset beam. Although, they should be drawn with different dimensions and the partition, as shown in Figures 4.21–4.23, respectively.

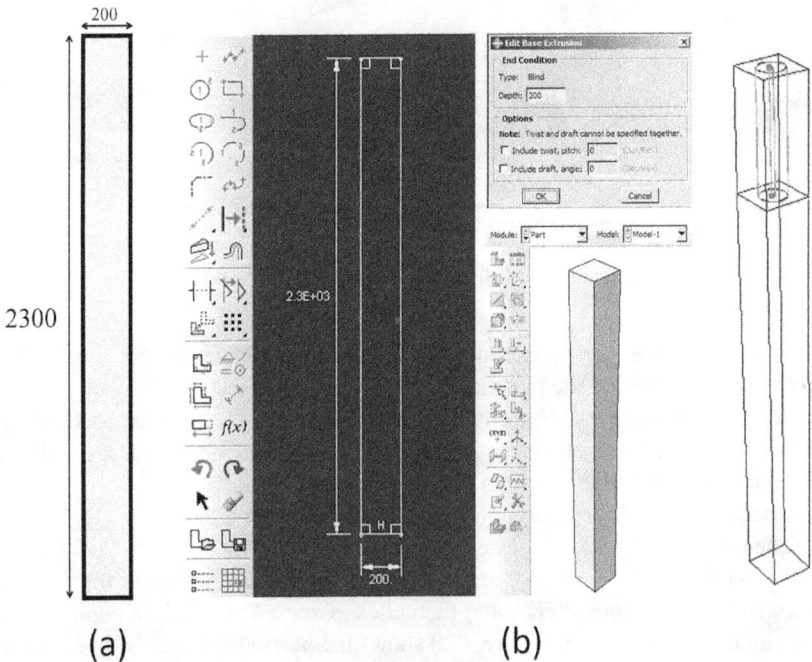

(a)   (b)

FIGURE 4.21   (a) Dimension of the precast column; (b) shaded and wireframe views of the precast column.

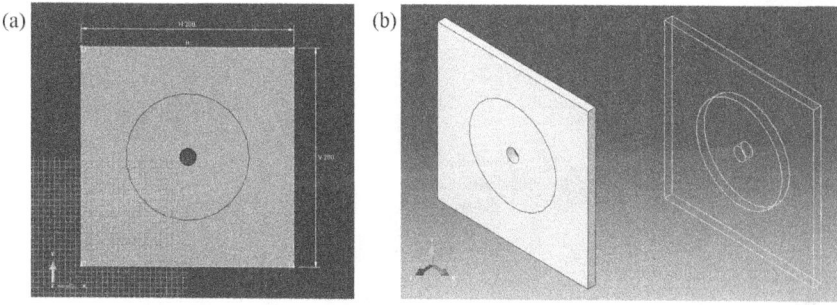

FIGURE 4.22 (a) Dimension of the rubber bearing pad; (b) shaded and wireframe views of the rubber bearing pad.

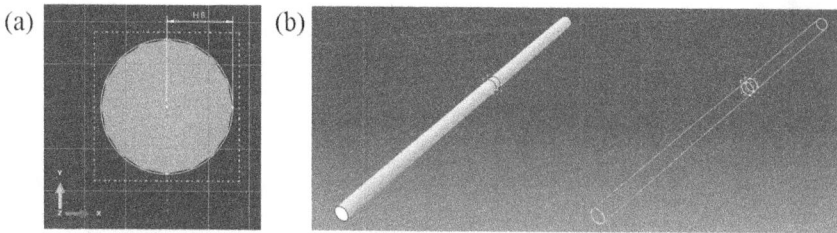

FIGURE 4.23 (a) Dimension of the dowel bar; (b) shaded view of the dowel bar.

Steel reinforcement:

The steel reinforcement cage in this model is assembled from the "Bottom bar", "Top bar", and "Shear Link" parts. Hence, these parts need to be created individually.

4.4.1.6 Create a New Part

From the menu bar, select "Part → Create" to create a new part.

The "Create Part" dialog box appears. Use the "Create Part" dialog box to name the part; to choose its modeling space, type, and base feature; and to set the approximate size. The name of the part may be edited once it has been created, but the modeling space, type, or base feature cannot be changed.

Name the part "Bottom Bar". Choose "3D", "Deformable", and "Wire" from the base feature (see Figure 4.24).

Enter an "Approximate size" as "10000". The value entered in the approximate size text field at the bottom of the dialog box sets the approximate size of the new part.

Click "Continue" to exit the "Create Part" dialog box.

4.4.1.7 Define Bottom Bar Section with Dimension

Use the "Create Lines: Connected" tool located in the upper right corner of the sketcher toolbox to begin drawing the geometry of the "Bottom Bar", as shown in Figure 4.25. The user can pick a starting corner for the "Bottom Bar" at the

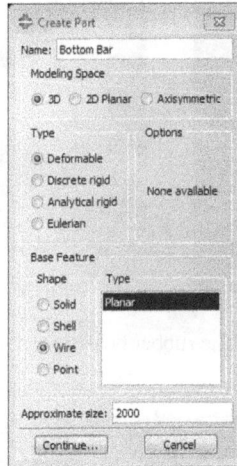

FIGURE 4.24   Create bottom bar part.

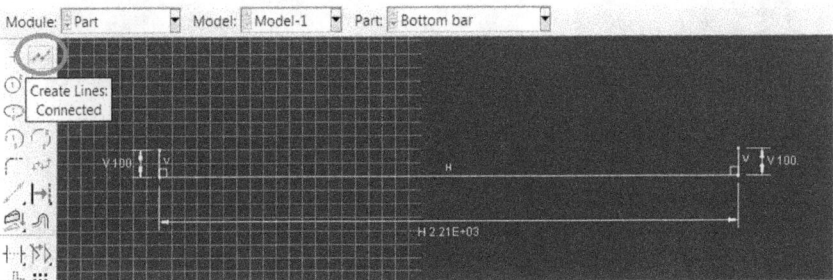

FIGURE 4.25   Draw and add dimensions of the bottom bar sketch.

viewport or enter the "X" and "Y" coordinates. Then create a line with known coordinates, or alternatively, the user can also define the dimension of the geometry by clicking on the add dimension tool. Once finished sketching the section for the dimension, right click and click on "Cancel Procedure" to exit the sketcher. Click "Done" to complete creating the "Bottom Bar" part.

Repeat the same steps in creating the "Top bar" and "Shear link" parts, with their geometry as shown in Figures 4.26 and 4.27, respectively.

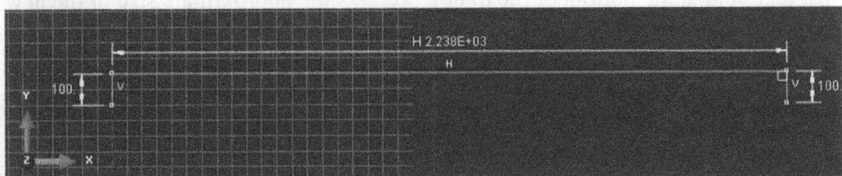

FIGURE 4.26   Dimensions of the top bar.

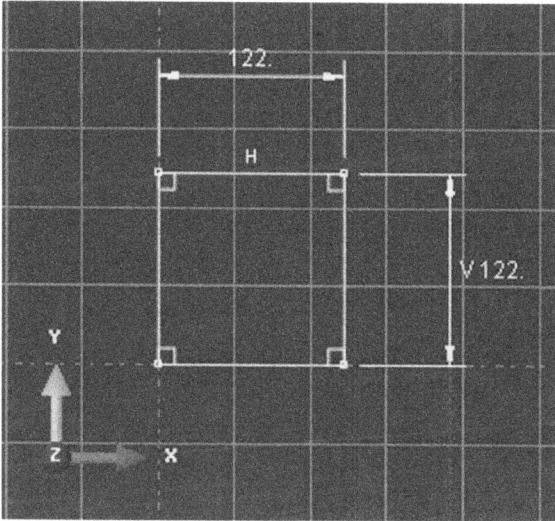

FIGURE 4.27 Dimension of the shear link.

4.4.2 PROPERTY MODULE

In this module, the material properties for the analysis are defined and assigned to the available parts.

4.4.2.1 Material Properties

The precast reinforced concrete frame being modeled here consists of concrete, high tensile steel, mild steel, and rubber materials. The elastic and plastic behaviors of the "Concrete" material are defined according to the steps described below:

In the "Module" list located under the toolbar, select "Property" to enter the "Property" module. The cursor changes to an hourglass while the "Property" module loads.

From the menu bar, select "Material → Create" to create a new material.

The "Edit Material" dialog box appears, as shown in Figure 4.28.

Name the material "Concrete".

From the material editor's menu bar, click "Mechanical → Elasticity → Elastic". The software displays the "Elastic" data form.

Enter the value of "26,600" for "Young's Modulus" and "0.2" for "Poisson's Ratio" in the respective cells. Use the [Tab] button or move the cursor to a new cell and click to move between cells.

Click "General → Density" to enter a "Mass Density" as "2400E-9".

To demonstrate the nonlinear material behavior of concrete in the software, it is necessary to include the concrete damaged plasticity.

In the next step, click on "Mechanical → Plasticity → Concrete Damaged Plasticity".

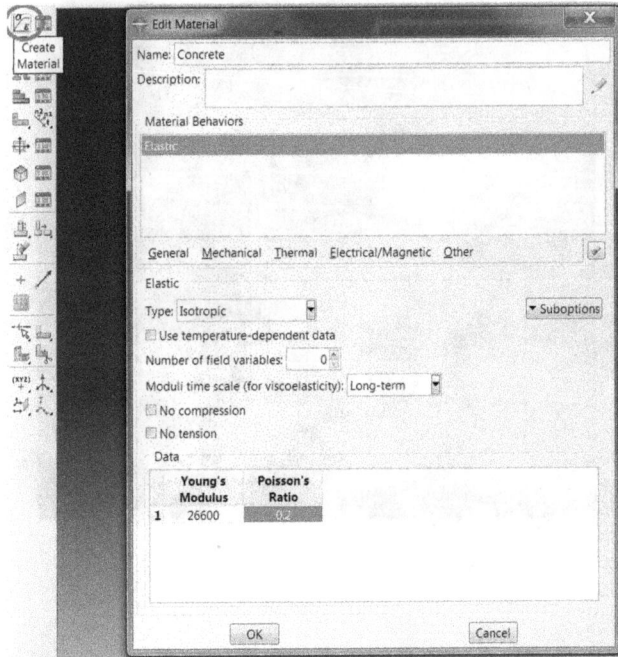

FIGURE 4.28   Create concrete material properties.

Enter the values for each text field (Dilation angle, Eccentricity, f_{b0}/f_c, K, and Viscosity Parameter) as stated previously in Table 4.3.

Click the "Compressive Behavior" tab and enter or copy and paste the yield stress and inelastic strain values from Table 4.3 into the cells.

Then, click the "Sub Options → Compressive Damage", as shown in Figure 4.29. A "Sub Option Editor' for the "Concrete Compression Damage" dialog box appears.

Enter or copy and paste the compressive damage parameter and inelastic strain values from Table 4.3 into the cells. Click "OK" to exit the dialog box.

Next, click the "Tension Behavior" tab, enter or copy and paste the yield stress and cracking strain values from Table 4.3 into the cells.

Then, click "Sub Options → Tension Damage" to open "Sub Option Editor" for the "Concrete Tension Damage" dialog box.

Next, enter or copy and paste the tension damage parameter and cracking strain values from Table 4.3 into the cells. Click "OK" to exit the dialog box.

Click "OK" to exit the material editor.

The properties of "Rubber", "High Tensile Steel", and "Mild Steel" materials are created with similar steps as mentioned above by referring to the values listed in Tables 4.4 and 4.5. Whereas only elastic ("Mechanical → Elasticity → Elastic") properties are defined for "Rubber" and for "High Tensile Steel", and "Mild Steel" is needed to define both elastic and plastic ("Mechanical → Plasticity → Plastic") properties.

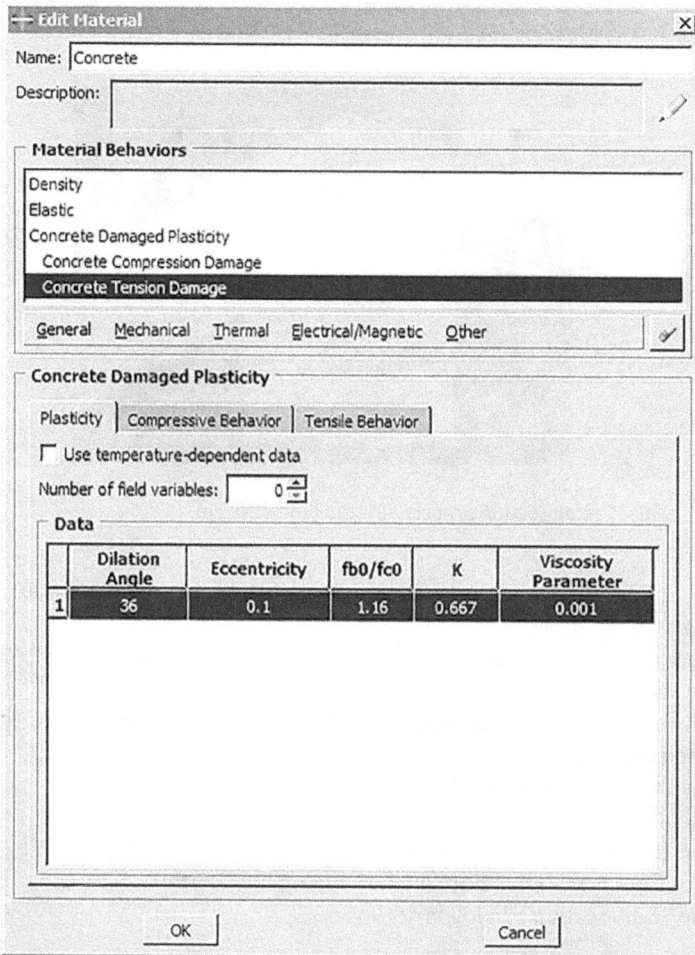

FIGURE 4.29   Enter the parameters for the concrete damaged plasticity model.

4.4.2.2   Section Properties

From the main menu bar, select "Section → Create". The "Create Section" dialog box appears.

In the "Create Section" dialog box:

Name the section "Precast Beam".

In the "Category" list, select "Solid".

In the "Type" list, select "Homogeneous" (see Figure 4.30).

Click "Continue". The "Edit Section" dialog box appears.

In the "Edit Section" dialog box:

Click the arrow next to the "Material" text box and scroll through the "Material" to see a list of available materials and select the required material. To define each material, it is necessary to perform the same process as the "Create Section".

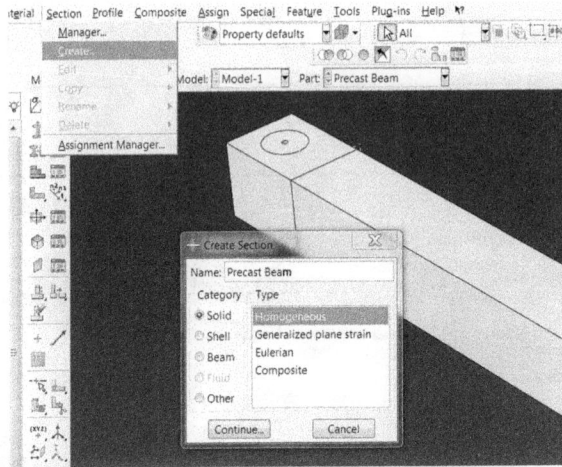

FIGURE 4.30 Create section property for the precast beam.

Next, in the "Plane stress/strain thickness" field, accept the default. Click "OK" (see Figure 4.31).

Repeat the same steps as mentioned above for the "Precast Column", "Rubber Bearing Pad", and "Dowel Bar" sections.

From the main menu bar, select "Section → Create". The "Create Section" dialog box appears.

In the "Create Section" dialog box:

Name the section "R8."

In the "Category" list, select "Beam".

Click "Continue". The "Edit Section" dialog box appears.

In the "Edit Section" dialog box:

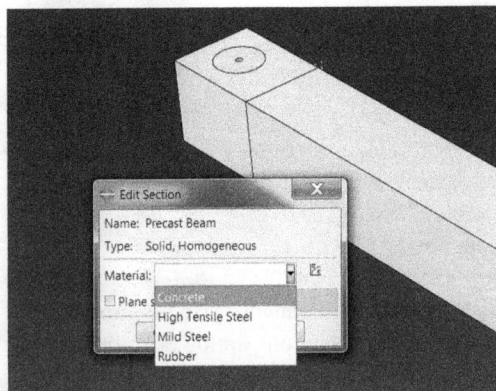

FIGURE 4.31 Solid section of concrete.

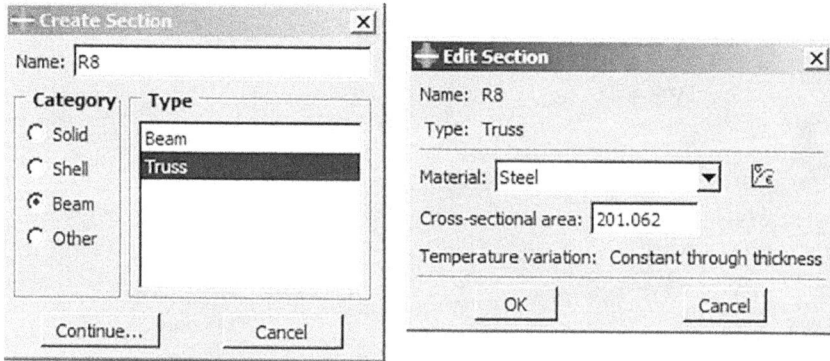

FIGURE 4.32　Create the section property for shear link.

Click the arrow next to the "Material" text box and scroll through the materials to see a list of available materials and select "Steel" as material. To define each material, it is necessary to follow the same process for "Create Section" and then determine the material by clicking on the "Continue" button.

Insert the value of the "Cross-sectional area" as 50 for the "shear link". Click "OK" (see Figure 4.32).

Repeat the same steps as mentioned above for "T12" and "H20" of the steel reinforcement.

4.4.2.3　Section Assignments

Next, assign the defined section to the corresponding part. The "Assign menu" is used in the property module to assign the section "Precast Beam" to the beam. To assign the section to the beam, perform the following steps:

From the main menu bar, select "Assign → Section". The software displays prompts in the prompt area to guide the user throughout the procedure.

Alternatively, expand the menu under "Beam" and double click on the "Section Assignments".

Select the entire part as the region to which the section will be applied.

Click and hold down the left mouse button in the upper left corner of the viewport.

Drag the mouse pointer to create a box around the beam.

Release the left mouse button. The software highlights the entire beam.

Right click on the viewport or click "Done" in the prompt area to accept the selected geometry. The "Assign Section" dialog box appears.

In "Section", scroll to "Precast Beam" and click "OK" as shown in Figure 4.33.

The "Part" changes color to green once the section is assigned, as shown in Figure 4.34.

The "Precast Column", "Dowel Bar", "Rubber Bearing Pad", and all steel reinforcements are assigned material to the part using the same steps as described above.

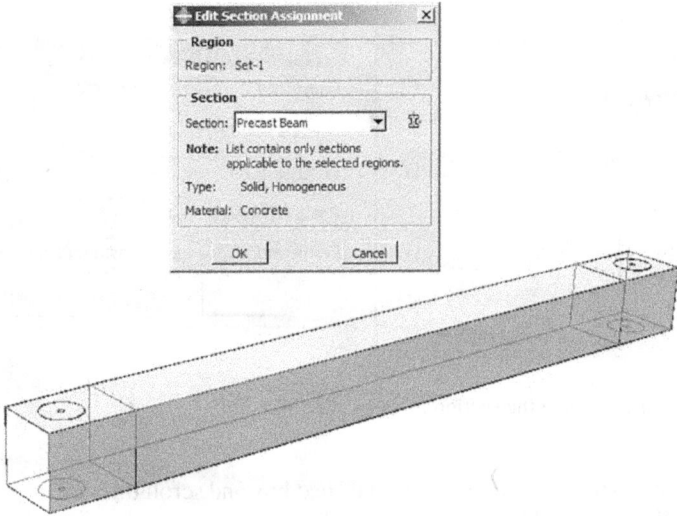

FIGURE 4.33 Section assignment to the concrete beam.

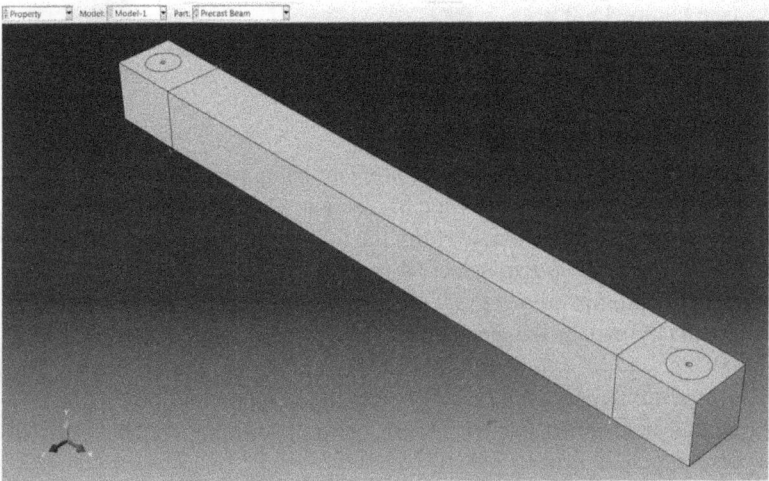

FIGURE 4.34 The beam assigned section.

4.4.3 ASSEMBLY MODULE

In the "Module" list located under the toolbar, click "Assembly" to enter the "Assembly" module. The cursor changes to an hourglass while the 'Assembly" module loads. From the main menu bar, select "Instance → Create". The "Create Instance" dialog box appears.

Assemble for the beam's steel reinforcement cage:

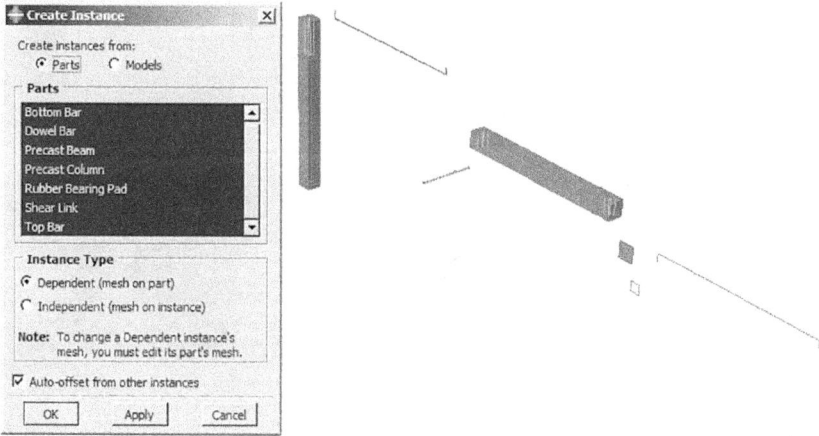

FIGURE 4.35 Insert all parts to assembly.

In the opened window, under the "Instance Type" box, choose "Dependent (mesh on the part)".

The first step, the instance part, is to bring all parts needed for the beam reinforcement together, which have been created previously. Toggle the "Auto-offset from other instances" on to space out the parts among each other, as shown in Figure 4.35.

Next, click "Apply" to bring the part in and select the remaining parts, click "OK" to bring the part in and close the Instance dialog box.

The user needs to Translate Instance to move the element laterally or vertically to the required position, Rotate Instance to rotate the element to the required direction, and Linear Pattern to define instance patterns as shown in Figure 4.36–4.38.

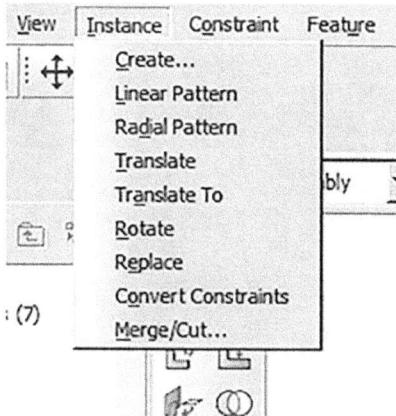

FIGURE 4.36 Translate, rotate, and linear pattern tools.

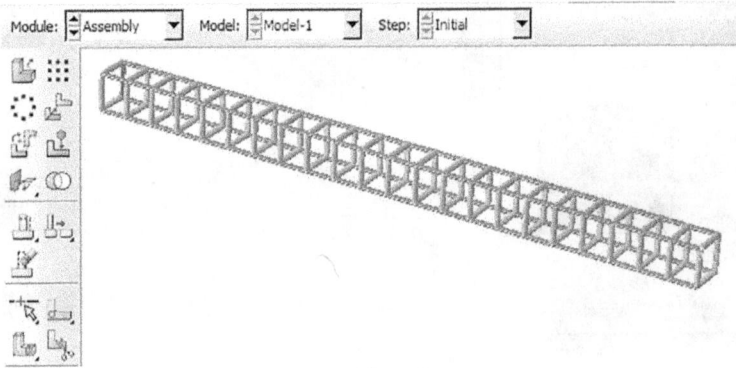

FIGURE 4.37   Define the steel reinforcement of the beam.

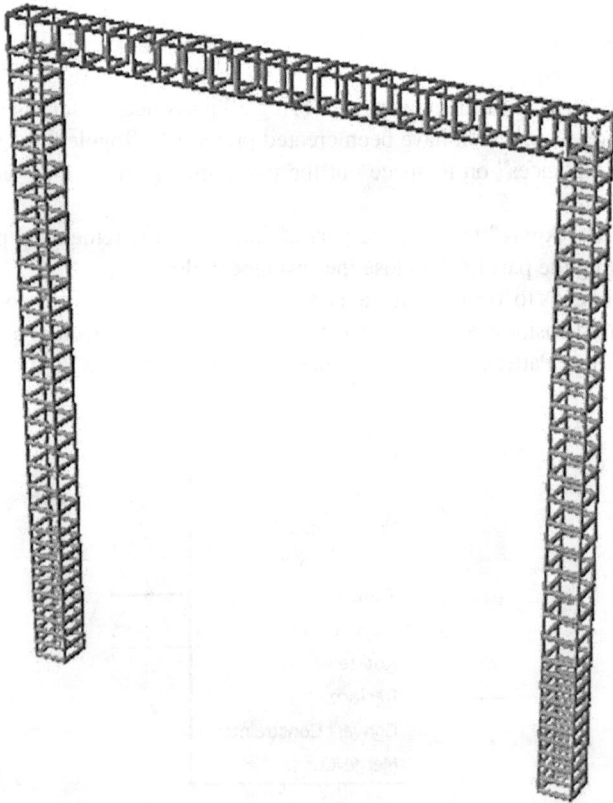

FIGURE 4.38   Define the steel reinforcement for the beam and columns.

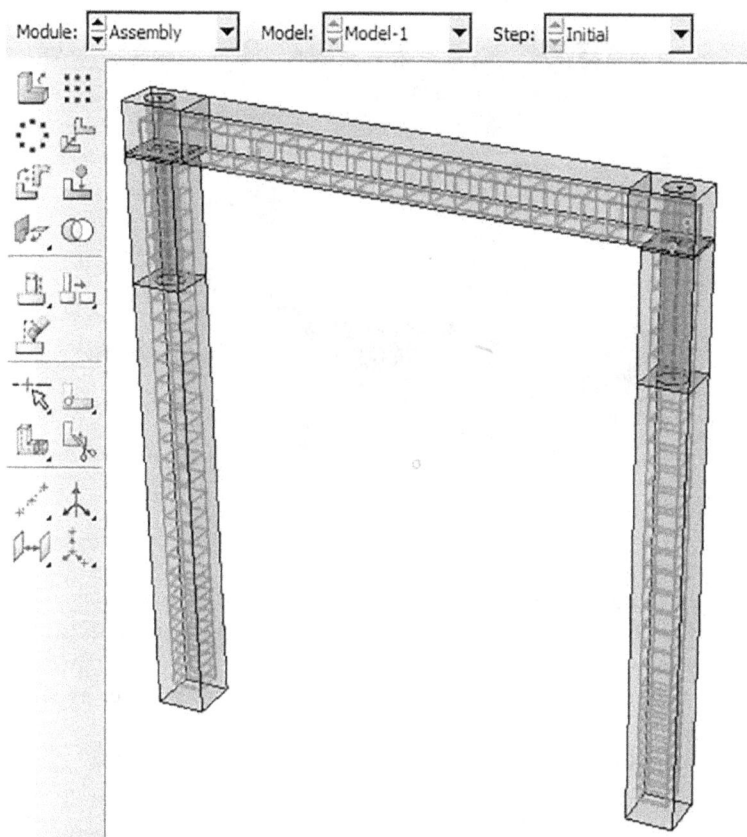

FIGURE 4.39 The precast frame model (translucency mode activated).

Other parts included beam, column, rubber plate, and dowel should be added to define the final assembly as shown in Figure 4.39.

4.4.4 STEP MODULE

In the module list located under the toolbar, click "Step" to enter the "Step" module.

From the main menu bar, select "Step → Create" to create a step. The "Create Step" dialog box appears with a list of all general procedures and a default step named "Step-1".

Change the step name to "Gravity".

Select "General" as the "Procedure type".

Scroll through the available list, select "Static, General", and click on "Continue" (see Figure 4.40).

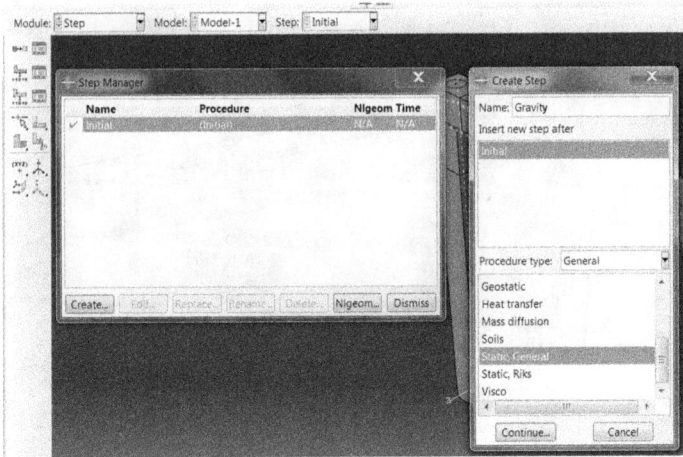

FIGURE 4.40   Create Gravity step.

Next, the "Edit Step" dialog box appears.

Under the "Basic" tab, select "Nlgeom" to include nonlinear effects of geometry in the analysis.

Under the "Incrementation" tab, change the increment size, as shown in Figure 4.41.

Click "OK" to create the step and to exit the Edit Step dialog box.

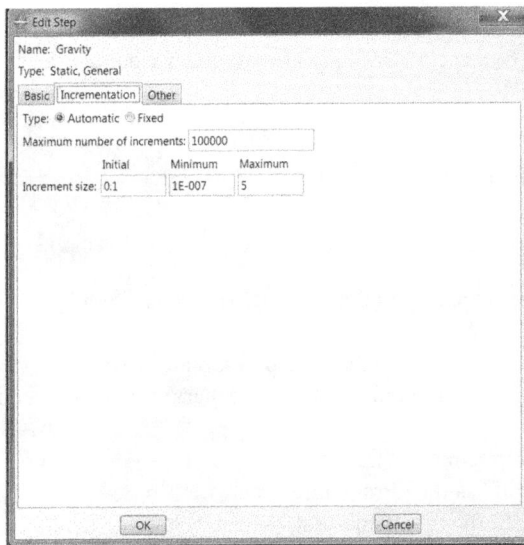

FIGURE 4.41   Gravity step settings.

Repeat the same steps mentioned above for Permanent Load. Cyclic Load step will be created by selecting the "Static, General" and consider 347 as Time period same incrementation as the previous step.

Create amplitude:

Under the Model tree, double click on "Amplitudes".

The "Create Amplitude" dialog box appears.

Set the name of the amplitude as "Cyclic". Select the "Tabular" from "Type" and click "Continue". Input the incremental amplitude data listed in Table 4.5. Click "OK" to exit from the dialog box as shown in Figure 4.42.

TABLE 4.5
Amplitude of Displacement-Based Cyclic Load

Time (s)	Amplitude	Time (s)	Amplitude (mm)	Time (s)	Amplitude (mm)	Time (s)	Amplitude (mm)
0	0	22	0	86	0	198	0
0.5	2	23	−10	88.5	25	202	−40
1	0	24	0	91	0	206	0
1.5	−2	25.5	15	93.5	−25	210	40
2	0	27	0	96	0	214	0
2.5	2	28.5	−15	99	30	218	−40
3	0	30	0	102	0	222	0
3.5	−2	31.5	15	105	−30	226.5	45
4	0	33	0	108	0	231	0
4.5	2	34.5	−15	111	30	235.5	−45
5	0	36	0	114	0	240	0
5.5	−2	37.5	15	117	−30	244.5	45
6	0	39	0	120	0	249	0
6.5	6	40.5	−15	123	30	253.5	−45
7	0	42	0	126	0	258	0
7.5	−6	44	20	129	−30	262.5	45
8	0	46	0	132	0	267	0
8.5	6	48	−20	135.5	35	271.5	−45
9	0	50	0	1.139	0	276	0
9.5	−6	52	20	142.5	−35	281	50
10	0	54	0	146	0	286	0
10.5	6	56	−20	149.5	35	291	−50
11	0	58	0	153	0	296	0
11.5	−6	60	20	156.5	−35	301	50
12	0	62	0	160	0	306	0
13	10	64	−20	163.5	35	311	−50
14	0	66	0	167	0	316	0

(Continued)

TABLE 4.5 (*Continued*)
Amplitude of Displacement-Based Cyclic Load

Time (s)	Amplitude	Time (s)	Amplitude (mm)	Time (s)	Amplitude (mm)	Time (s)	Amplitude (mm)
15	−10	68.5	25	170.5	−35	321	50
16	0	71	0	174	0	326	0
17	10	73.5	−25	178	40	331	−50
18	0	76	0	182	0	336	0
19	−10	78.5	25	186	−40	341.5	55
20	0	81	0	190	0	347	0
21	10	83.5	−25	194	40		

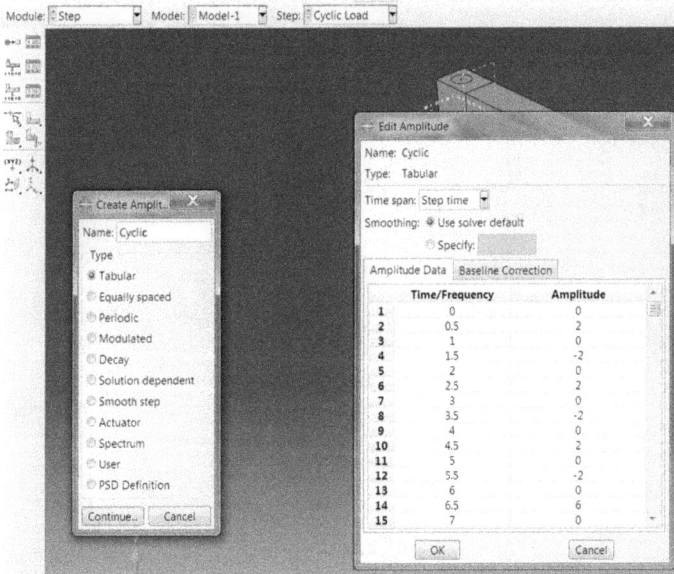

FIGURE 4.42 Create amplitude.

To obtain the required outputs at the visualization module, it is necessary to request the field outputs and history outputs in the "Step" module. In this study, the user followed the default settings for both the field outputs and history outputs.

4.4.5 INTERACTION MODULE

This module is used to define various interactions within the model or interactions between regions of the model and surroundings. Analysis constraints can also be applied between two similar and different materials such as the "Tie" connection,

"Rigid", "Embedded Region", and "Shell to Solid Coupling". Before assigning the type of interaction to the connecting surfaces, the interaction property needs to be created. Here, the interaction property between the concrete and rubber bearing surfaces is created. The interactions used in this example are a weld connection between the concrete and the dowel bar. To demonstrate the condition of the weld connection, the tie constraints in the software will be implemented in this model. Another type of constraint utilized in this study is the embedded region, which has been used to represent the relationship between the reinforcements and the concrete.

4.4.5.1 Surface-to-Surface Contact Interaction

A surface-to-surface contact defines the interactions between specific surfaces in a model. Certain interaction behaviors can be defined only by using the surface-to-surface contact. Here, the user can define surface-to-surface contact in any step, including the initial step. The user can also define contact between edges of wire or between the faces of a solid or shell. Certain connectivity restrictions apply to contact surfaces depending on the type of contact formulation. The user can obtain contact data for a specific surface-to-surface contact interaction by using the field and history output request editors in the "Step" module.

To define surface-to-surface contact for analysis, perform the following steps:

In the module list located under the toolbar, click "Interaction" in order to enter the "Interaction" module.

Select "Create Interaction Property" from the toolbox. The "Create Interaction Property" dialog box appears with a list of all types of interaction properties and a default interaction property name.

Name the interaction property "Concrete-Rubber", as shown in Figure 4.43.

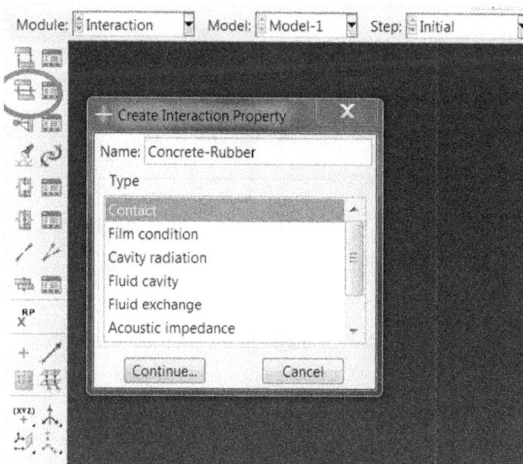

FIGURE 4.43 Create and define the interaction property.

Select "Contact" as the "Type" of interaction property and click on "Continue". The "Edit Contact Property" dialog box appears. Under the "Mechanical" tab, select "Tangential Behavior".

Click the arrow next to the "Friction formulation" field and then select "Penalty" to use a stiffness (penalty) method that permits some relative motion of the surfaces (an elastic slip) when they should be sticking. While the surfaces are sticking (i.e., $\overline{\tau} < \overline{\tau}_{crit}$), the magnitude of sliding is limited to this elastic slip. The software will continually adjust the magnitude of the penalty constraint to enforce this condition. Then, specify the friction coefficient with a value of 0.5.

Keep the default setting in the "Shear Stress" and "Elastic Slip" tabs, as shown in Figure 4.44.

Click "OK" to complete the definition of contact property.

Create the "Dowel-Rubber" interaction property with the steps described above but set its "Friction Coefficient" as 0.3.

The interaction between the concrete and rubber bearing pad will be created in the following steps.

In the "Interaction" module, click "Create Interaction" from the toolbox. The "Create Interaction" dialog box appears.

Name the interaction "Conc-Rub", as shown in Figure 4.45. Select "Initial" as the step in which the interaction will be created. Choose "Surface-to-surface contact (Standard)" from the "Type" of interaction.

Click "Continue" to proceed to define the master and slave surfaces of the contact instances. The master surface shall be made from a harder material and the slave surface shall be made from the softer material.

In the prompt area, choose the "Master type: Surface". There are two ways to select the surface.

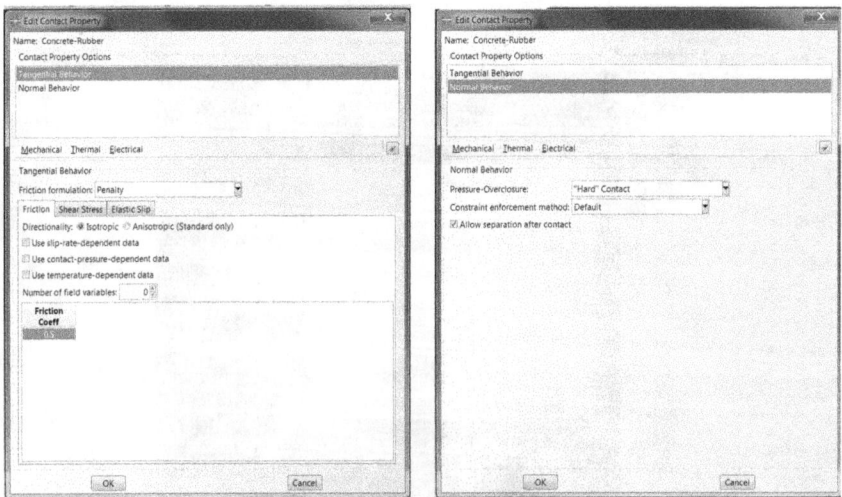

FIGURE 4.44 Edit concrete–rubber contact properties.

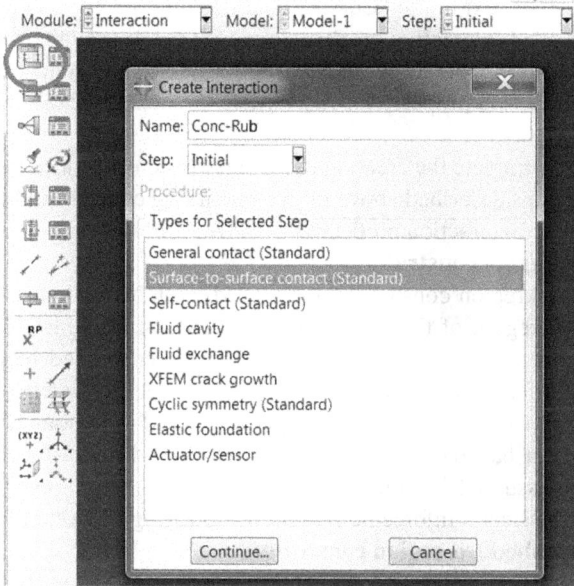

FIGURE 4.45 Create surface-to-surface contact interaction.

Use the existing surface to define the region. On the right side of the prompt area, click "Surfaces". Select an existing surface name from the "Region Selection" dialog box that appears and click "Continue".

Use the mouse to select a region in the viewport. Click the mouse button 2 to indicate that the user has finished selecting the region. Certain connectivity restrictions apply to contact surfaces depending on the type of contact formulation. The main surface of the column is selected as the master surface. Then click Done.

The master surface that the user selects is highlighted in red in the viewport.

In the prompt area, choose the "Slave type: Surface".

Select "Surface" if the user wants to select a surface.

Select "Node Region" if the user wants to select a region to create a contact node set.

Use one of the same methods described earlier to select the slave surface or region. The slave surface or region that the user selected becomes highlighted in magenta in the viewport. The left edge surface of the beam is selected as the slave surface. Click "Done". The "Edit Interaction" dialog box appears.

The "Switch Surfaces" option allows the user to interchange the master and slave surface selections without needing to start over. However, it is available only when both master and slave regions are from the same type; either surfaces or both node-based regions.

Use the default settings for the finite sliding formulation and surface-to-surface discretization method. No adjustment for the slave is required.

Click the arrow next to the "Contact interaction property" text box and scroll through the contact interaction property to view a list of available properties and to select the required contact interaction property. Select "Concrete-Rubber" as the contact interaction property for the surface-to-surface contact interaction, as shown in Figure 4.46.

Click OK to complete the creation of this interaction and close the dialog box.

Repeat the steps described above for the remaining contact surfaces with their respective contact interaction property.

Embedded region constraint:

An embedded region constraint can be used to embed a region of the model within a "host" region of the model or within the entire model. An embedded region constraint can be created by specifying the embedded region, the host region, a weight factor round-off tolerance, and an absolute exterior tolerance or fractional exterior tolerance. In this example, the total rigid connection between the reinforcement bar and its surrounding concrete with nonslip reinforcement bar approach is assumed. Therefore, the "Embedded Region" is used to model the total fixity of the steel reinforcement to the surrounding concrete in this model.

Create an embedded region constraint:

In the "Interaction" module, select "Create Constraint" from the toolbox. The "Create Constraint" dialog box appears with a list of all types of constraints.

Next, name the constraint as "Embedded Region".

Select "Embedded region" as the "Type" of constraint and click on "Continue" (see Figure 4.47).

Select the beam reinforcement in the viewport. Click the mouse button 2 to indicate that the user has finished the selection. The region that the user selects becomes highlighted in red in the viewport.

Select the host region. In the prompt area, select one of the following methods:

Select "Region" if the user wants to select a region in the viewport or use an existing set to define the region. Use one of the methods described in the previous step to select the host region. The precast beam is selected as the host region. The region that the user selects becomes highlighted in magenta in the viewport.

Whole Model: if the user wants to embed the embedded region within the whole model.

The Edit "Constraint" dialog box appears.

In the constraint editor, specify values for the tolerance parameters. If the user specifies values for both exterior tolerance parameters, the software will use the smaller values of the two tolerances.

Weight factor round-off tolerance: The user can specify a small value below which the weighting factors will be zeroed out. The default value is 10^{-6}.

Absolute exterior tolerance: The user can specify the absolute value in which a node on the embedded region may lie outside the host region.

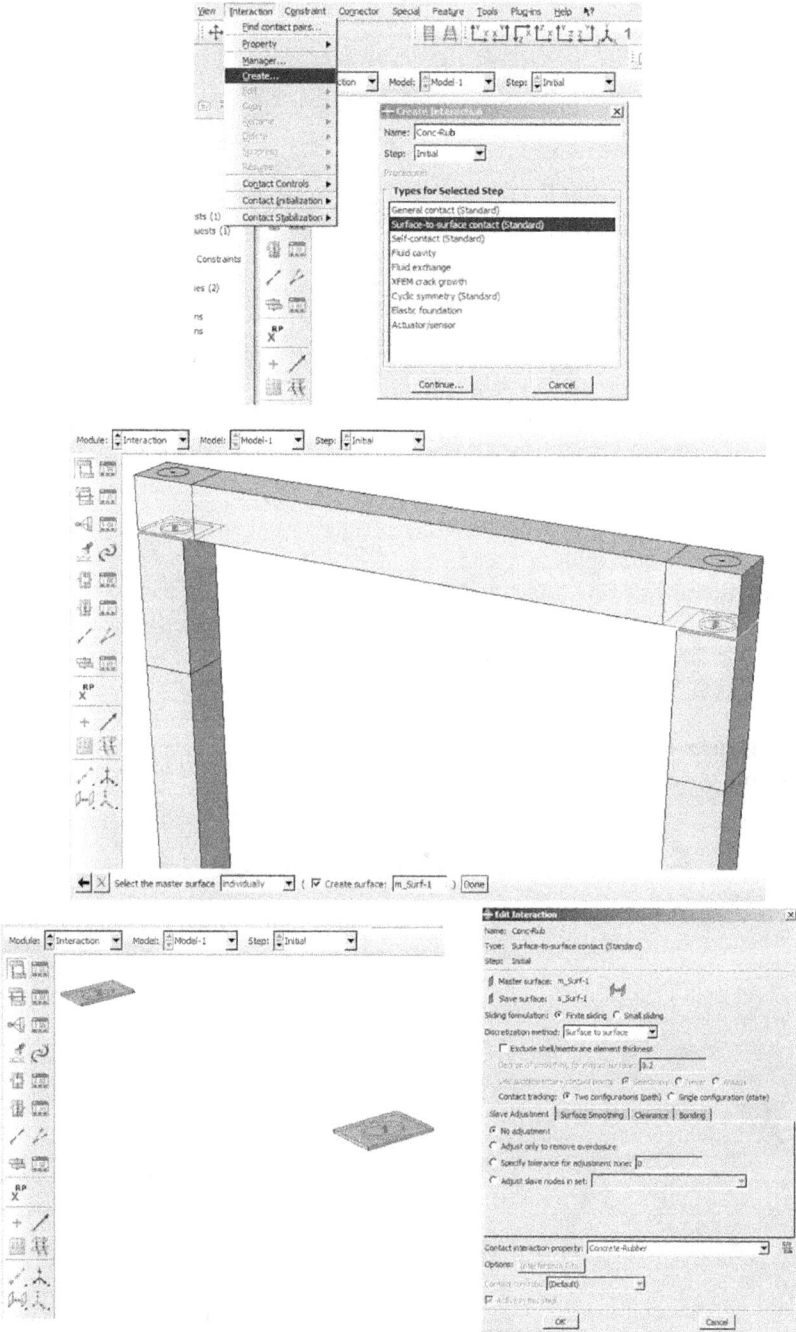

FIGURE 4.46 Edit the contact interaction.

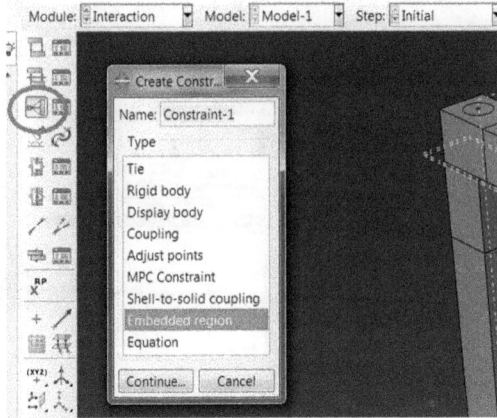

FIGURE 4.47   Embedded region constraint.

If this parameter is omitted or has a value of 0.0, the fractional exterior tolerance will apply.

Fractional exterior tolerance: The user can specify the fractional value by which a node on the embedded region may lie outside the host region. The fractional value is based on the average element size within the host region. The default value is 0.05.

Use the default setting and click "OK" to define the constraint definition and to exit from the editor (see Figure 4.48).

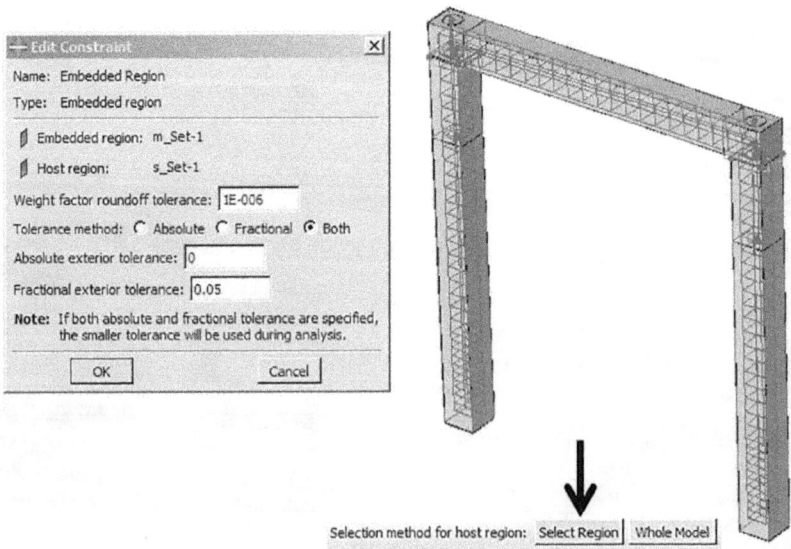

FIGURE 4.48   Edit constraint for the embedded region.

Repeat the same steps described above to make an "Embedded Region" constraint between the two precast columns and their reinforcements.

4.4.5.2 Tie Constraint

In the "Interaction" module, click "Create Constraint". The "Create Constraint" dialog box appears with a list of all types of constraints.

Name the constraint as "Tie". Select "Tie" as the "Type" of constraint and click on "Continue", as shown in Figure 4.49.

In the prompt area, choose the master type: Surface.

Select regions for the master surface and click "Done". There are two ways to select the surface:

The user can use the existing surface to define the region. On the right side of the prompt area, click Surfaces. Then, select an existing surface name from the Region Selection dialog box that appears and click "Continue".

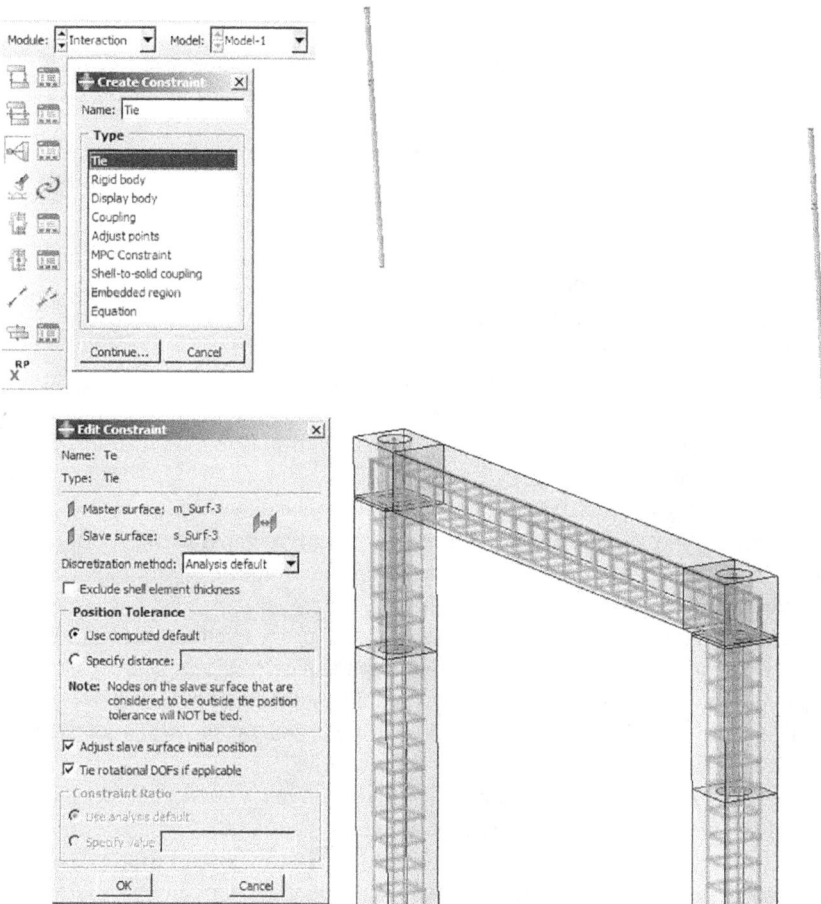

FIGURE 4.49   Create tie constraint for the concrete–dowel connection.

The user can use the mouse to select a region in the viewport. The circumference surface of the dowel bars is selected as the master surface.

In the prompt area, choose the slave type: "Surface". The inner contact surface of the holes on the concrete beam and concrete column is selected as the slave regions. Select the regions for the slave surface and click Done.

The "Switch Surfaces" option allows the user to interchange the master and slave surface selections without having to start over. However, it is available only when both master and slave regions are from the same type; either surfaces or both node-based regions. Accept the default.

Click "OK" to save the constraint definition and to exit from the editor.

4.4.6   LOAD CONDITION MODULE

In this module, the user can create the loading that needs to be assigned to the elements or parts of the structure. The prescribed conditions, such as loads and boundary conditions, are step-dependent, which therefore means that the user needs to specify the steps in which they become active. Since the steps in the analysis have been defined, then the Load module can be used to define the prescribed conditions.

Apply the boundary condition to the column base:

In the module list located under the toolbar, click "Load" to enter the "Load" module.

From the main menu bar, select "BC → Create". The "Create Boundary Condition" dialog box appears.

In the "Create Boundary Condition" dialog box:

Name the boundary condition "Fix End".

From the list of steps, select "Initial" as the step in which the boundary condition will be activated. All the mechanical boundary conditions specified in the Initial step must have zero magnitudes. This condition is enforced automatically by the software.

In the "Category" list, accept "Mechanical" as the default category selection.

In the "Types" list, select "Symmetry/Axisymmetry/Encastre", and click "Continue". The software displays prompts in the prompt area to guide the user throughout the procedure (see Figure 4.50).

To apply a prescribed condition to a region, the user can either select the region directly in the viewport or apply the condition to an existing set. Sets are a convenient tool that can be used to manage large and complicated models.

There are several boundary conditions identified by the software such as pinned and fixed.

In the viewport, select the base of the column. This is the region to which the boundary condition will be applied, as shown in Figure 4.51.

Right click on the viewport or click "Done" in the prompt area to indicate the end of selecting the regions. The "Edit Boundary Condition" dialog box appears.

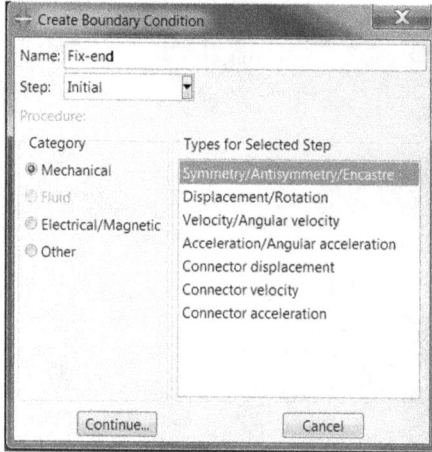

FIGURE 4.50 Create the fixed-end boundary condition.

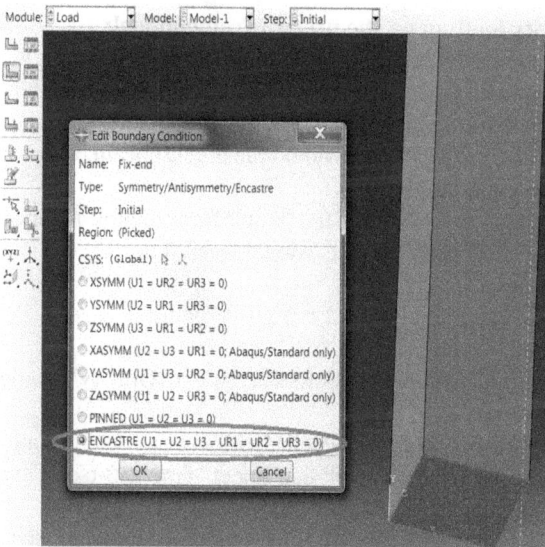

FIGURE 4.51 Select boundary regions and edit the boundary condition.

Select "ENCASTRE" since all the horizontal and vertical degrees of freedom need to be constrained as a fixed support. Click "OK" to create the boundary condition and close the dialog box as shown in Figure 4.51.

Click "OK" to create the boundary condition and to close the dialog box.

The boundary conditions at the column base will be illustrated, as shown in Figure 4.52.

FIGURE 4.52   Fixed-end boundary condition.

Apply gravity loading on the precast frame model:

To analyze the performance of a structure under the cyclic loading, the self-weight of the structure is necessary to be included in the dynamic analysis. Hence, the precast frame in this model will be assigned its gravity loading and a permanent uniformly distributed load on the precast beam, in predefined analysis steps (created in the Step Module).

To apply the gravity load to the model, perform the following steps:

Double click on "Loads" in the "Model tree". The "Create Load" dialog box appears.

In the "Create Load' dialog box:

Name the load "Gravity".

Scroll through the available list of steps, select "Gravity" as the step in which the load will be exerted.

In the Category list, accept "Mechanical" as the default category selection. In the "Types for the selected step" list, select "Gravity" (see Figure 4.53). Click "Continue".

The software displays prompts in the prompt area to guide the user throughout the required procedure. The user is requested to select a region to which the load will be applied. As with boundary conditions, the region to which the load will be applied can be selected either directly in the viewport or from a list of existing sets. As before, select the region directly in the viewport.

In the viewport, select the whole precast frame model where the load will be applied.

Click on the viewport or click "Done" in the prompt area to finish selecting the regions. The "Edit Load" dialog box appears.

In the dialog box:

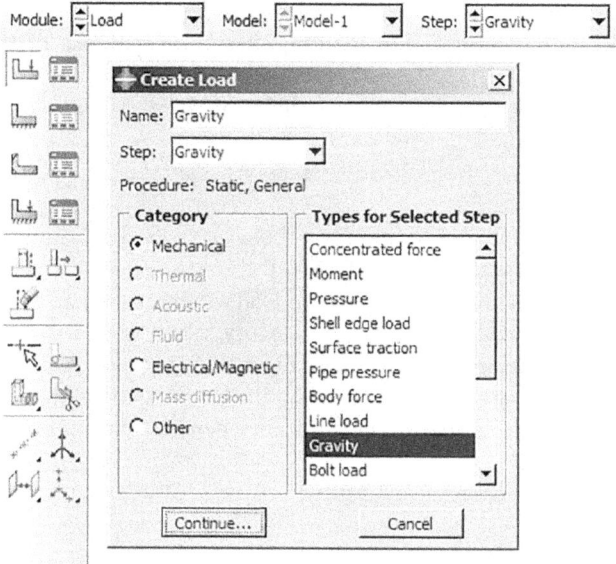

FIGURE 4.53 Create gravity loading.

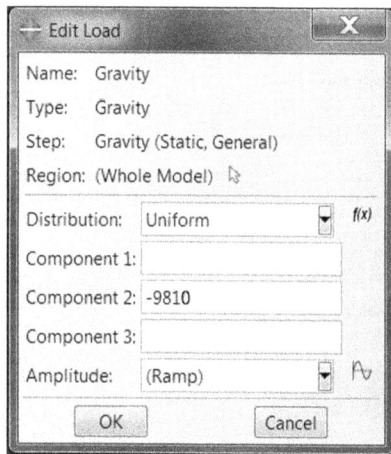

FIGURE 4.54 Input the load value for gravity load.

Enter a magnitude of "–9810" as the gravity load in the "Component 2" text field. The negative sign in front of the value indicates the load applied is a downward force as the gravitational force is acting downwards, as shown in Figure 4.54. Accept the remaining default settings; applied force "Region" as the "Whole Model" and "Amplitude" in "Ramp" mode.

Click "OK" to create the load and to close the dialog box.

Apply uniformly distributed loading on the top surface of the precast beam:

To apply the 1000 kg of the total load that is uniformly distributed on the precast beam, perform the following steps:

Click "Create Load" from the toolbox upon which the "Create Load" dialog box appears.

In the "Create Load" dialog box:

Name the load "Permanent Load". Scroll through the available list of steps, select "Cyclic Load" as the step in which the load will be exerted. In the "Category" list, accept "Mechanical" as the default category selection. In the "Types for the Selected Step" list, select "Pressure" (see Figure 4.55).

Click "Continue" to select the regions for the boundary condition. In the viewport, select the top surface of the precast beam where the load will be applied (see Figure 4.56).

Click on the viewport or click "Done" in the prompt area to finish selecting the regions. The "Edit Load" dialog box appears.

In the dialog box:

Enter a value of "10791 N" of "Total force" in "Distribution" (see Figure 4.57). Click "OK" to create the load and to close the dialog box. The uniform distributed load (in total force mode) will appear on the top surface of the precast beam.

Apply cyclic loading to the precast frame model:

Click "Create Boundary Condition" from the toolbox. The "Create Boundary Condition" dialog box appears.

In the "Create Boundary Condition" dialog box:

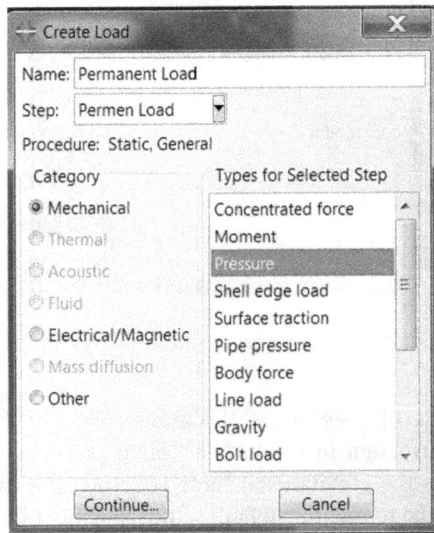

FIGURE 4.55   Create a permanent load on the precast beam.

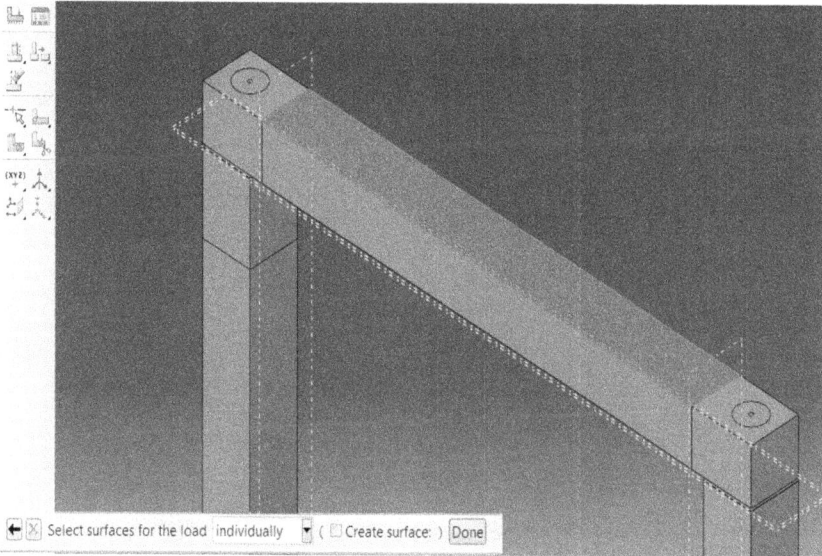

FIGURE 4.56 Select the surface to apply pressure.

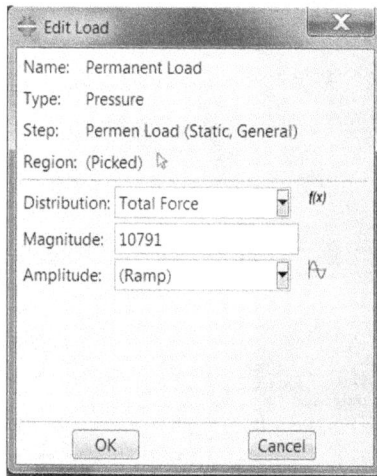

FIGURE 4.57 Input the load value.

Name the load "Cyclic". Scroll through the available list of steps, select "Cyclic load" as the step in which the load will be exerted. In the "Category" list, accept "Mechanical" as the default category selection.

In the "Types for the Selected Step" list, select "Displacement/Rotation" (see Figure 4.58).

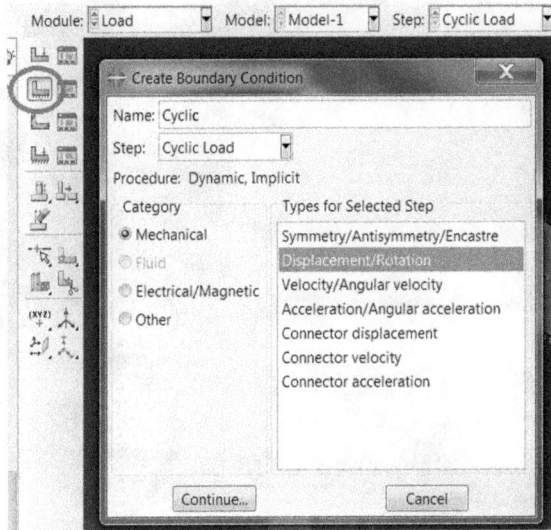

FIGURE 4.58   Create a cyclic load on the frame.

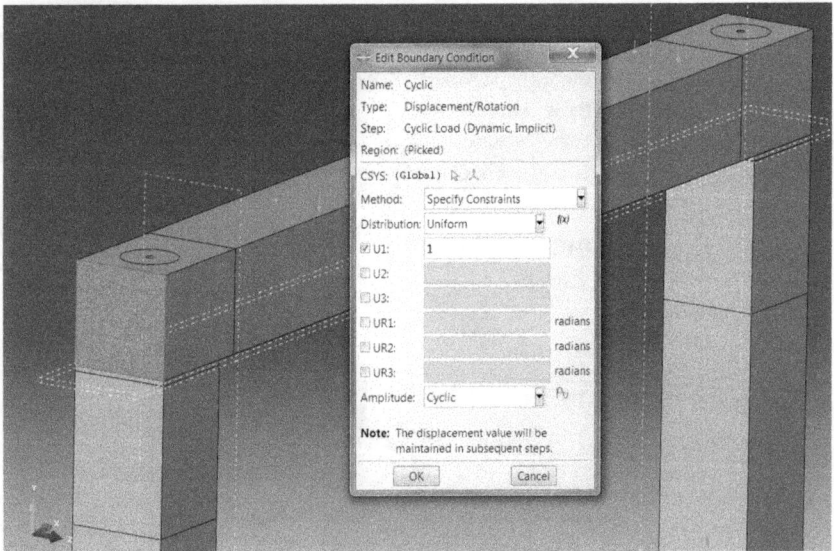

FIGURE 4.59   Select the surfaces to apply the cyclic load and edit the boundary condition.

Click "Continue" to select the regions for the boundary condition. In the viewport, select the upper-left side surface of the precast column where the displacement-based load will be applied (see Figure 4.59).

Click on the viewport or click "Done" in the prompt area to finish selecting the regions. The "Edit Boundary Condition" dialog box appears.

In the dialog box:

Accept the default setting in "Method" and "Distribution".

Check "U1" and enter "1" as the multiplication factor to the amplitude created in the "Step" module, as shown in Figure 4.59.

Choose "Cyclic" (created in Step Module) from the "Amplitude". Click "OK" to save the load and to close the dialog box. The cyclic load will appear on the selected precast column surface.

Repeat the steps above for another cyclic load to be applied on the upper right of the precast column. This is required because the precast frame will be simultaneously pushed and pulled at both sides of the precast column.

4.4.7 Mesh Module

In the module list located under the toolbar, click "Mesh" to enter the "Mesh" module. At the context bar, click "Part", to unclick the "Assembly". Select "Bottom Bar" from the list of parts.

From the main menu bar, select "Mesh → Element Type".

In the viewport, select "Bottom Bar" as the region to be assigned an element type.

In the prompt area, click "Done". The "Element Type" dialog box appears, as shown in Figure 4.60.

In the dialog box, select the following:

- "Standard" as the "Element Library" selection (the default).
- "Linear" as the "Geometric Order" (the default).
- The "Truss" as the "Family" of elements.

Then, element size should be defined by seed parts. By seeding, the software places the nodes of the mesh at the seeds whenever possible.

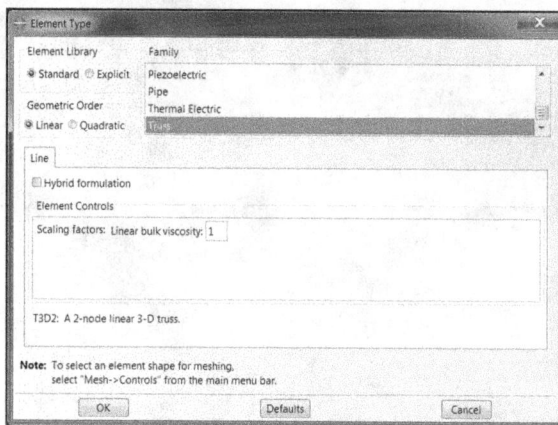

FIGURE 4.60 Select the element type for beam reinforcement.

Seed and mesh the beam reinforcement instance:

From the main menu bar, select "Seed → Part" to seed the part instance.

Alternatively, select the "Seed Part" on the upper-left corner of the meshing toolbox. The "Global Seeds" dialog box will appear, as shown in Figure 4.61.

Type the appropriate value for the "Approximate global size" of the mesh elements, e.g., "50", in this example.

Click "OK" to accept the seeding.

From the main menu bar, select "Mesh → Part" and in the prompt area click "Yes".

Once meshed, the part changes color to blue, and the meshed geometry is shown in Figure 4.62.

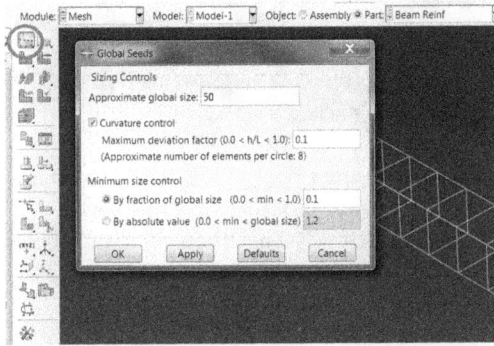

FIGURE 4.61 Assign the approximate global size for the mesh elements.

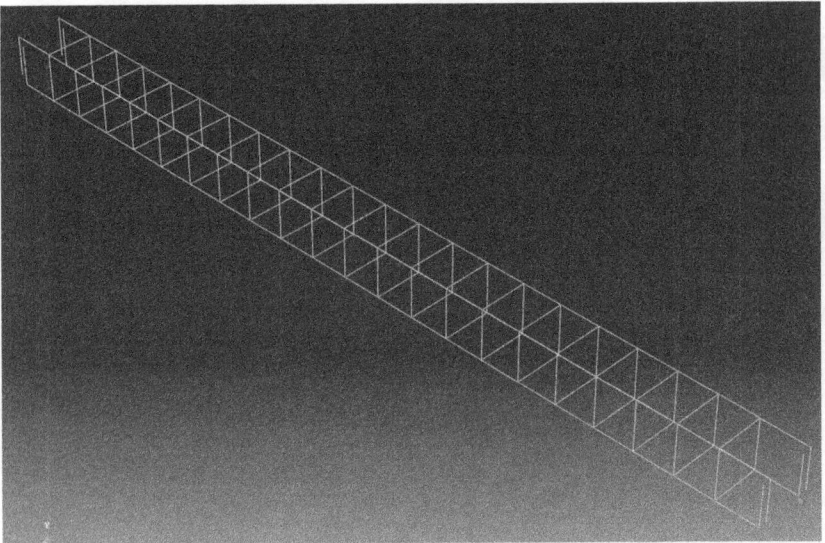

FIGURE 4.62 Meshing on beam reinforcement is done.

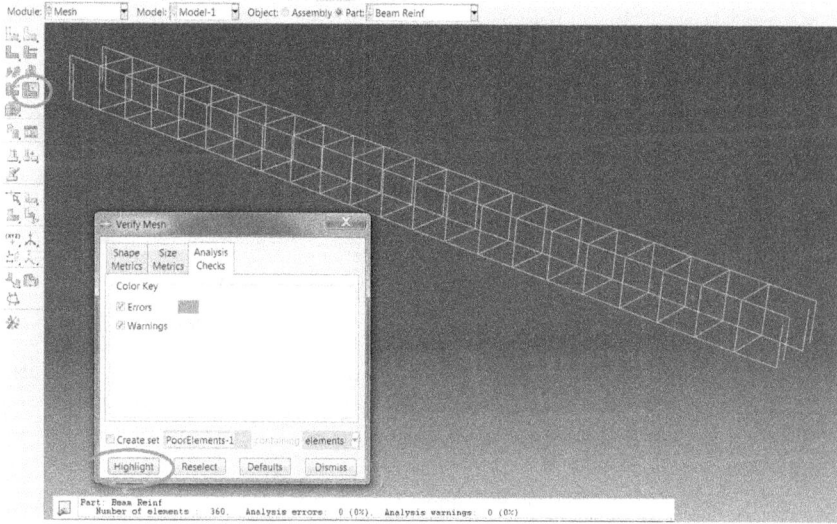

FIGURE 4.63 Verify the mesh.

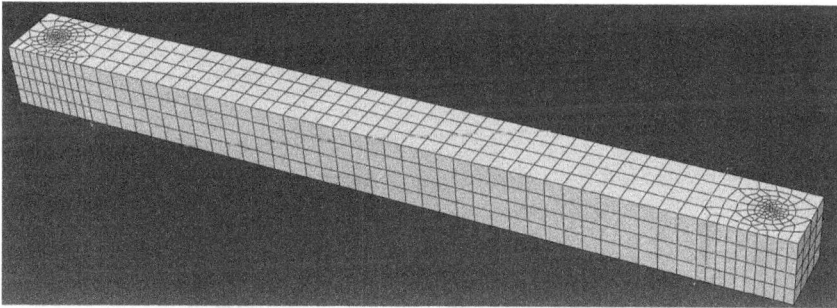

FIGURE 4.64 Whole meshed model.

The user can verify the model meshing by clicking on the "Verify Mesh" icon, and the message will appear in the text messages at the bottom of the screen. The software highlights any elements that fail in the mesh quality tests and displays the number of elements tested along with the number of errors and warnings in the message area (see Figure 4.63).

Repeat the same steps described above for other parts. Figure 4.64 shows the meshed model.

4.4.8 HISTORY OUTPUT DEFINITION

For ease of selection, the needed group of instances or nodes is recorded/presented in the output; however, the considered group/nodes can be created as a homogeneous set using the Set toolset.

4.4.8.1 Create Sets of Nodes

From the main menu bar, select "Tool → Set → Create" and the "Create set" dialog box appears.

In the "Create set" dialog box name the set "Displacement" (see Figure 4.65). Select "Node" from the "Type" of set and click "Continue".

At the viewport, select the node on the mid-span and mid-height of the precast beam (see Figure 4.66). Click "Done" to complete the node assignation to the set.

Repeat the same steps described above to create a "Reaction" set at the column bases.

4.4.8.2 History Output for the Sets

To obtain the required outputs at the visualization module, it is necessary to request the history outputs in the "Step" module. The steps are illustrated below as follows:

Double click on "History Output Requests" in the Model tree to create "History Output Requests".

The "Create History" dialog box appears.

Name the history output "Displacement"; the select step is "Cyclic Load" (see Figure 4.67). This is because only the cyclic force–displacement data are interested in being obtained in this example.

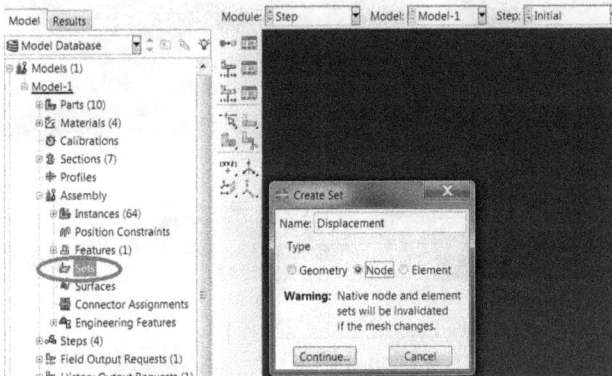

FIGURE 4.65   Create a displacement node set.

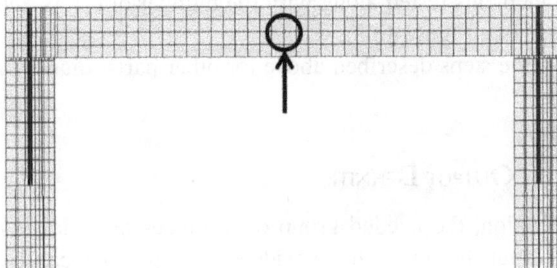

FIGURE 4.66   Select the displacement node on beam.

Click "Continue". The "Edit History Output Request" dialog box appears.

At the "Domain", select "Set".

Click the field next to the domain and scroll through the Set to view a list of available sets and to select the required "Set" as shown.

Expand the "Displacement/Velocity/Acceleration" and toggle on the "U1" under "U, Translations and rotations" (see Figure 4.67).

Click "OK" to exit from the dialog box.

Repeat the same procedure for the "Reaction" (base shear) by requesting the reaction force in the x-direction, "RF1" under the "Forces/Reactions" (see Figure 4.68).

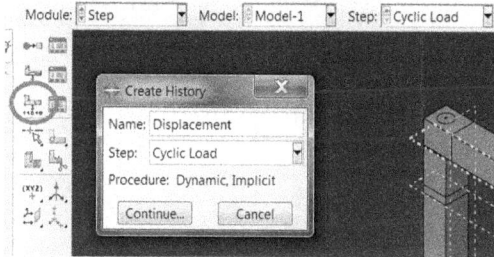

FIGURE 4.67 Create history output for displacement.

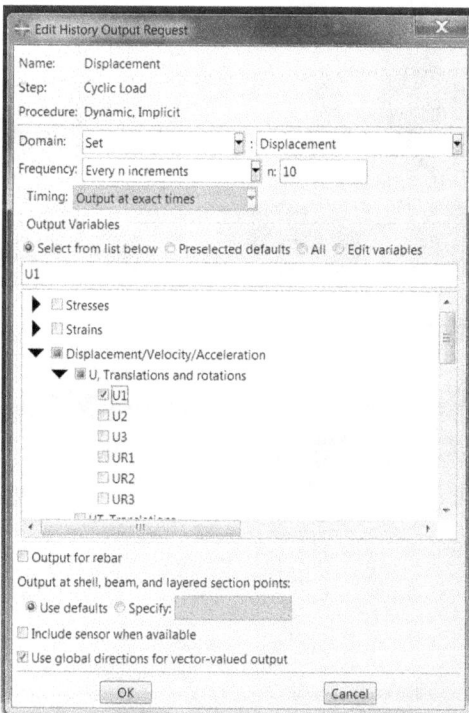

FIGURE 4.68 Create history output for reaction force.

4.5 ANALYSIS: JOB MODULE

The "Job" module can be used to create and manage the analysis of jobs and submit them for analysis.

4.5.1 CREATE AN ANALYSIS JOB: JOB-1

Double click on "Jobs" in the Model tree. Name the "job" and click "Continue" (see Figure 4.69).

The "Edit Job" dialog box appears. The user can describe the job in the "Description" field. In the Submission tabbed page, select "Full Analysis" from the "Job Type".

Click "OK" to accept all other default job settings in the job editor and to close the dialog box.

Click on "Submit" to start checking the input file and running the analysis (see Figure 4.70).

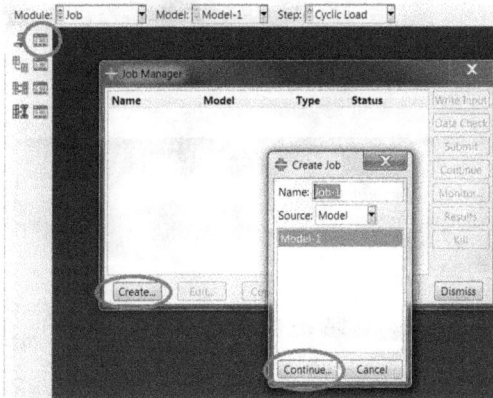

FIGURE 4.69   Create job.

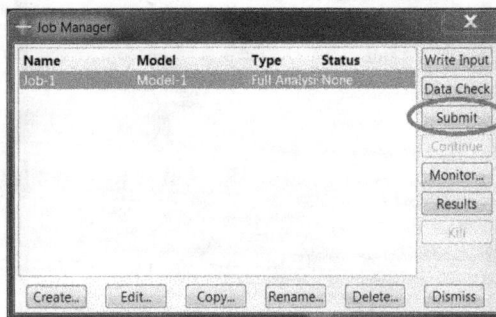

FIGURE 4.70   Job submission.

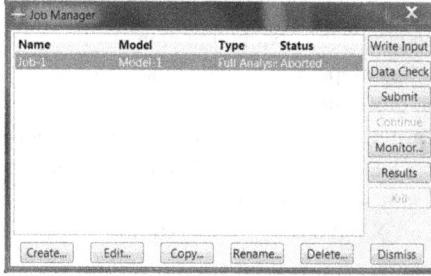

FIGURE 4.71 Select result to enter the Visualization module.

Once the running of analysis has been completed, right click on the job and click "Results" to enter the "Visualization" module (see Figure 4.71).

4.6 VISUALIZATION MODULE

In the "Visualization" module, the user can view and check results and export plots.

4.6.1 VIEW THE RESULTS OF THE ANALYSIS

Double click the "XYData" from the "Result tree", as shown in Figure 4.72.

The "Create XY Data" dialog box appears. Click "Continue". The "History Output" dialog box appears.

Select only the displacement results (see Figure 4.72) and click "Save As", then the "Save XY Data As" dialog box appears. Name the data as "Displacement" and choose "as is" from "Save Operation" and click "OK" to save the graph.

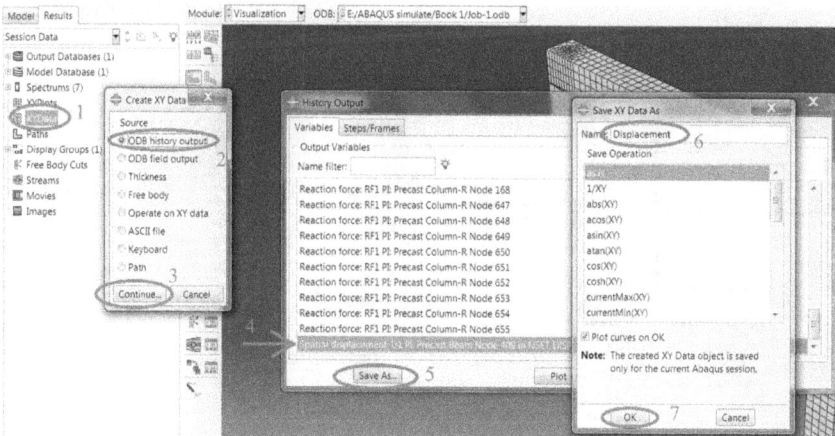

FIGURE 4.72 Extracting displacement history output.

Select all the reaction results (see Figure 4.73) and click "Save As", the "Save XY Data As" dialog box appears. Name the data as "Reaction", and choose "sum ((XY,XY,…))" from "Save Operation" and click "OK" to save the graph.

Click "Dismiss" to close the "History Output" dialog box.

Combine the obtained different (X,Y) data into a single graph:

Double click the "XYData" from the "Result tree", as shown in Figure 4.74. The "Create XY Data" dialog box appears.

Choose "Operate on XY data". Click "Continue". The "Operate" on the "XY Data" dialog box appears.

Select "combine (X,X)" from the "Operators" list.

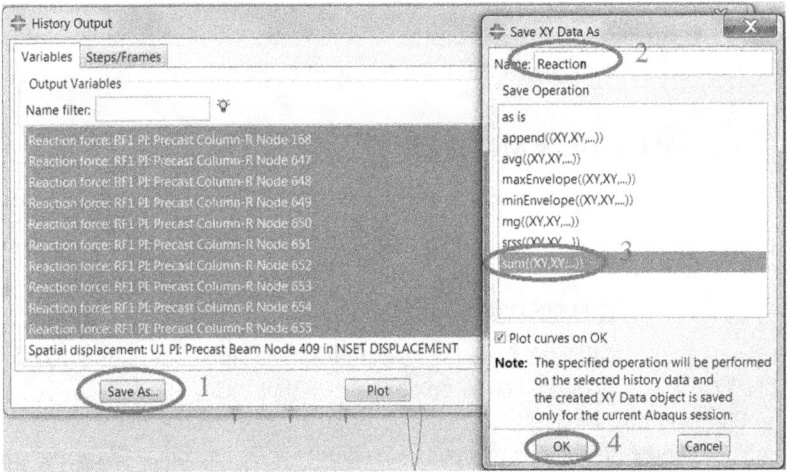

FIGURE 4.73   Extracting reaction force history output summation.

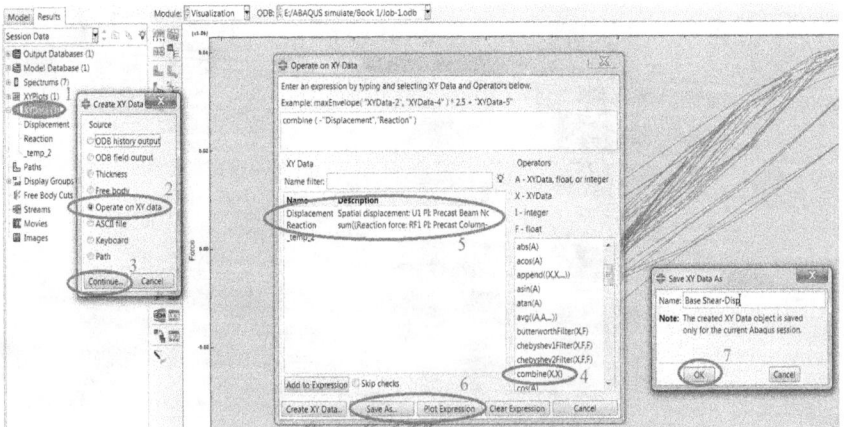

FIGURE 4.74   Combine the displacement and force graphs into a single graph.

Click "Save as" to save the graph by giving it a name.

A base shear vs. displacement graph is plotted.

All the results will be extracted from this module and are further explained in the next section.

4.7   CONTOUR PLOTS

Right click on all concrete parts underneath "Cyclic" in the Results tree and click "Replace", then click "Plot contours on deform shape". The stress contour illustrated as shown in Figure 4.75

Perform the same for steel reinforcement, dowel bars, and rubber bearing pads as shown in Figures 4.76–4.78, respectively.

To view the magnitude of plastic strain contour distribution, right click all concrete parts and select "Replace". Then select "PEMAG" as a primary variable as shown in Figure 4.79.

Perform the same for steel reinforcement and dowel bars as shown in Figures 4.80 and 4.81, respectively.

To view tensile and compressive damage of concrete parts, replace them in the viewport, then select "DAMAGET" and "DAMAGEC" as the "Primary" variable as shown in Figures 4.82 and 4.83, respectively.

To view total strain in the whole model, replace all parts in the viewport, then select "E" as the "Primary" variable and "Max. Principal" as invariant as shown in Figure 4.84.

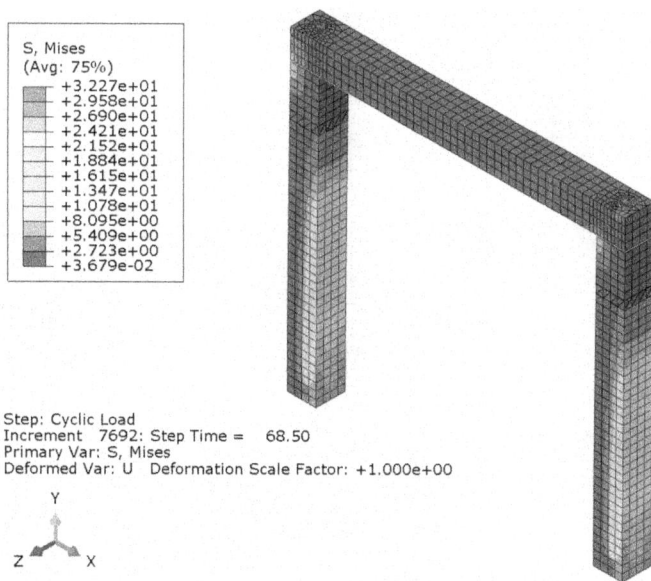

FIGURE 4.75   Von Mises stresses contour distribution for precast concrete frame.

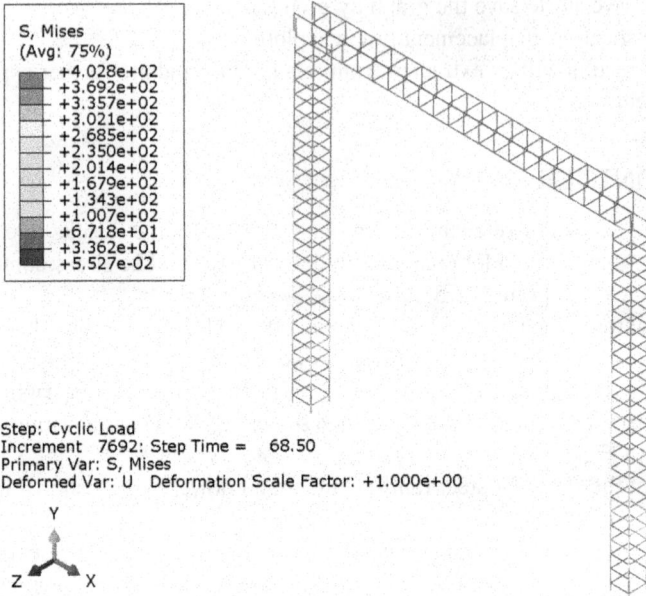

S, Mises
(Avg: 75%)
+4.028e+02
+3.692e+02
+3.357e+02
+3.021e+02
+2.685e+02
+2.350e+02
+2.014e+02
+1.679e+02
+1.343e+02
+1.007e+02
+6.718e+01
+3.362e+01
+5.527e-02

Step: Cyclic Load
Increment 7692: Step Time =   68.50
Primary Var: S, Mises
Deformed Var: U Deformation Scale Factor: +1.000e+00

Y

Z   X

FIGURE 4.76   Von Mises stresses contour distribution for steel reinforcement.

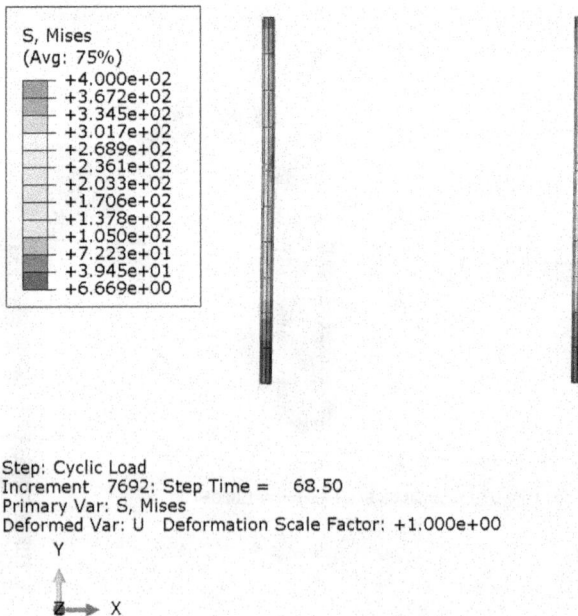

S, Mises
(Avg: 75%)
+4.000e+02
+3.672e+02
+3.345e+02
+3.017e+02
+2.689e+02
+2.361e+02
+2.033e+02
+1.706e+02
+1.378e+02
+1.050e+02
+7.223e+01
+3.945e+01
+6.669e+00

Step: Cyclic Load
Increment 7692: Step Time =   68.50
Primary Var: S, Mises
Deformed Var: U Deformation Scale Factor: +1.000e+00

Y

Z   X

FIGURE 4.77   Von Mises stresses contour distribution for dowel bars.

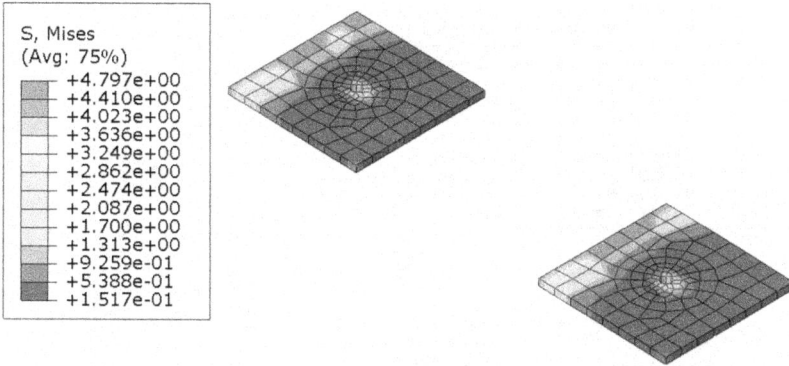

Step: Cyclic Load
Increment 7692: Step Time =   68.50
Primary Var: S, Mises
Deformed Var: U Deformation Scale Factor: +1.000e+00

FIGURE 4.78　Von Mises stresses contour distribution for rubber bearing pads.

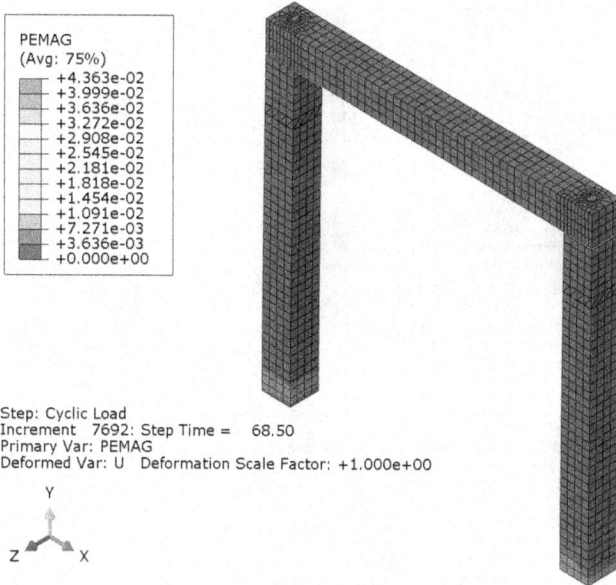

Step: Cyclic Load
Increment 7692: Step Time =   68.50
Primary Var: PEMAG
Deformed Var: U Deformation Scale Factor: +1.000e+00

FIGURE 4.79　Plastic strain contour distribution for precast concrete frame.

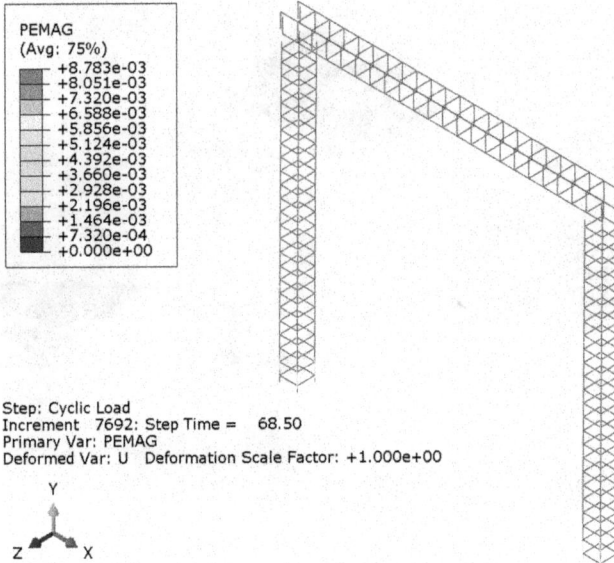

FIGURE 4.80 Plastic strain contour distribution for steel reinforcement.

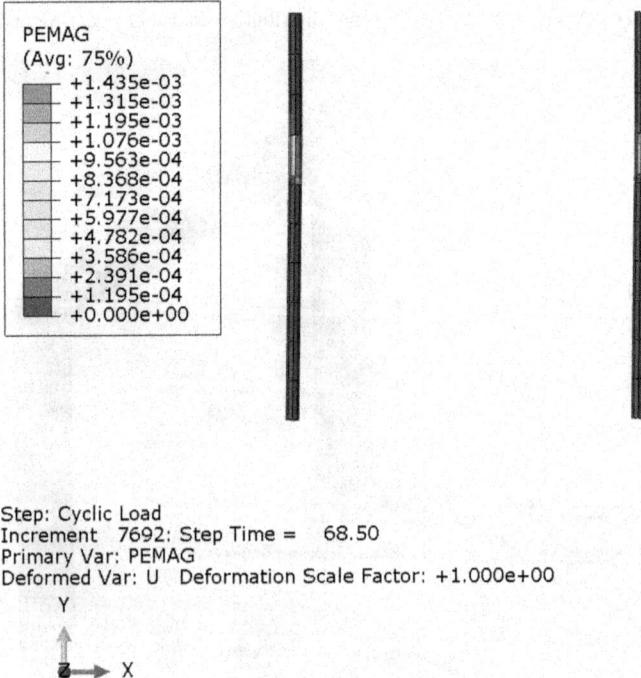

FIGURE 4.81 Plastic strain contour distribution for dowel bars.

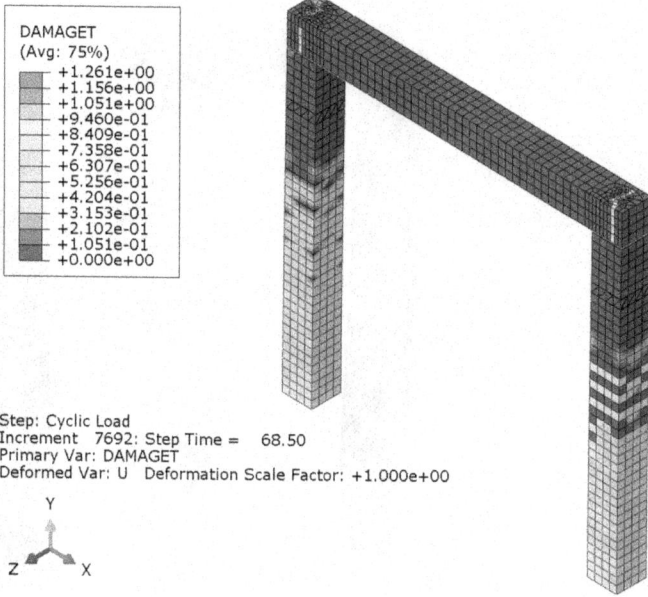

FIGURE 4.82 Concrete tensile damage contour distribution.

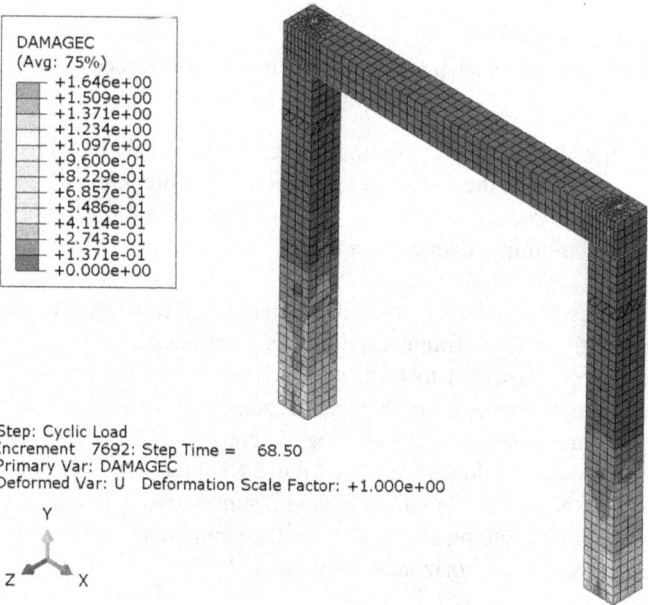

FIGURE 4.83 Concrete compressive damage contour distribution.

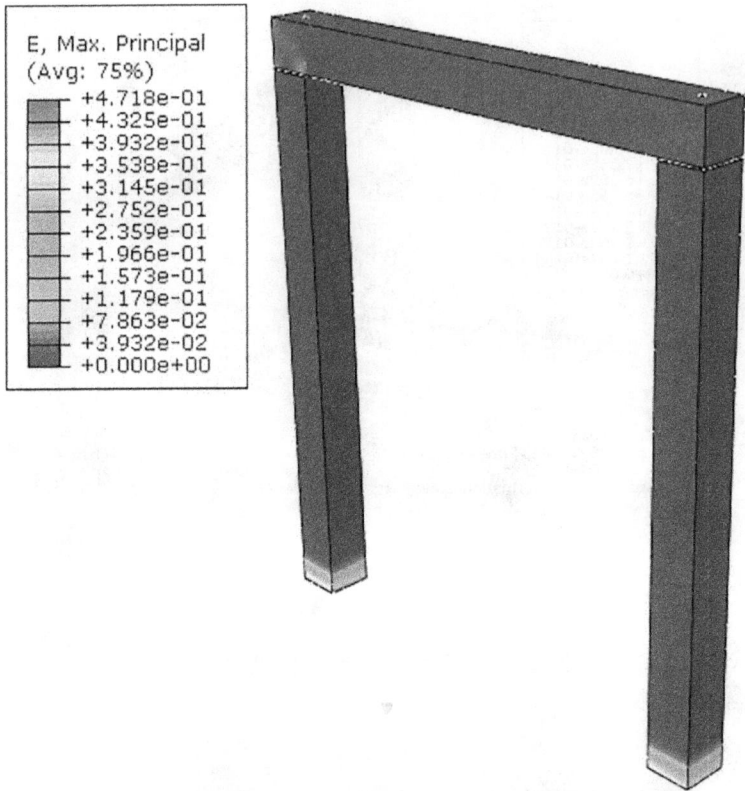

E, Max. Principal
(Avg: 75%)
+4.718e-01
+4.325e-01
+3.932e-01
+3.538e-01
+3.145e-01
+2.752e-01
+2.359e-01
+1.966e-01
+1.573e-01
+1.179e-01
+7.863e-02
+3.932e-02
+0.000e+00

FIGURE 4.84   Total maximum principal strain in the whole model.

To plot the displacements through time, double click on "Displacement" underneath XYData in the Result tree as shown in Figure 4.85.

To plot the "Force–Displacement" graph, double click on the "Plot" which is saved by combining displacement and force history output as shown in Figure 4.86.

From Figures 4.75 and 4.82, both column bases and concrete around the dowel connection of the precast frame significantly encounter high principal stresses. Hence, cracks are expected to form and propagate at those locations. Besides, the dowel bars also encounter high principal stress concentration within a 25 mm depth of the embedment length from the concrete-rubber surface. The attained 400 MPa stresses in the dowel bars (see Figure 4.77) show the dowel bars yielded at 25 mm of displacement demand. Table 4.6 summarizes the maximum principal stresses and absolute plastic strains for the components of the precast frame model during the 25 mm horizontal displacement.

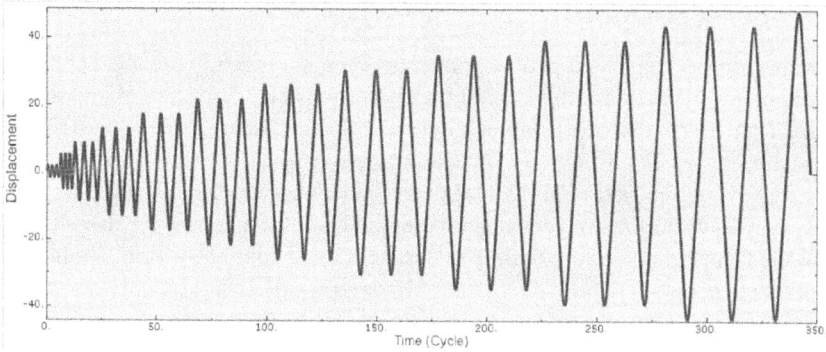

FIGURE 4.85   Displacement vs. time graph.

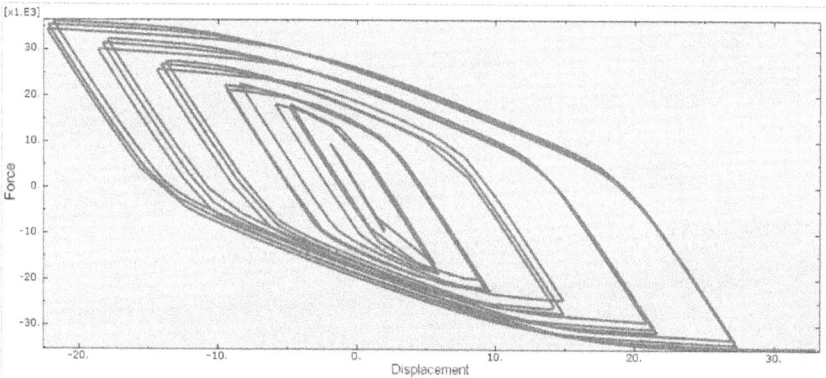

FIGURE 4.86   Force vs. displacement hysteresis and envelop curves.

TABLE 4.6
Von Mises Stresses and Magnitude of Plastic Strain

Components	Von Mises Stresses (MPa)	Magnitude of Plastic Strains
Concrete frame	27.8	9.88×10^{-2}
Steel reinforcement	400.6	9.78×10^{-4}
Dowel bars	322.4	2.943×10^{-4}
Rubber bearing pad	7.8	–

4.8 CONCLUSION

A single dowel-connected precast frame has been numerically modeled and analyzed under the action of horizontal cyclic loading. The deformation and force–displacement curves of the model were obtained. Overall, the precast frame model achieved its ultimate force capacity. However, its strength began to deteriorate beyond the 40 mm horizontal displacement demand, causing the numerical complexity to the convergence problem. The numerical results also showed that the formation of cracks mainly occurred at the column bases and around the dowel connection.

5 Simulation of the Preloaded Bolt Connection under Cyclic Loading

5.1 INTRODUCTION

Nowadays, implementing bolted joints for beam–column connections in the structures has increased dramatically due to their advantages such as rapid construction process and high quality of welding parts.

In this chapter, the simulation process of the preloaded bolt connection under cyclic load by the implicit method using ABAQUS/standard finite element package has been explained step by step. The displacement control analysis has been conducted and then analysis results have been discussed by focusing on the energy dissipation capacity of the considered connection in the plastic range.

5.2 PROBLEM DESCRIPTION

The model contains the column, beam, L-plate, and bolts, and a schematic representation of the model is shown in Figure 5.1.

Both the beam and column sections are considered as "IPE400" and the bolts and nuts are "M16". All dimensions of the components are shown in Figures 5.2 and 5.3.

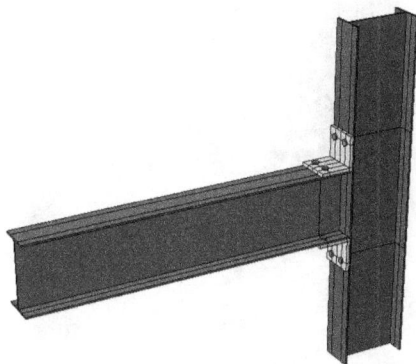

FIGURE 5.1 Bolt connection assembly.

DOI: 10.1201/9781003219491-5

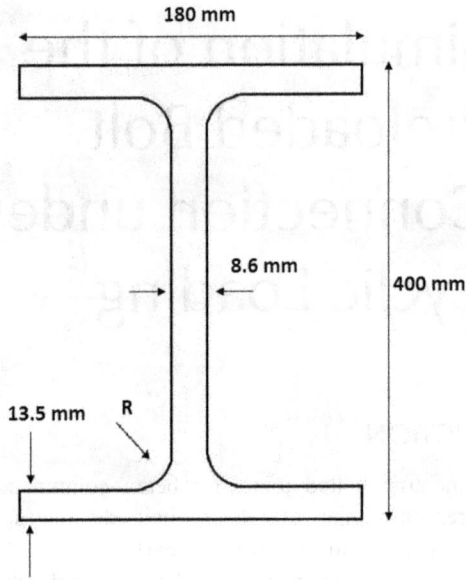

FIGURE 5.2   IPE400 dimensions.

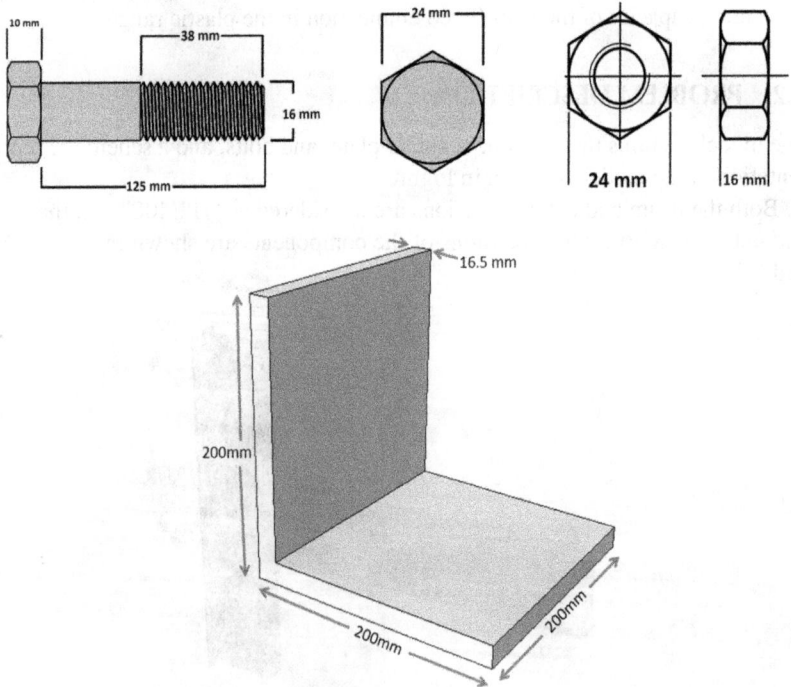

FIGURE 5.3   Bolt, Nut, and L-plate dimensions.

Both the beam and the column are 2 m in length. For the simplification of the model, the corner with "R" radius has been ignored.

All small fillets and treading are ignored for the bolts and nuts.

5.3 OBJECTIVES

1. To simulate preloaded bolt connections using the finite element method.
2. To implement cyclic incremental loading method for the time-dependent problems.
3. To investigate load–displacement and plastic energy dissipation in preloaded bolt connections.

5.4 MODELING

Run the software from the start menu and then close the "Start Session" dialog box (Figure 5.4).

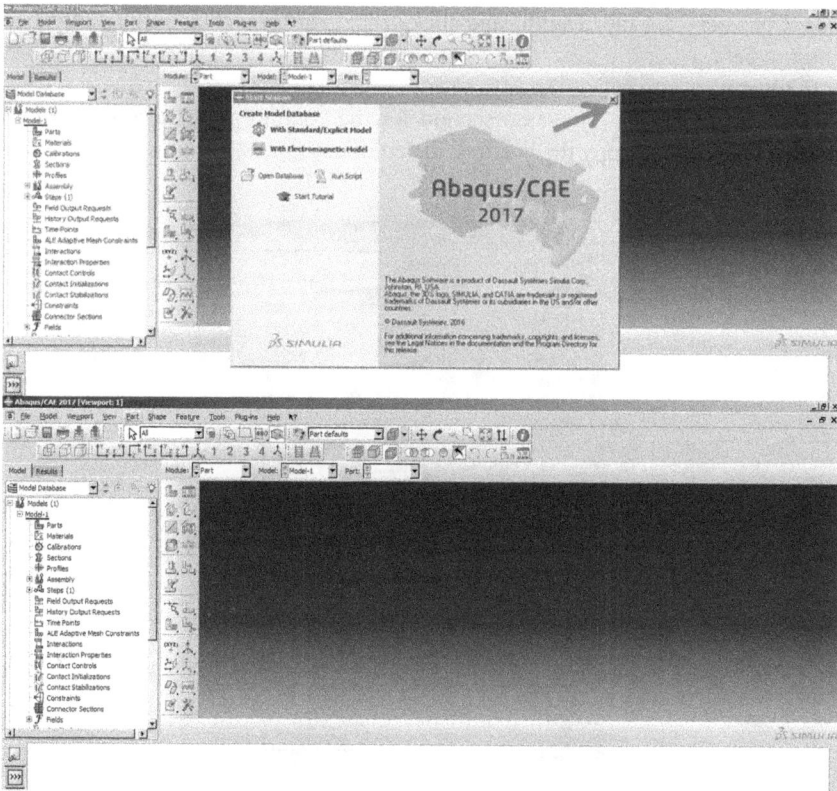

FIGURE 5.4   Run ABAQUS/CAE.

5.4.1 Part Module

In the first step, the geometry of the considered model must be defined. In this example, the geometry contains the beam, column, bolt, and L-plate and their geometry should be considered as three dimensional and deformable. Moreover, they are assumed to be solid with simple geometries for modeling. The drawing method, in all components, is considered as "Extrusion".

Beam:

Double click on "Parts" in the "Model tree" and modify the "Create Part" dialog box, as shown in Figure 5.5. The sketcher approximate size is considered to be 2. Click "Continue" to open sketcher.

Draw a profile similar to the "IPE" sections by using the "Create Lines; Connected" tool (Figure 5.6).

Select "Add Constraint" and make corresponding lines of equal length, as shown in red (Figure 5.7).

Select "Add Dimension" and determine the dimensions according to Figure 5.2. If the drawing changes significantly, correct it using the "Drag entity" tool. The dimensions of drawing are shown in Figure 5.8.

Click the middle mouse button twice to open the "Edit base extrusion" dialog box. Consider 2 for the "Depth" and click "Ok" to apply and close the dialog box (Figure 5.9).

Column:

For simplicity, use the same section for the column as there is no need to redraw the section. Copy the beam as the column.

FIGURE 5.5 Creating the beam.

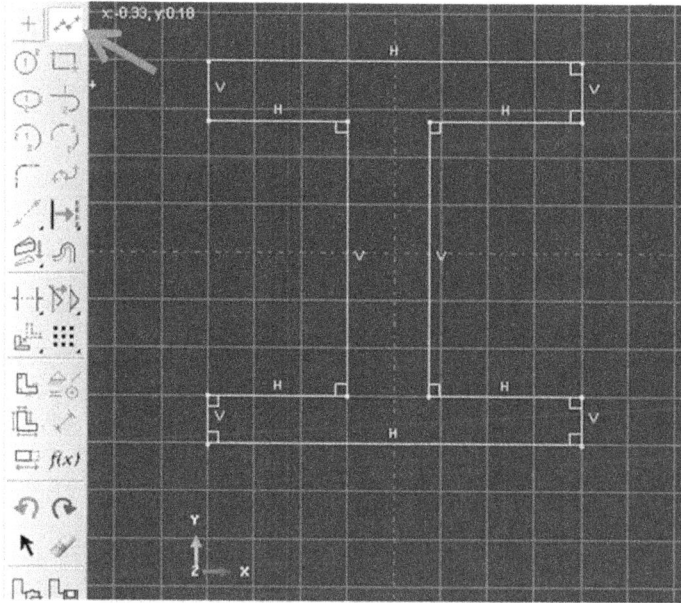

FIGURE 5.6 Draw the IPE section sketch.

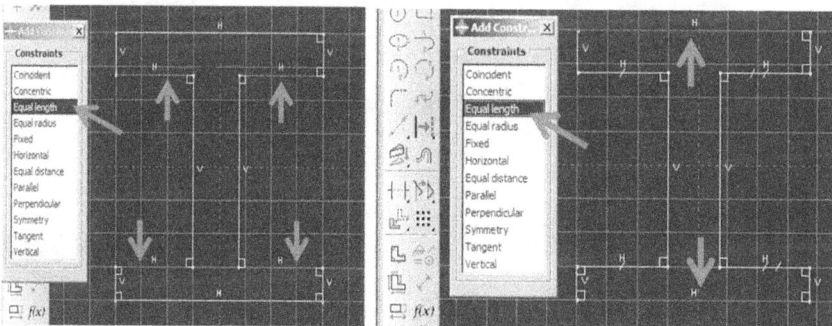

FIGURE 5.7 Equal length constraints.

Right click on "Beam" in the "Model tree" and select "Copy", and then rename it as "Column" but do not change any other settings. Then click "Ok" (Figure 5.10).

Bolt and nut:

In the next step, the bolt and nut need to be drawn.

Double click on "Part" in the "Model tree" and change the dialog box as shown in Figure 5.11 and then click "Continue".

Draw a hexagon: This must be drawn by "Create Lines; Connected" and then using the equal length tool. All six lines should be equal (Figure 5.12).

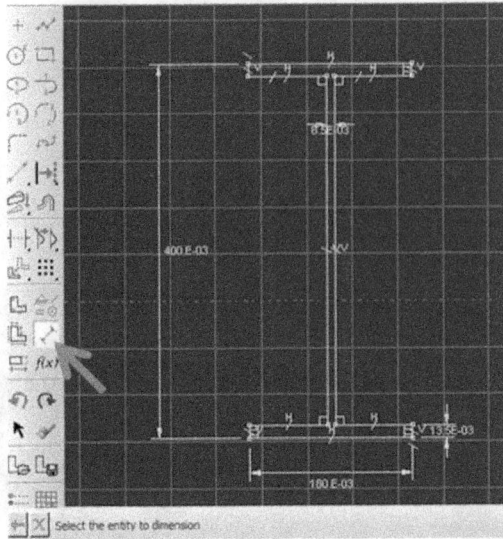

FIGURE 5.8   Add dimensions to drawing.

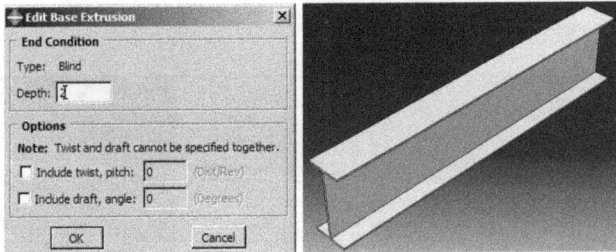

FIGURE 5.9   Extrusion of the drawing.

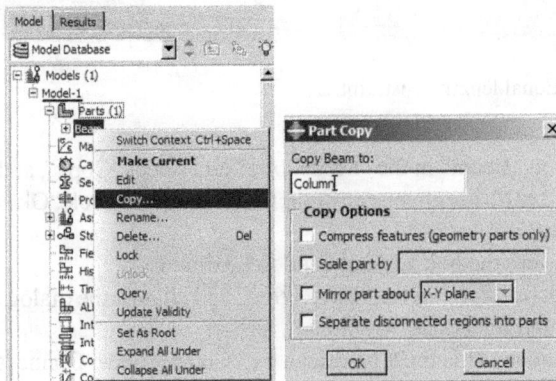

FIGURE 5.10   Copy of the beam as a column.

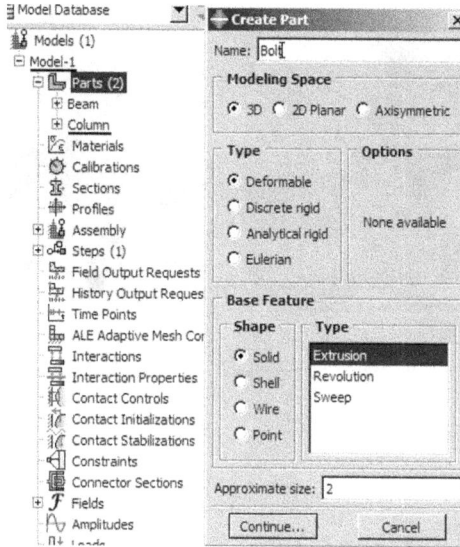

FIGURE 5.11 Create bolt and nut.

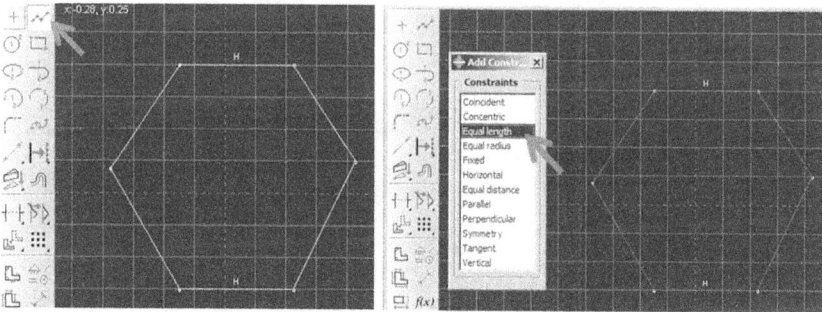

FIGURE 5.12 Draw the bolt head.

Select "Add dimension" and set the angles between two lines at "120°". This needs to be done for only two corners. Finally, according to Figure 5.2, set the distance between the two lines at 24 mm (Figure 5.13).

Click the middle mouse button twice to open the "Edit base extrusion" dialog box. Consider 0.01 for "Depth" and then click "Ok" to apply and close the dialog box (Figure 5.14).

In the next step, the body of the bolt needs to be drawn. Select "Shape → solid → extrude" and then click the front surface and choose an edge to open sketcher (Figure 5.15).

Draw a circle: Click "Add dimension" to define 0.008 for the radius.

Note: Since only 3 cm of bolt length is used in the connection, the length of the bolt (i.e., between bolt and nut) is considered as 3 cm. Click on the middle mouse

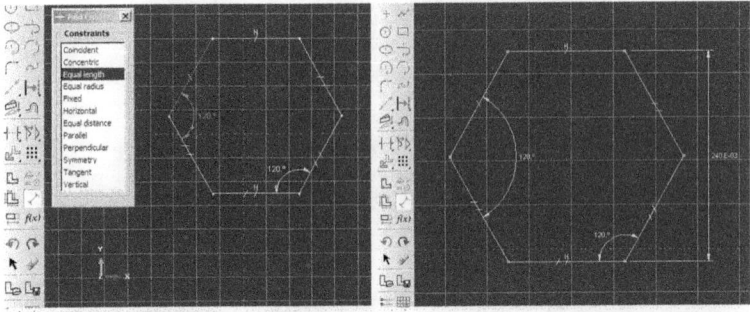

FIGURE 5.13   Add dimensions to the bolt head.

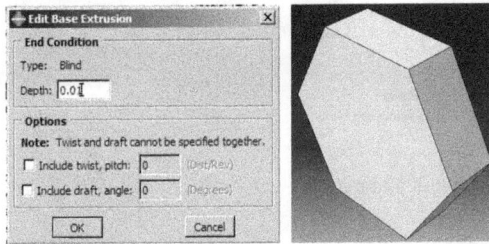

FIGURE 5.14   Extruding bolt head.

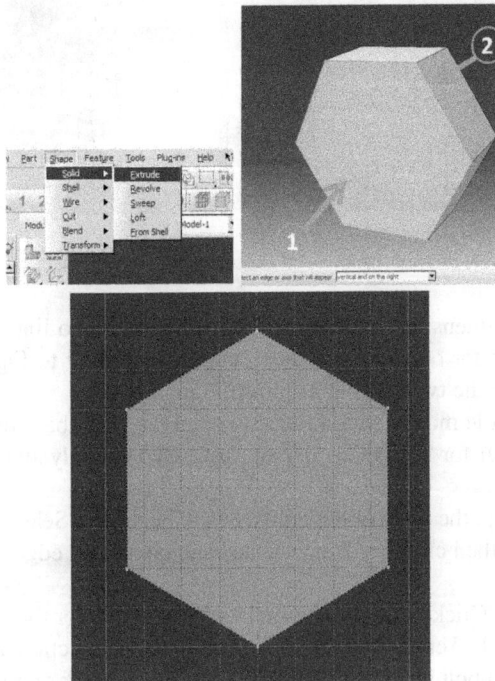

FIGURE 5.15   Bolt body: Open sketcher.

button twice to open "Edit base extrusion", and consider 0.03 for the depth and click "Ok" to apply and close the dialog box (Figure 5.16).

Since there is no movement between the bolt and the nut, for the simplicity of modeling, they can be modeled and viewed as connected.

Select the "Shape → Solid → Extrude", and then on the end of the bolt surface as the surface draw sketch, select the circumferential edge as an edge to open sketcher (Figure 5.17).

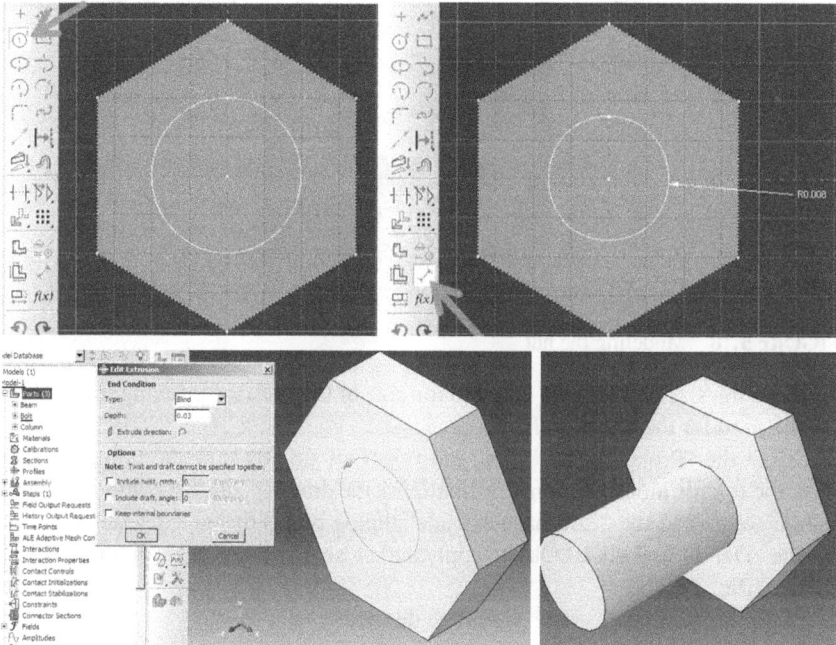

FIGURE 5.16   Drawing bolt body.

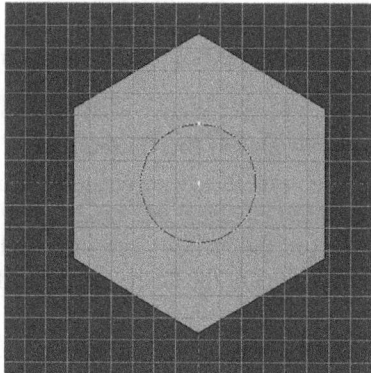

FIGURE 5.17   Open sketcher to draw the nut.

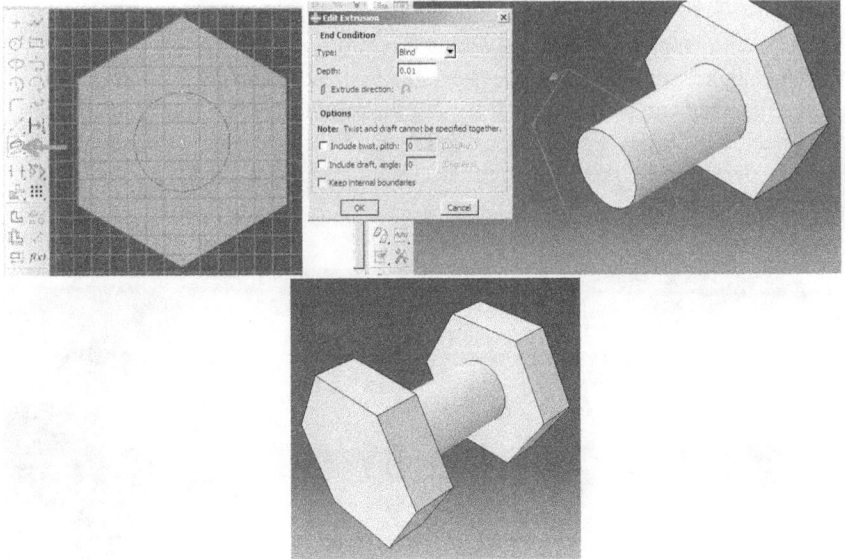

FIGURE 5.18 Modeling the nut.

Since the size of the nut is equal to the size of the bolt, the drawing for the bolt can be used for the nut as well.

Select the "Project edges" tool, and select all six edges of the bolt head and click the middle mouse button. For finalizing the drawing, click the middle mouse button twice, and the "Edit base extrude" dialog box is displayed. Consider 0.01 for the "Depth" and click "Ok" to apply and close the dialog box (Figure 5.18).

L-plate:

In this section, the L-plate should be defined.

Double click on "Parts" in the "Model tree" to change the "Create Part" in the dialog box (Figure 5.19) and click "Continue". Draw an L-section by "Create Lines: Connected" and then define the dimensions by using "Add dimension", as shown in Figure 5.19.

Click the middle mouse button twice to open the "Edit base extrusion" dialog box. Consider 0.2 as the "Depth" and click "Ok" to apply and close the dialog box (Figure 5.20).

5.4.1.1 Modifying Parts for Use in the Connection

In this step, it is necessary to correct each component for use in the connection.

The beam needs to consider the bolt holes at the end of the beam.

Double click on beam from the "Model tree", then select "Shape → Cut → Extrude", then select the top wing of the beam surface and click an edge to open the sketch area. The dimensions are shown in Figure 5.21. Finally, exit the sketch and cut the circles through all parts.

Note: The holes are applied on both the top and bottom wings (Figure 5.21).

Due to the connection geometry, four similar holes should be defined in the column. So, repeat the same procedure, as for the beam, in order to create the

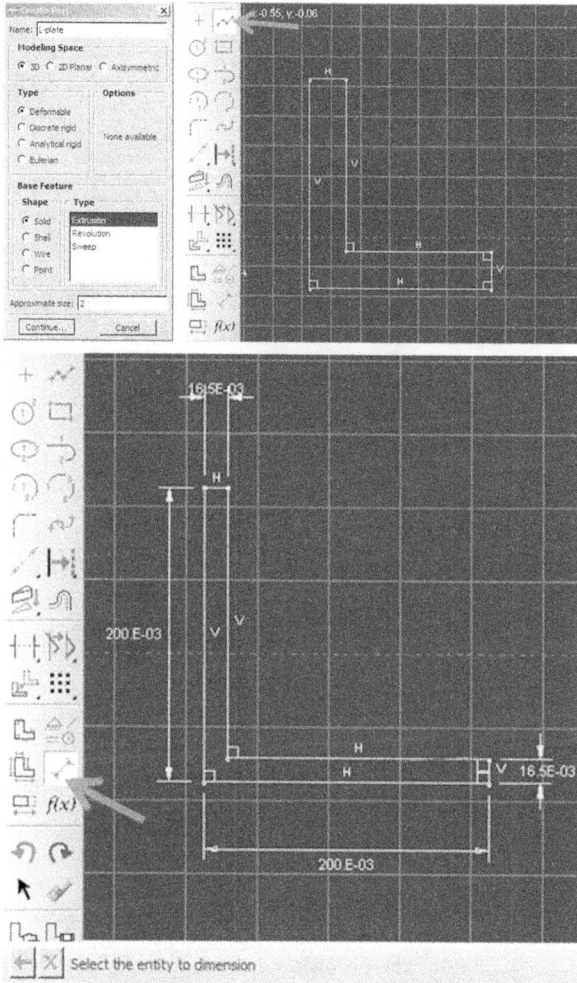

FIGURE 5.19 Drawing the L-plate.

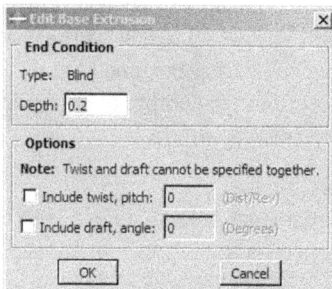

FIGURE 5.20 Modeling the L-plate.

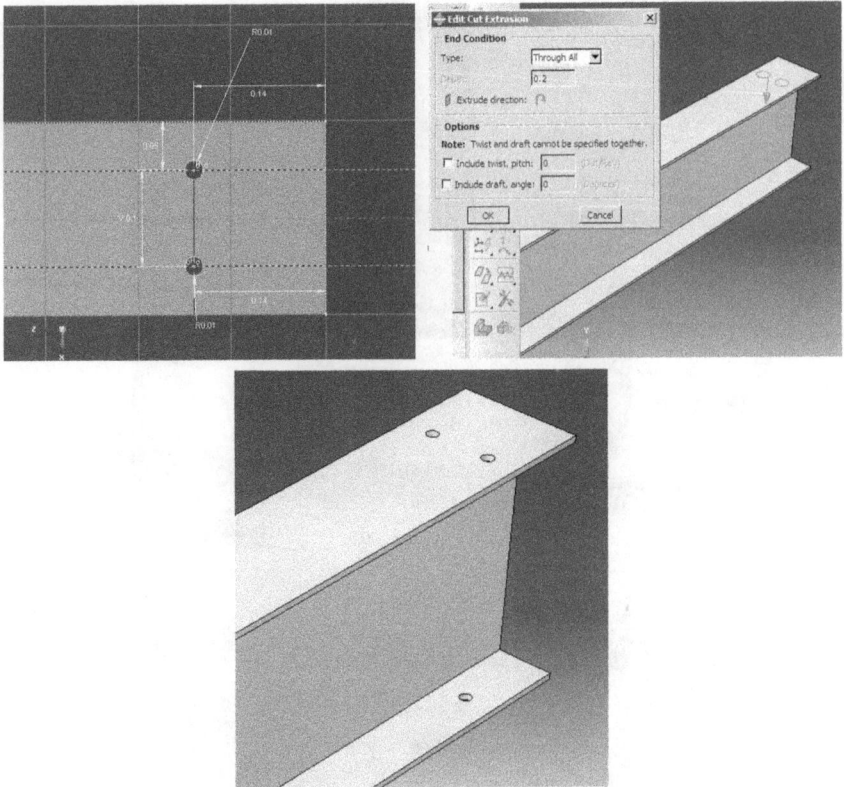

FIGURE 5.21   Creating the connection holes for the beam.

holes, as shown in Figure 5.22. However, the holes should be created just on one wing.

Double click on column in the "Model tree", then select "Shape → Cut→ Extrude" and choose the upper flange of the column and the right edge as the vertical edge (Figure 5.22).

The L-plate should be modified by creating holes in both areas.

Double click on L-plate from the Model Tree, and then select the "Shape → Cut → Extrude" option and choose one area and an edge to open sketcher. Draw two circles and add dimensions according to Figure 5.23. Click on the mouse middle button twice to open the "Edit cut extrusion" dialog box and click "Ok" to apply and close the dialog box. Repeat the procedure for another area (Figure 5.23).

5.4.2 MATERIAL PROPERTIES

In this step, material models should be defined. Here, all components are considered to be steel with the mechanical properties as proposed in Table 5.1.

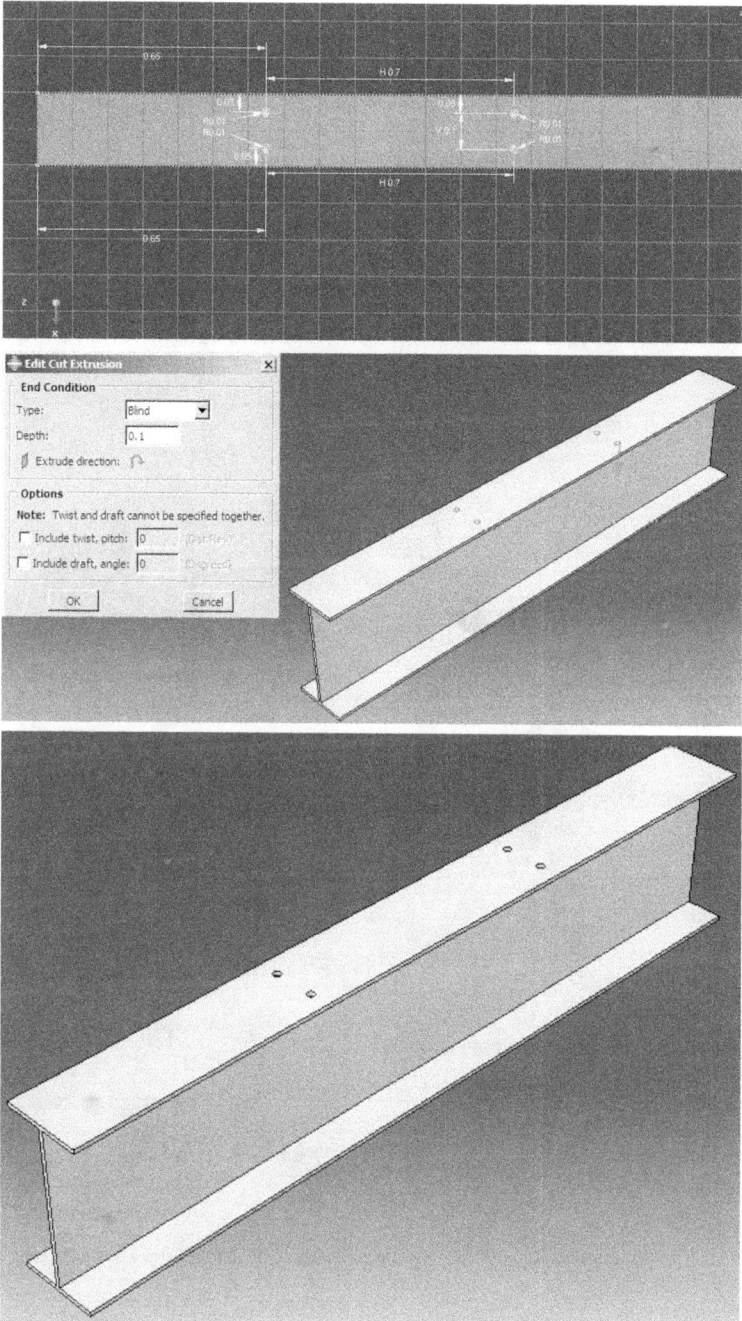

FIGURE 5.22   Creating the connection holes for the column.

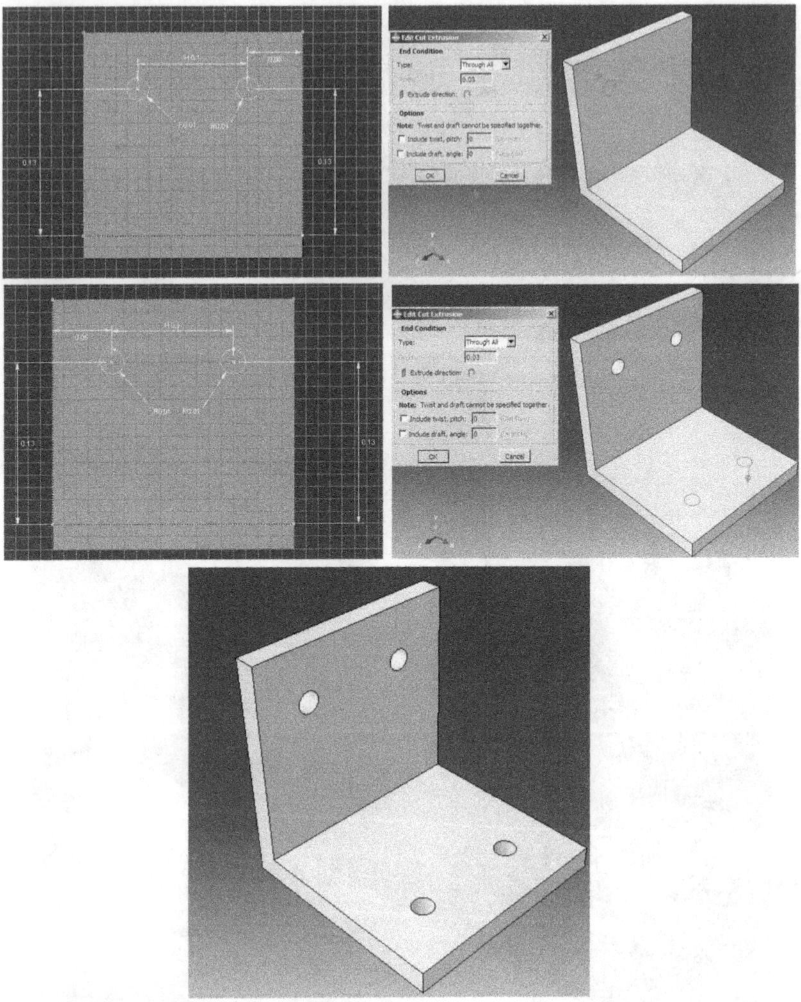

FIGURE 5.23 Creating the connection holes in the L-plate.

TABLE 5.1
Mechanical Properties for All Components

	Elasticity		Plasticity	
	Young's Modulus	Poisson's Ratio	Yield Stress	Plastic Strain
Property	(GPa)		(MPa)	
Value	200	0.3	300	0
			380	0.01
			470	0.02

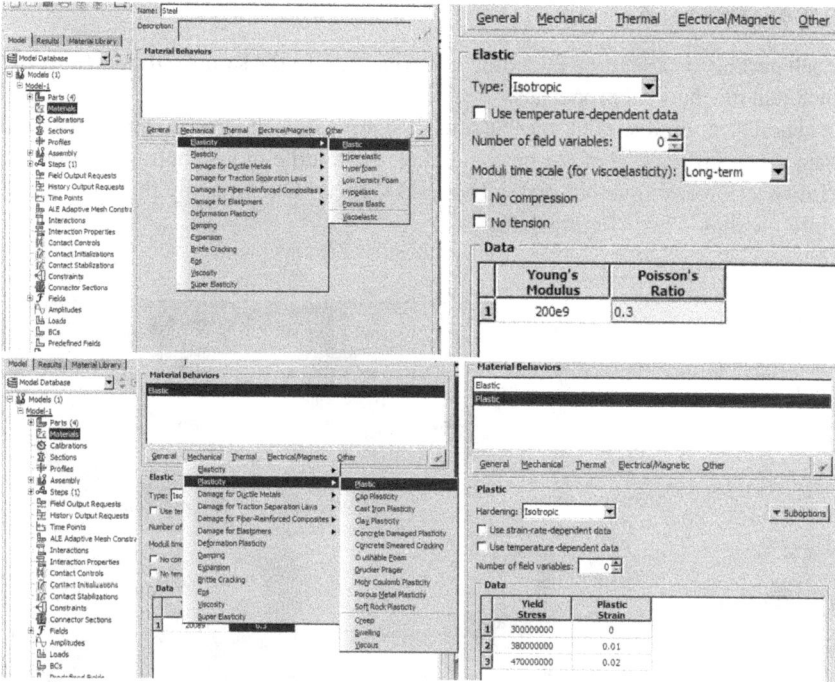

FIGURE 5.24 Defining the material model.

Double click on "Materials" in the "Model tree" and define all properties, as shown in Figure 5.24.

5.4.3 Section Properties

In the sample, the material was assumed homogenous and considered as such in the section definition.

Double click on the "Sections" in the "Model tree" and select "Solid, Homogenous" and click "Continue". On the subsequent dialog box, select the material defined recently (Figure 5.25).

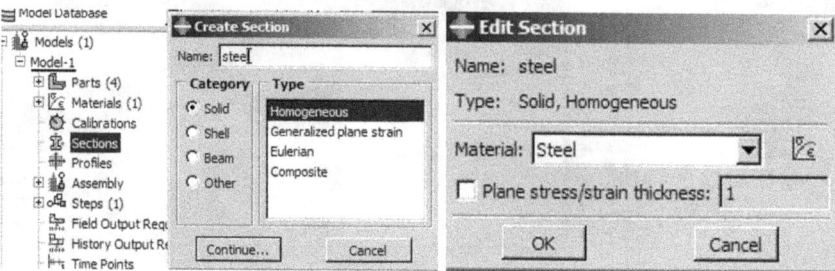

FIGURE 5.25 Defining the section.

5.4.4 Section Assignment

Each part needs a section, including all properties. In the sample, the section only included mechanical properties.

Therefore, double click on "Section assignments" underneath each part, and then select all regions of the part and click the middle mouse button to open the "Edit section assignment" dialog box. Next, click "Ok" to apply the section and close the dialog box (Figure 5.26).

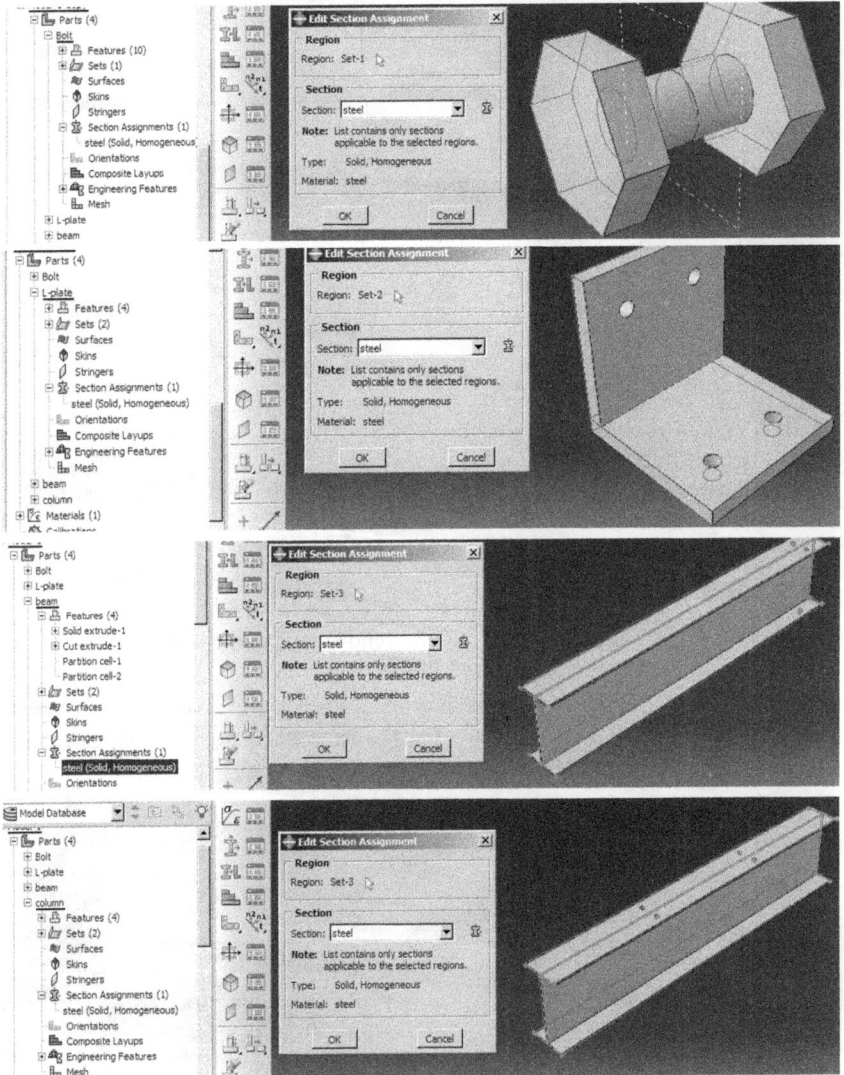

FIGURE 5.26   Assigning the section.

5.4.5 Assembly Module

In the next step, all parts need to be correctly combined.

Double click on Instances underneath "Assembly" in the "Model tree" to open the corresponding dialog box. Then, click on beam and "Ok" to add a beam in the assembly (Figure 5.27).

In order to display the parts in the assembly, choose "Parts" from the "Color map" section and change the color in this level (Figure 5.28).

Double click on "Instances" to add the column. Next, select the "Auto-offset from other instances" for a better view. In this case, along with the beam, the column will exist and click "OK" (Figure 5.29).

To make the assembly more accurate and easier, geometric datum points using the datum tool should be defined in the middle of the column top wing and middle of the beam flange.

Select "Tools → Datum" and then choose "Midway between two points" as a point datum "Type", as shown in Figure 5.30.

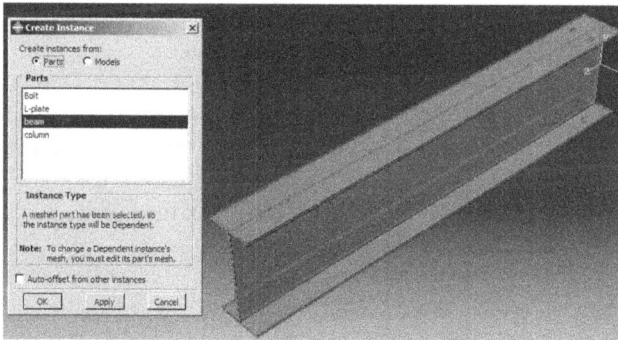

FIGURE 5.27 Adding a beam in assembly.

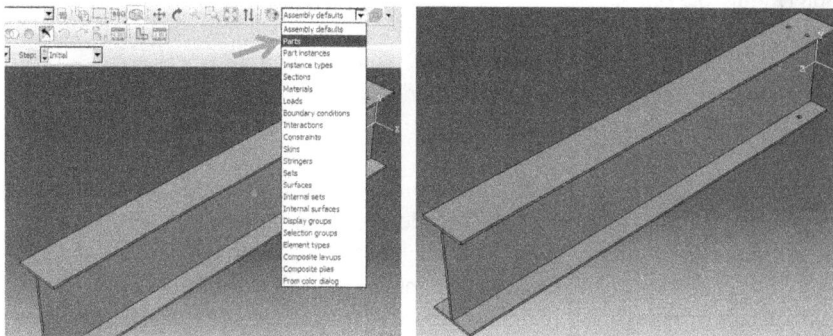

FIGURE 5.28 Recoloring instances in the assembly to display better.

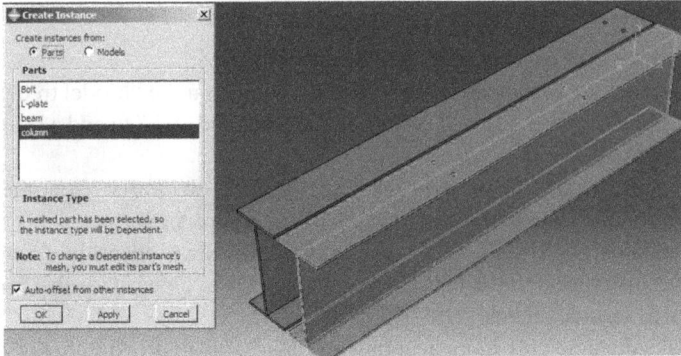

FIGURE 5.29   Adding the column in the assembly.

A 1 cm distance is assumed between the beam and the column. Therefore, offset a datum point from a point in the middle of the beam flange according to (0, 0, –0.01).

Select "Offset" from a point in the "Create datum" dialog box and consider the offset of 0.01 off from the beam's datum (Figure 5.31).

In the next step, it is necessary to rotate the column.

Select "Instance → Rotate". Then, choose column instance from the instance to rotate and then click the middle mouse button. Next, choose the points, as shown in Figure 5.32, that define the axis of rotation. Click the middle mouse button to accept 90° as the angle of rotation. Then, click the middle mouse button to accept the new position (Figure 5.32).

Select "Instance → Translate" and choose the column instance and click the middle mouse button. Then, select the points created earlier, as shown in Figure 5.33, to reposition properly.

In the next step, the L-plate should be added to the assembly.

Double on "Instances" underneath the assembly and select L-plate to add in the assembly, then rotate 90° and then translate it so that it is in the correct position (Figure 5.34).

Perform the same process to generate another L-plate into the assembly, then rotate it 90° twice, and finally translate it, as shown in Figure 5.35.

In the next step, the bolt should be added and positioned in the assembly. For the part, first a bolt will be imported and positioned; then using the linear pattern, other bolts will be added to the assembly.

Double click on Instances and select bolt. Select "Constraint → Coaxial", and then select bolt body and the L-plate hole inside, respectively. Note that arrows should be in the same direction, if not, click flip in command prompt and click "Ok" in the prompt for coaxial to be done. Select "Constraint → Face to Face" and select bolt head face and L-plate face, respectively. Again, observe the arrows to see if they are in the same direction. If not, select Flip in the prompt and click "Ok" to apply the changes, and consider the offset of 0 (Figure 5.36).

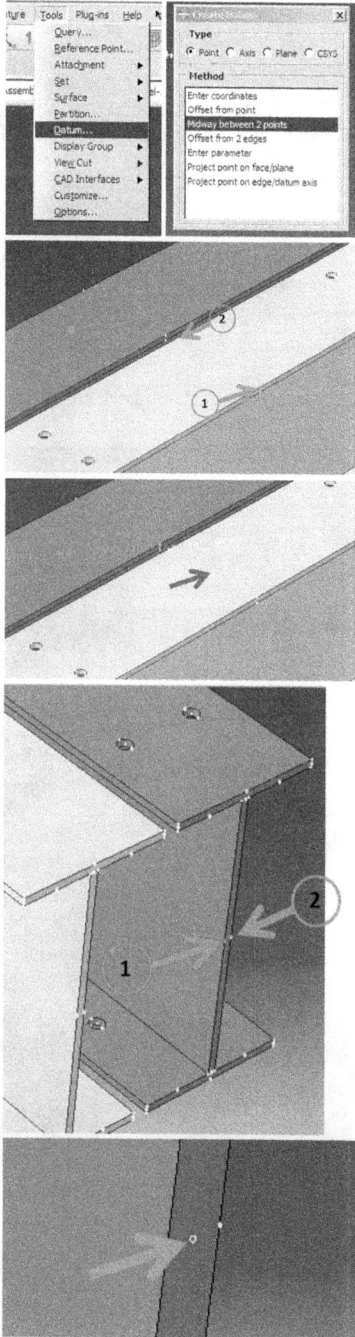

FIGURE 5.30 Defining the datum point in the middle of the beam flange and column top face.

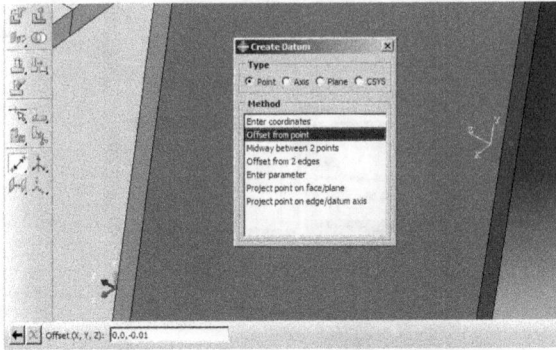

FIGURE 5.31   Defining a datum point of distance 1 cm from the beam to the column.

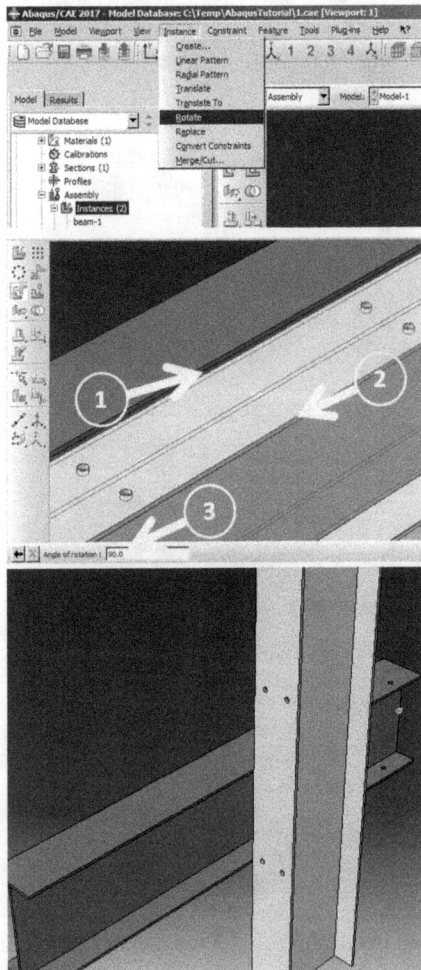

FIGURE 5.32   Rotating column.

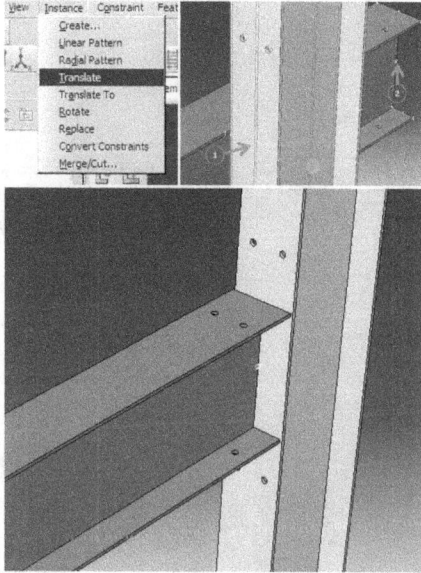

FIGURE 5.33 Translating column to beam.

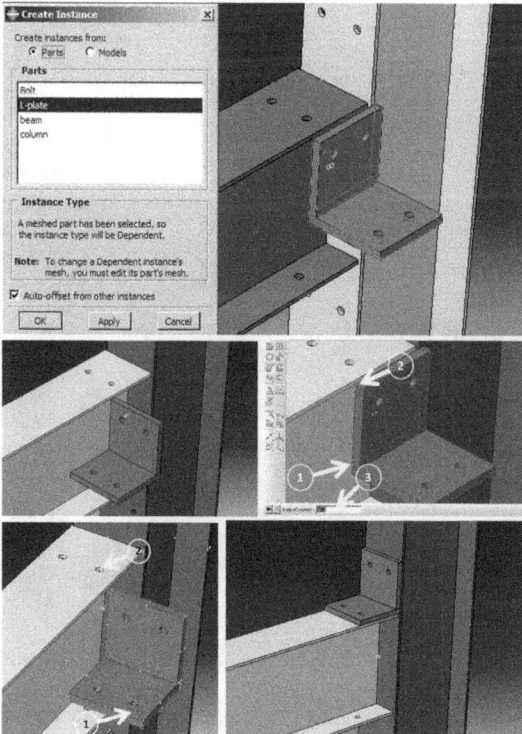

FIGURE 5.34 Adding and positioning the first L-plate in the assembly.

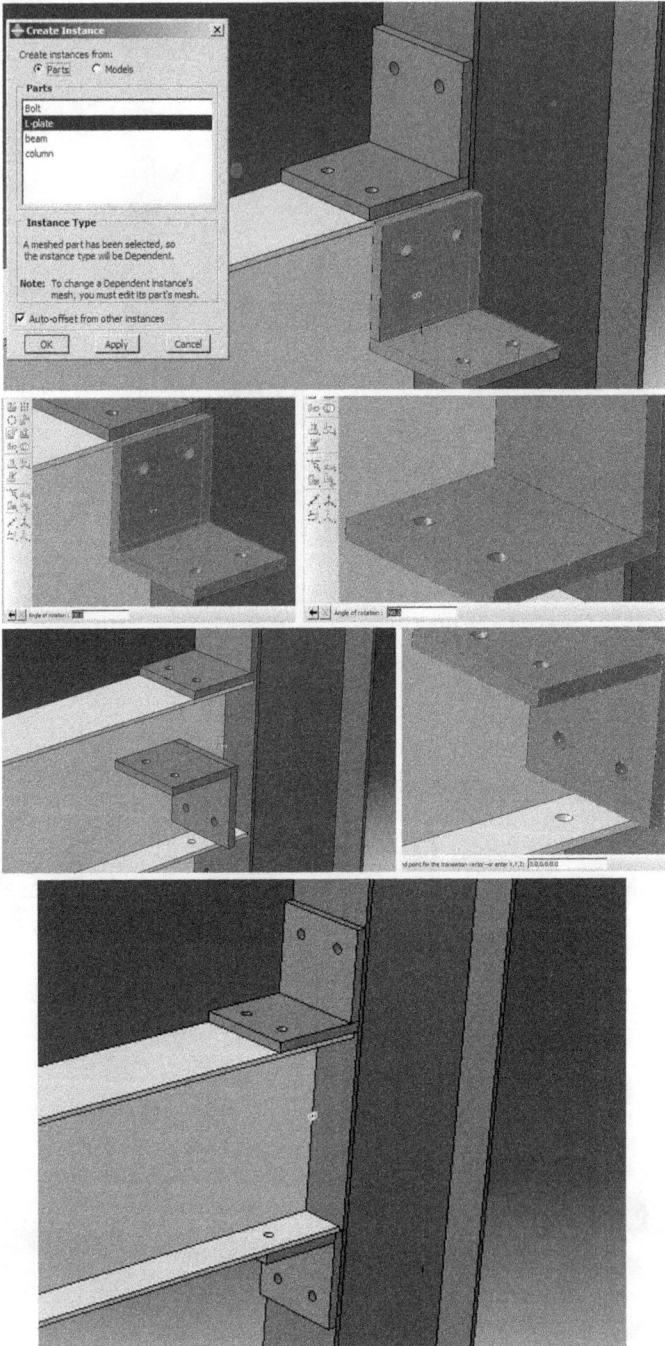

FIGURE 5.35   Adding and positioning the second L-plate in the assembly.

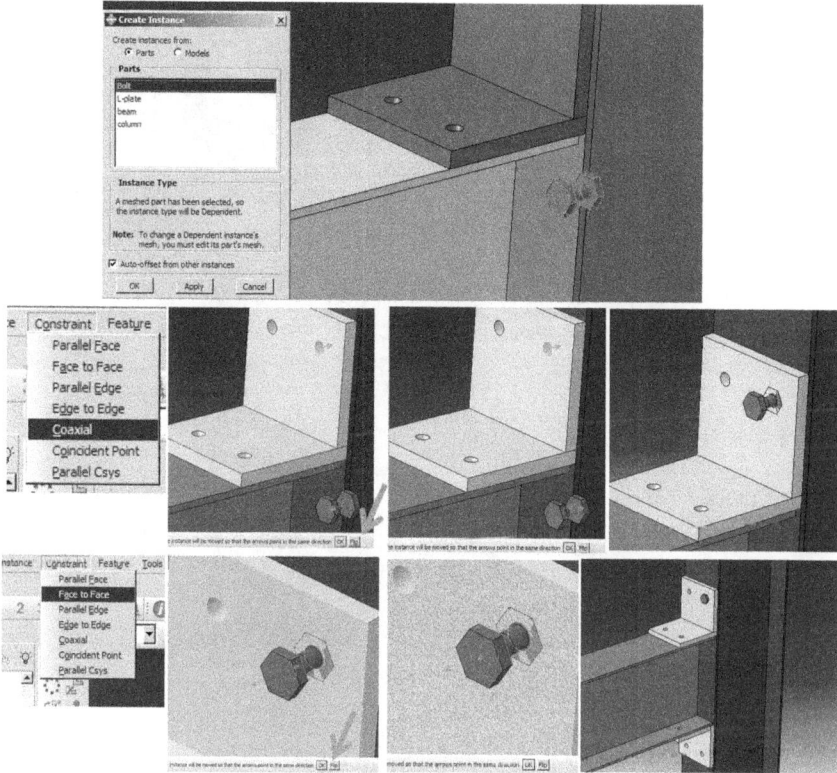

FIGURE 5.36   Adding and positioning the bolt.

At this point, the bolt should be a four-directional arrangement.

Select "Instance → Linear Pattern" and choose the bolt and click the middle mouse button to open the "Linear Pattern" dialog box. In "Direction 1", enter 0.1 and in "Direction 2" enter 0.7 for the offset. Check that the bolts are in the correct position; if not, click "Flip" in the corresponding direction (Figure 5.37).

Repeat the procedure and add a bolt once again via the "Create Instance" dialog box. The bolt should be rotated in order to be positioned vertically. Use "Coaxial" and "Face to Face" constraint to put the bolt in the correct position and then use "Linear Pattern" to generate a 2×2 array for the bolt as shown in Figure 5.38.

5.4.6   Meshing Module

At this point, all components should be meshed. If some complexities occur for generating the mesh, the user should simplify the component geometry using suitable tools.

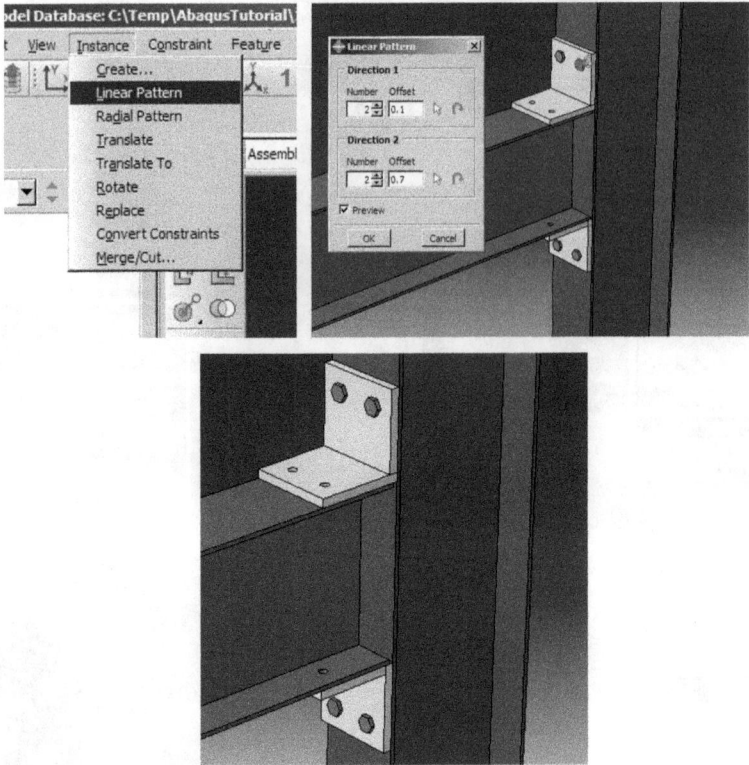

FIGURE 5.37   Linear pattern a bolt.

Beam:

Double click on "Mesh (empty)" underneath the part beam in the "Model tree". The beam color will be changed to orange color (Figure 5.39).

The orange color in the software means that the shape is not ready to mesh. To prepare the shape for meshing in the software, the shape should be partitioned into different parts.

To partition the beam, select "Tools → Partition", select "Cell" for the partition "Type", and use the "Define" cutting plane tool as the partition method (Figure 5.40).

Define the cutting plane by choosing point and normal that separate the flange from the wings. Select the point and a vertical edge, as shown in Figure 5.41, to perform this. Recoloring the top wing shows that it can be meshed, but the bottom wing remains. Repeat the same procedure for the bottom wing. When the beam is completely recolored to green and yellow, meshing can be defined (Figure 5.41).

At this point, the size of the elements should be defined. As the minimum size in the beam is about 0.02, this is suitable for consideration as an element size.

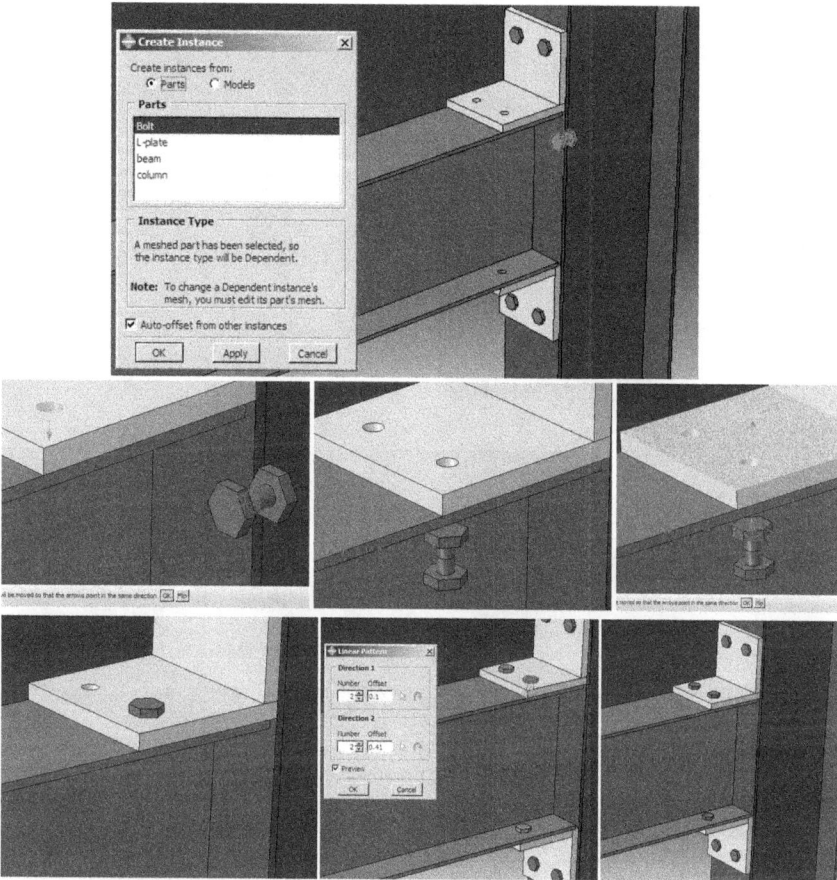

FIGURE 5.38 Adding and positioning other bolts.

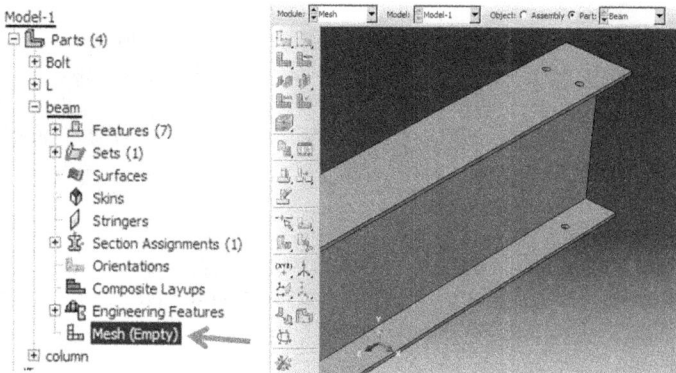

FIGURE 5.39 Importing the beam into the mesh module.

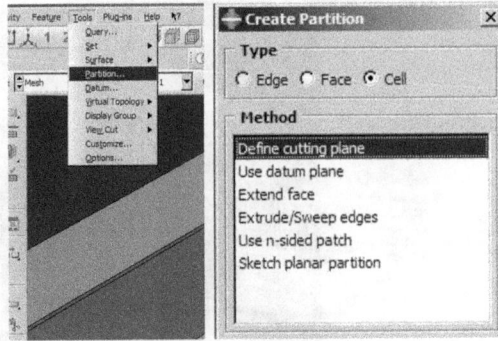

FIGURE 5.40   Defining type and method of partition.

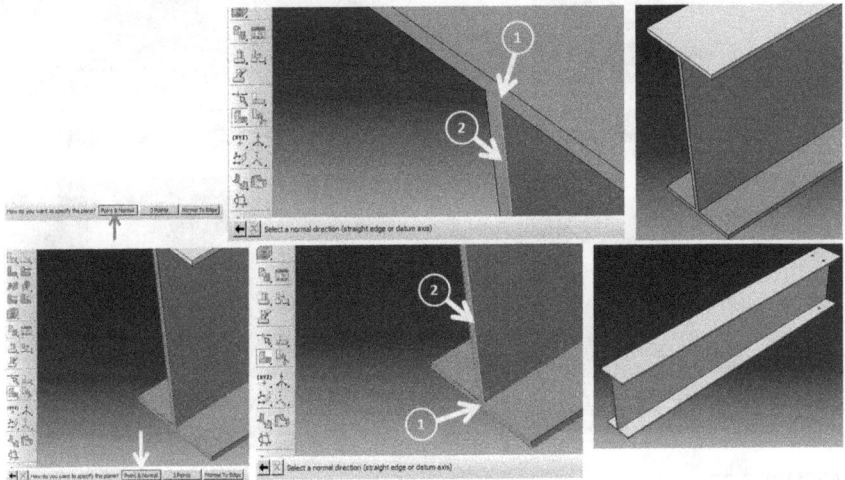

FIGURE 5.41   Partitioning the beam.

Select "Seed → Part" and open the "Global seeds" dialog box. Then, enter 0.02 for the "Approximate global size" and apply. Click "Apply" to see the nodes' approximate positions, and then click "Ok" (Figure 5.42).

Select "Mesh → Part" and click "Yes" in the bottom of the prompt area to generate mesh, as shown in Figure 5.43.

Column:

Double click on "Mesh (Empty)" underneath the column on the "Model tree" to open the "Mesh" module.

As shown, the column colored orange means that the part cannot be meshed. As the beam and column are nearly similar, the same procedure can be used for the column.

Select "Tools → Partition" and choose to define cutting plane as the method. Then select "Point and normal" for the "Method" of defining a cutting plane.

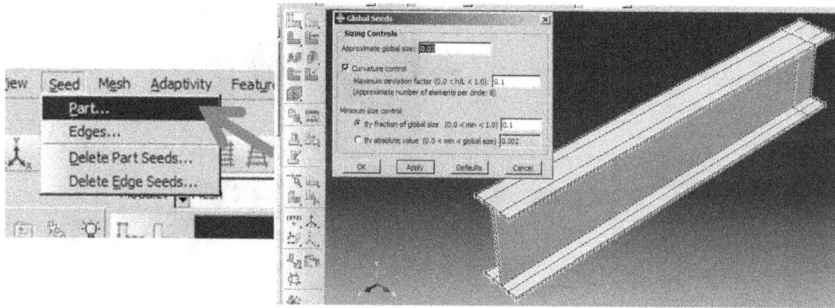

FIGURE 5.42   Defining the element size by seed.

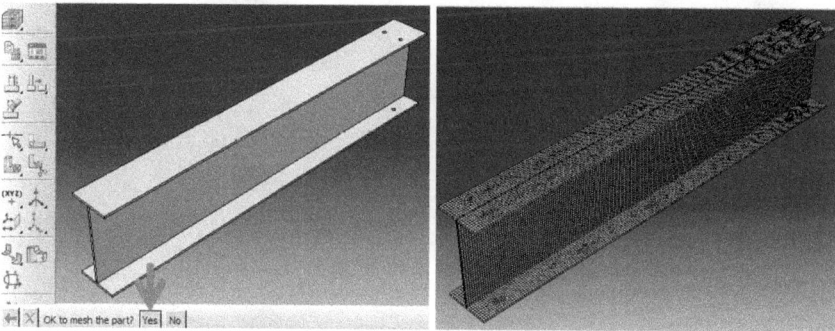

FIGURE 5.43   Meshing of the beam.

Then, select "Point and edge", as shown in Figure 5.44, and click "Create Partition". Column recoloring and could be assigned a mesh.

For meshing the column, the element size should first be defined.

Select "Seed → Part" to open the "Global seed part" dialog box. Similar to the beam, consider the column elements size of 0.02 and select Apply and then "Ok" to apply the changes and close the dialog box.

Select "Mesh → Part" and choose "Yes" from the prompt area to the meshing column, as shown in Figure 5.45.

Bolt:

Double click on "Mesh (Empty)" underneath the bolt to open the "Mesh" module. As can be seen, the bolt is colored orange, which means it cannot be meshed by default. So, partitions are needed to simplify the geometry.

Select "Tools → Partition" and choose to define the cutting plane as the partition method. The plane should be defined by "Point and Normal". Then, select the point and edge, as shown in Figure 5.46, to create a partition. Repeat the procedure for the other end. Finally, the bolt and the nut turn yellow.

In order to enable defining preload on the bolt, it needs to define the plane middle of the bolt.

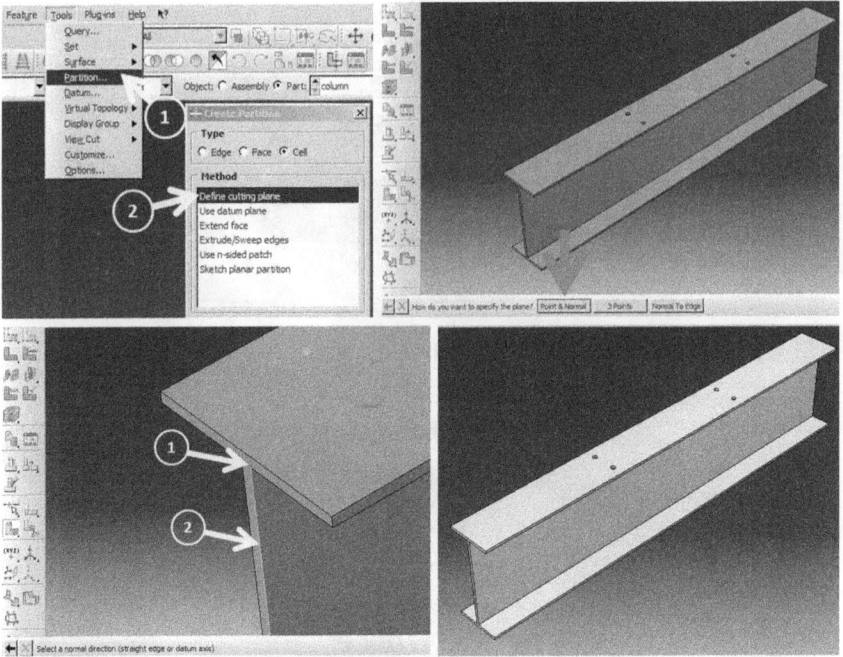

FIGURE 5.44   Partitioning the column.

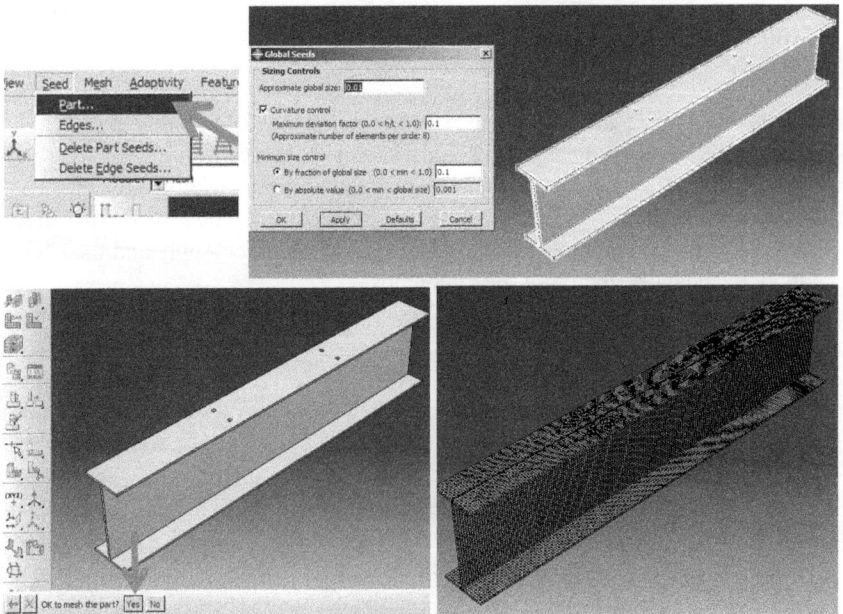

FIGURE 5.45   Seeding and meshing column.

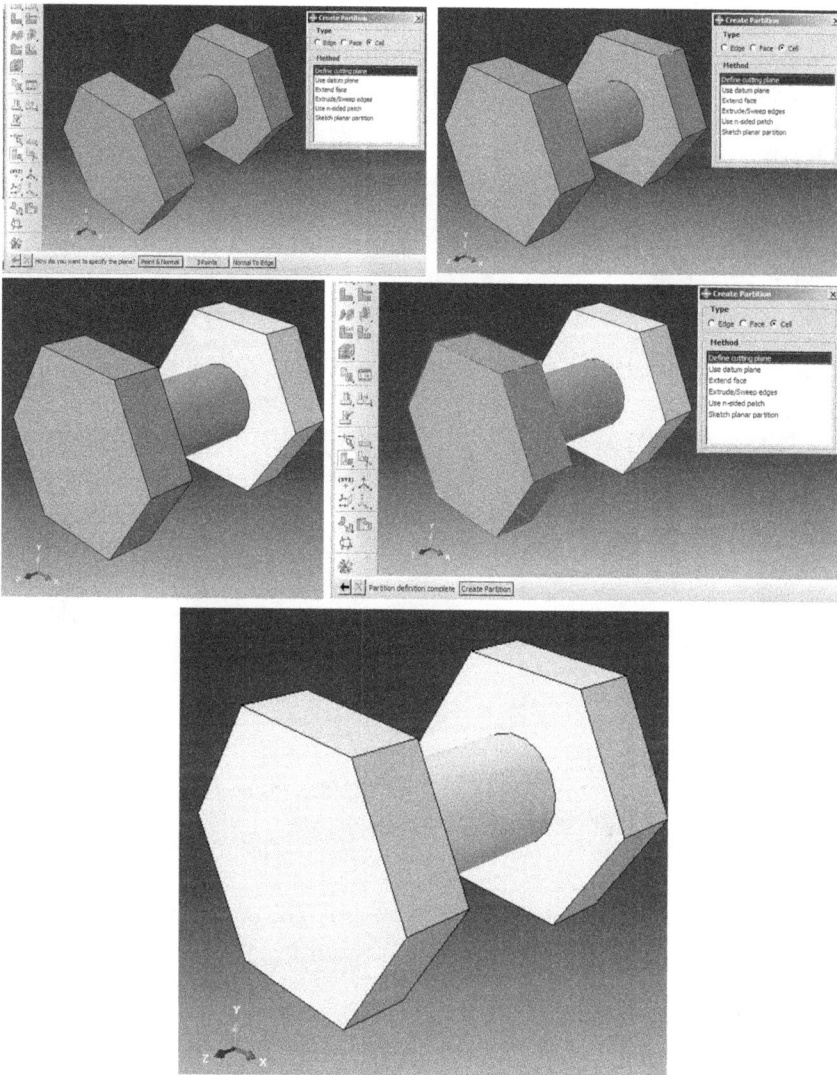

FIGURE 5.46 Partitioning the bolt by point and normal method.

Select "Tool → Datum" and choose the plane of the bolt head, consider an off-set in the positive direction z of 1.5 cm. This plane will be used to define another partition using a datum plane that separates the bolt into two parts, as shown in Figure 5.47.

Select "Use datum plane" as the "Method" in which the bolts body will be partitioned and select the bolt body. Then, select the plane and click "Create Partition" (Figure 5.48).

For meshing the bolt, the element size should first be defined.

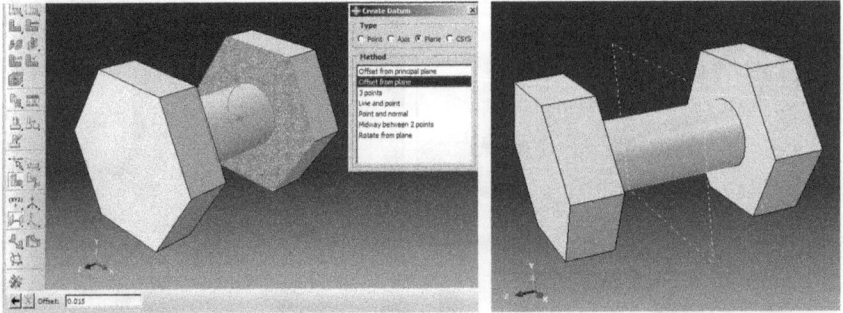

FIGURE 5.47   Defining a datum plane.

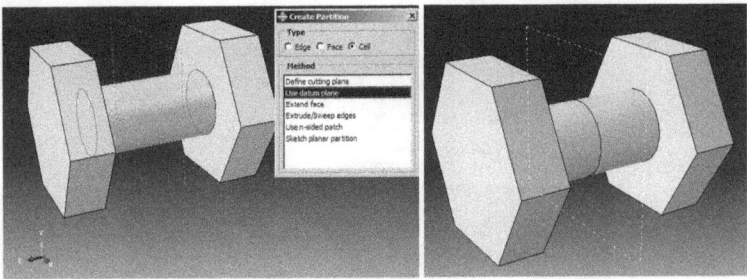

FIGURE 5.48   Partitioning the bolt by the datum and plane.

Select "Seed → Part" to open the "Global seed part" dialog box. Same as the other parts, consider the elements size of 0.02 and select "Apply" and "Ok" to apply the changes and close the dialog box.

Select "Mesh → Part" and choose "Yes" from the prompt area to mesh the column, as shown in Figure 5.49.

L-plate:

The last component to mesh is the L-plate.

Double click on "Mesh (Empty)" underneath the L-plate in the "Model tree" to open the "Mesh" module

Similar to the other parts, it is not able to mesh, so by partitioning the areas, it should be separated.

Select "Tools → Partition" and choose "Define" cutting plane from the partition method. The plane should be defined by "Point and Normal". Then, select "Point and edge", as shown in Figure 5.50, to create the partition.

For meshing the L-plate, the element size should first be defined.

Select "Seed → Part" to open the "Global seed part" dialog box. Consider elements of size 0.02 and select "Apply" and "Ok" to apply the changes and close the dialog box.

Select "Mesh → Part" and choose "Yes" from the prompt area to mesh the column, as shown in Figure 5.51.

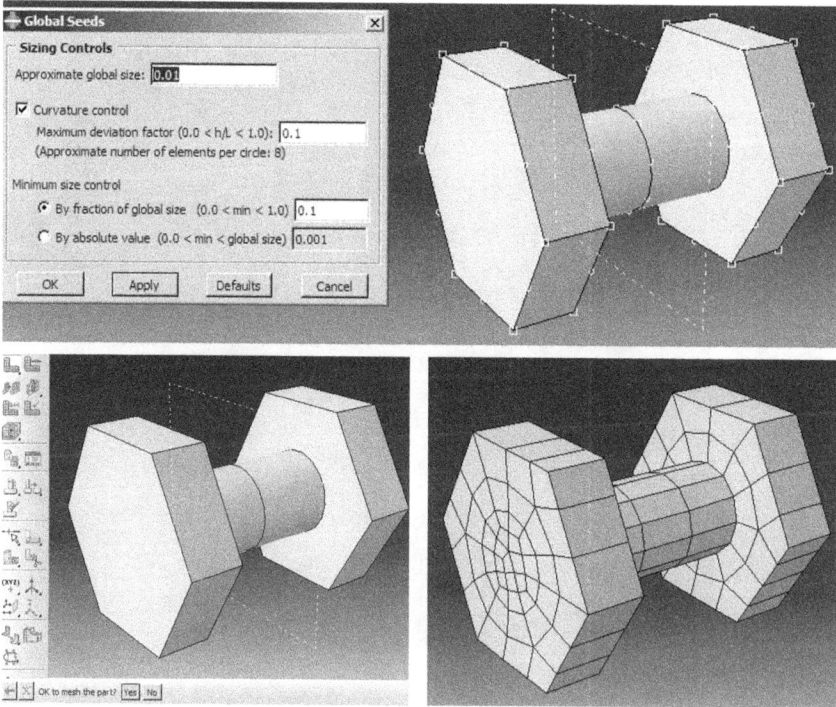

FIGURE 5.49 Seeding and meshing the bolt.

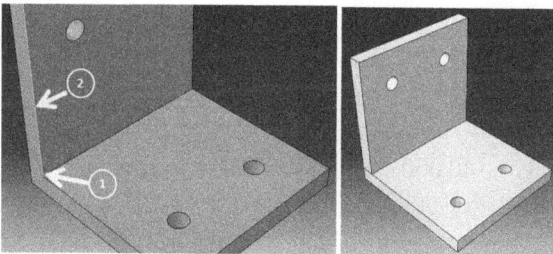

FIGURE 5.50 Partitioning the L-plate.

5.4.7 STEP MODULE

The next step needs to define the analysis steps. In the sample, analysis is divided into two steps. The first step is about applying the load bolts, statically called bolt load which is considered as "Static, General".

Double click on "Steps" in the "Model tree" and name it "Bolt_Load" and consider the deformation to be small. In the "Incrementation" tabbed page, set the "Maximum number of increments" equal to 1000 and assume the "Initial size" is 0.1, but do not change the rest (Figure 5.52).

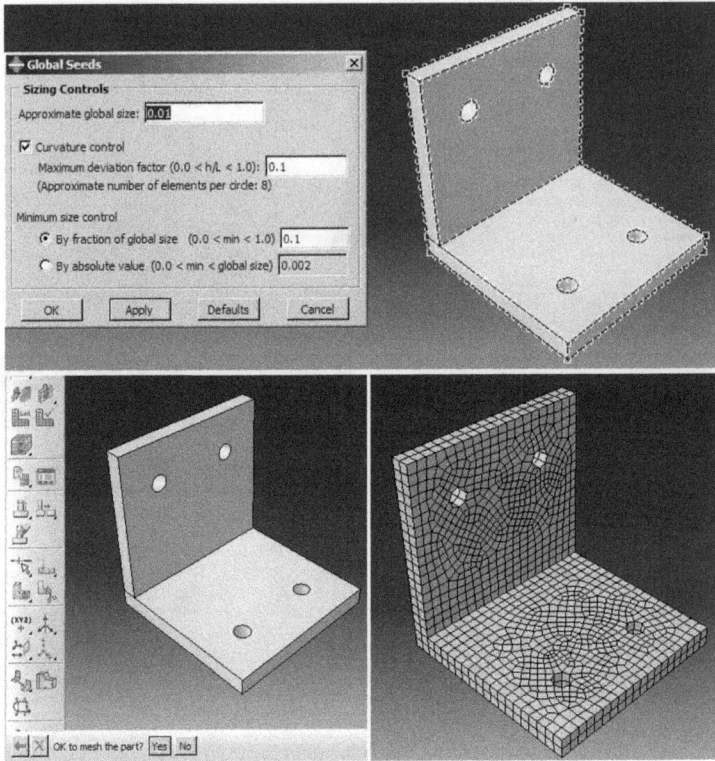

FIGURE 5.51   Seeding and meshing the L-plate.

Name the second step "Cyclic", enter 10 for "Time period" and consider nonlinear effects of geometry by toggling "Nlgeom" on. Increase the "Maximum number of increments" to 10,000, and set the "Initial", "Minimum", and "Maximum" increment size as 0.01, 0.00001, and 0.1, respectively (Figure 5.53).

5.4.8   INTERACTION MODULE

The next step needs to specify the contact between all interacted components, as many surfaces are in contact. For contact modeling in the entire model, a general contact definition should first be considered. This contact formulation considers a unique contact definition for all components. In this sample, a friction coefficient for all components is considered.

Double click on "Interactions" in the "Model tree" to open the "Create Interaction" dialog box. Select "General contact" as the "Type" and click "Continue". In the "Edit Interaction" dialog box, click "Create Interaction Property" to open the corresponding dialog box and then select "Contact" and click "Continue" to open the "Edit Contact Property". To define tangential

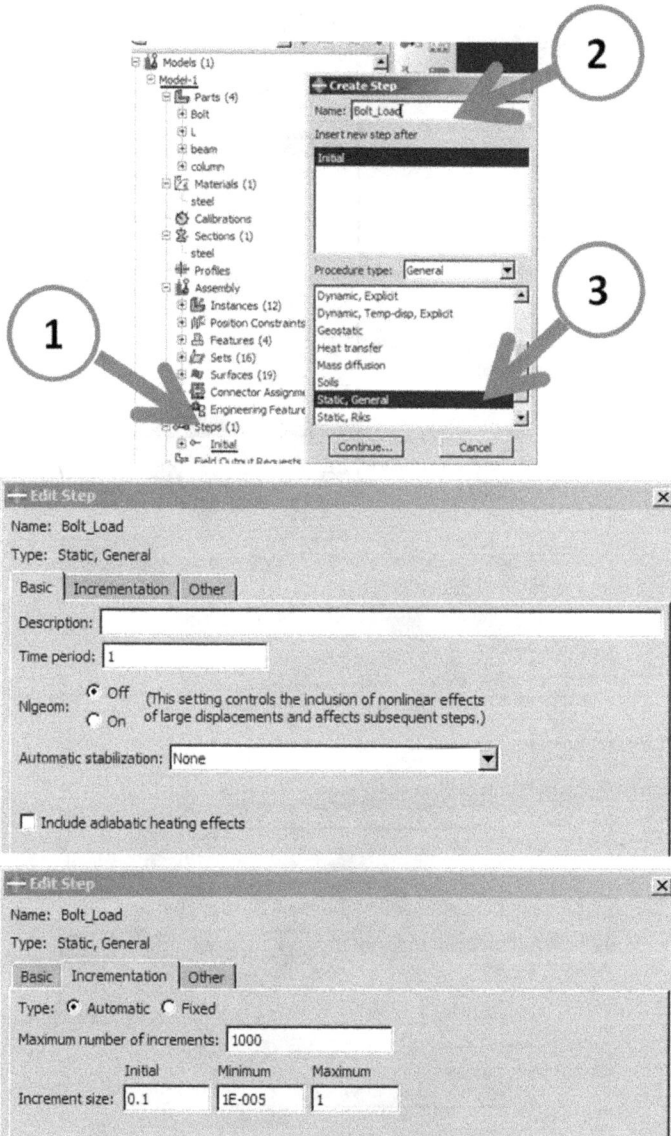

FIGURE 5.52 Defining the Bolt_Load step as the first step.

behavior, choose "Mechanical → Tangential behavior" and then select "Penalty" as a "Friction formulation" and enter 0.3 for a "Friction Coefficient".

To define normal behavior, select "Mechanical → Normal Behavior", but do not change the rest. To apply the contact properties, click "Ok" and close the dialog box. Also, select "IntProp-1" in the "Edit Interaction" dialog box (Figure 5.54).

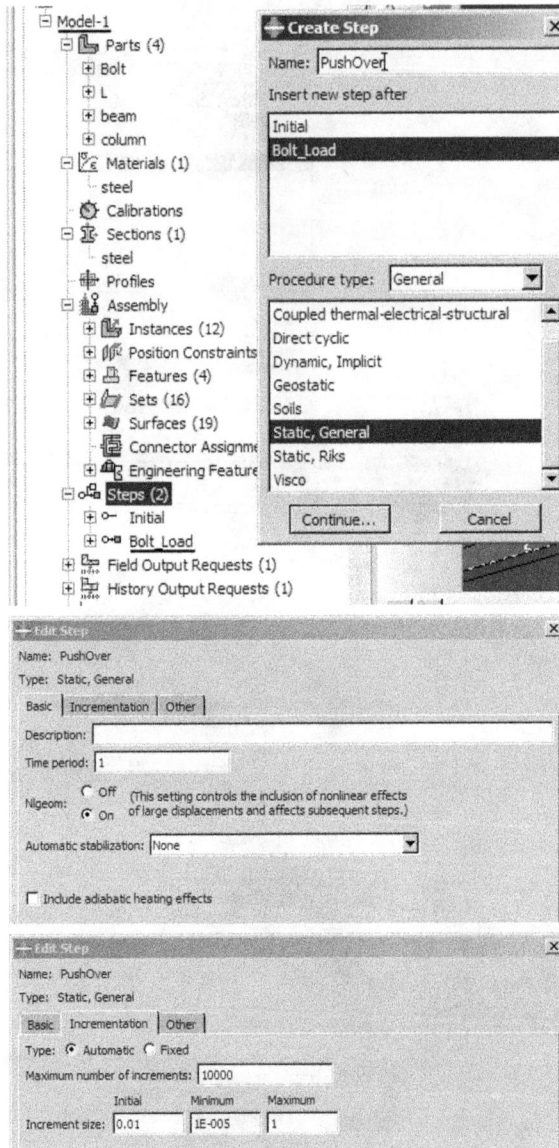

FIGURE 5.53 Defining the Cyclic analysis step as the second step.

5.4.8.1 Coupling Constraint

In the next step, the datum tool needs to be used to define the coupling constraint. A datum point should be defined as a "Reference point", and the front face of the beam should be coupled by the point. This point will be enforced displacement.

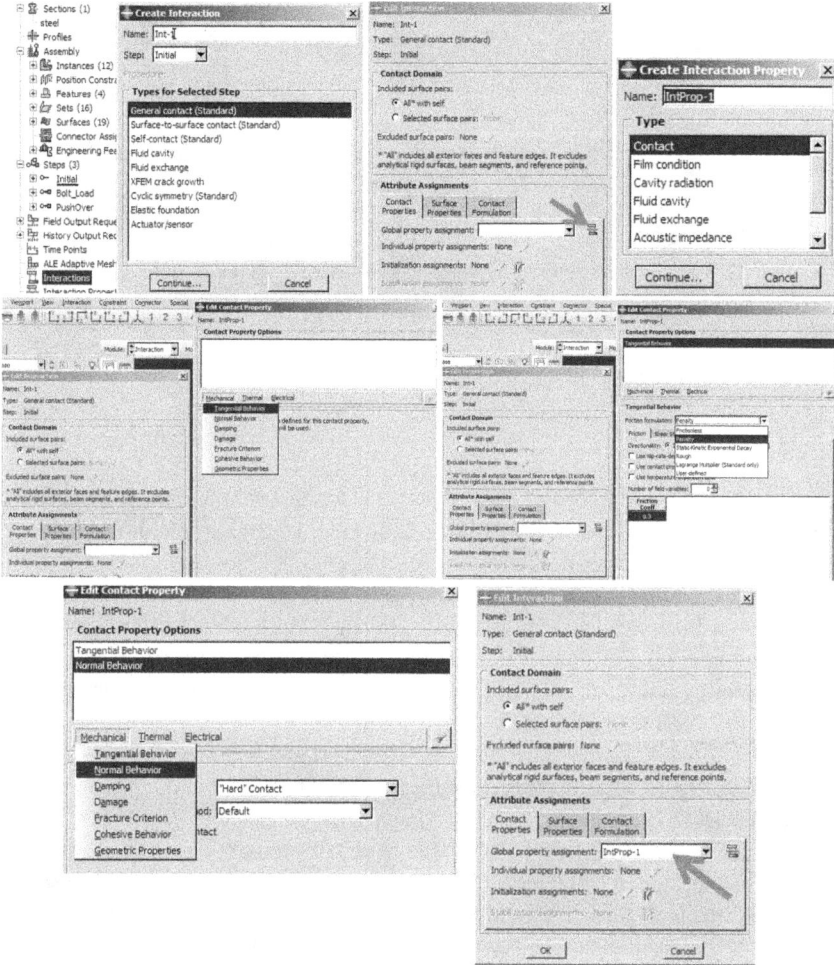

FIGURE 5.54 Defining general contact and contact interaction properties.

Open the "Datum" dialog box via "Tools → Datum" and select point from the "Type" and "Midway between 2 points" from the "Method". Then, select "Tools → Reference point" and choose the datum as defined (Figure 5.55).

The next step defines the coupling constraint.

Double click on "Constraints" in the "Model tree" and select "Coupling" and click "Continue". Choose the reference point as a coupling control point and click the middle mouse button and then select the front faces of the beam as the surfaces that the reference point should be coupled with. Next, click on the middle mouse button to open the "Edit Constraint" dialog box. Click "Ok" to apply and close the dialog box (Figure 5.56).

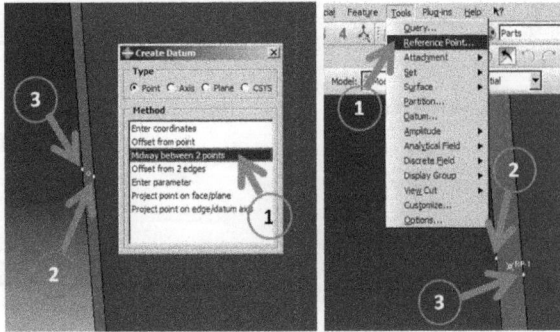

FIGURE 5.55   Defining datum point as the reference point.

5.4.9 LOAD MODULE

Next, bolt loads need to be defined. Eight (8) loads for eight bolts will be needed, which should be defined in the "Bolt_Load" step. In the sample, bolt loads are assumed to be 50 kN per bolt.

To make it simpler for bolt selections, select all parts, but the bolts from the assembly instance underneath as shown in the "Model tree" of the assembly should not be hidden (Figure 5.57).

For defining bolt loads, a datum axis needs to be defined and positioned in the middle of the bolt as well as in the correct direction.

Select "Tools→ Datum" and select "Axis of the cylinder" and choose bolt body as shown in Figure 5.58. Perform the same process for the remaining seven bolts.

Double click on "Loads" in the "Model tree" and select "Bolt load" from "Type" and "Bolt_Load" from "Step" and click "Continue". Then, select the middle internal surface of a bolt and click the middle mouse button to apply the selection. Next, the load surface should be determined. Select "Brown" as the side and click the axis which was recently defined to open in the "Edit load" dialog box. Enter 50,000 for the "Magnitude" and click "Ok" to apply and close the dialog box. Repeat for the other bolts (Figure 5.59).

All bolt loads should be changed in the "Cyclic" step so that instead of "Apply force", the bolt lengths should be fixed.

Right click on "Loads" in the "Model tree" and select "Manager" to open the corresponding dialog box. Next, double click on each load in the "Cyclic" column and change the "Loading method" to "Fix at current length". Follow the same process for all eight bolt loads, as shown in Figure 5.60.

5.4.10 BOUNDARY CONDITION

There are two types of boundary conditions in the considered model. The first is about clamping both ends of the column, and the second is a large enforced displacement that should be applied on the reference point on the beam.

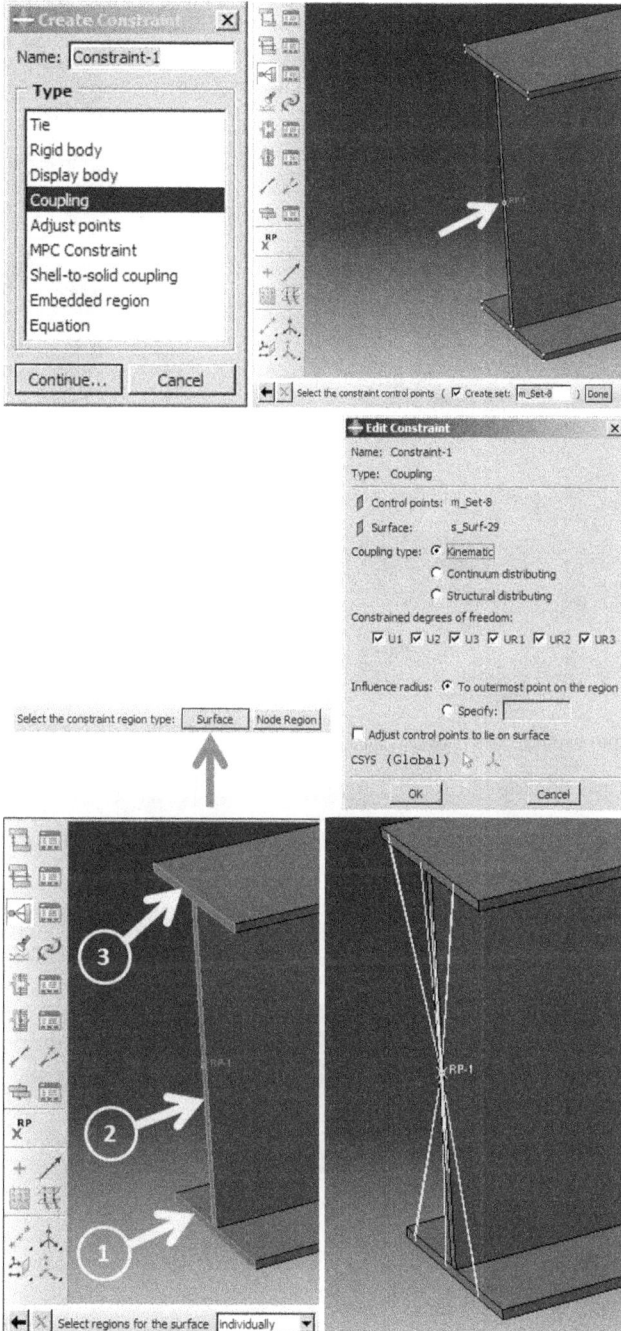

FIGURE 5.56   Coupling constraint.

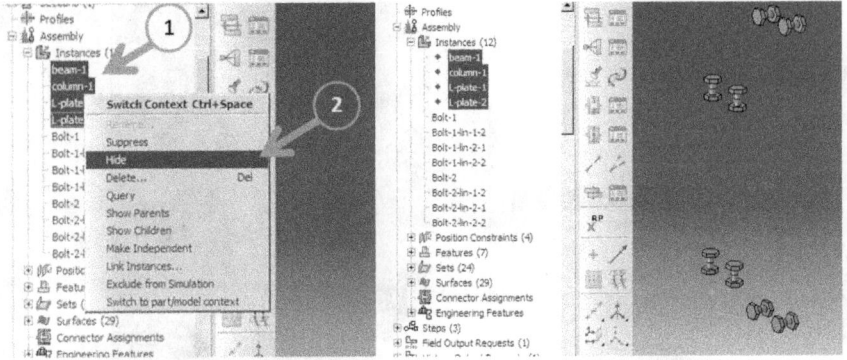

FIGURE 5.57 Hiding all parts except bolts.

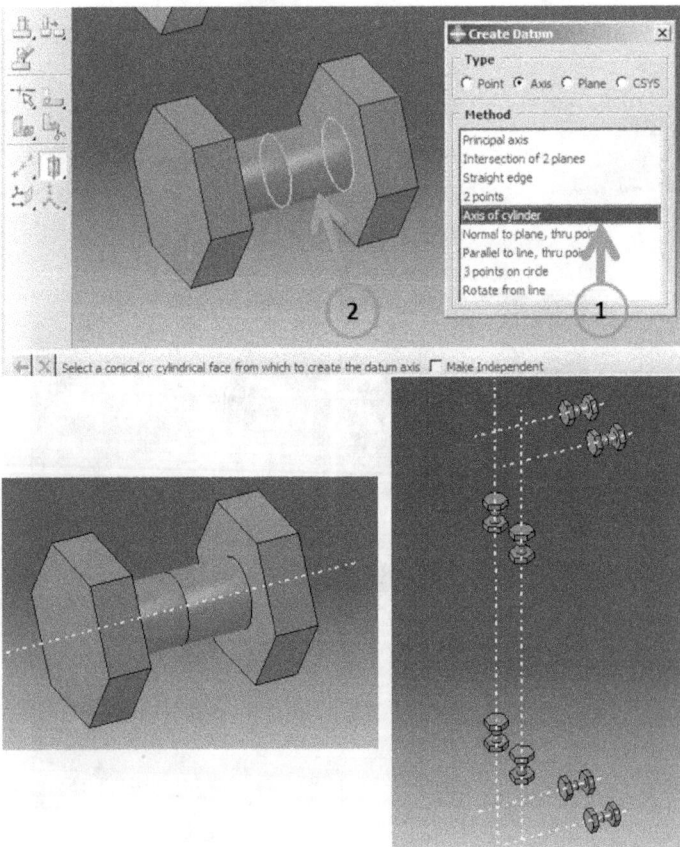

FIGURE 5.58 Defining the axis for the bolts.

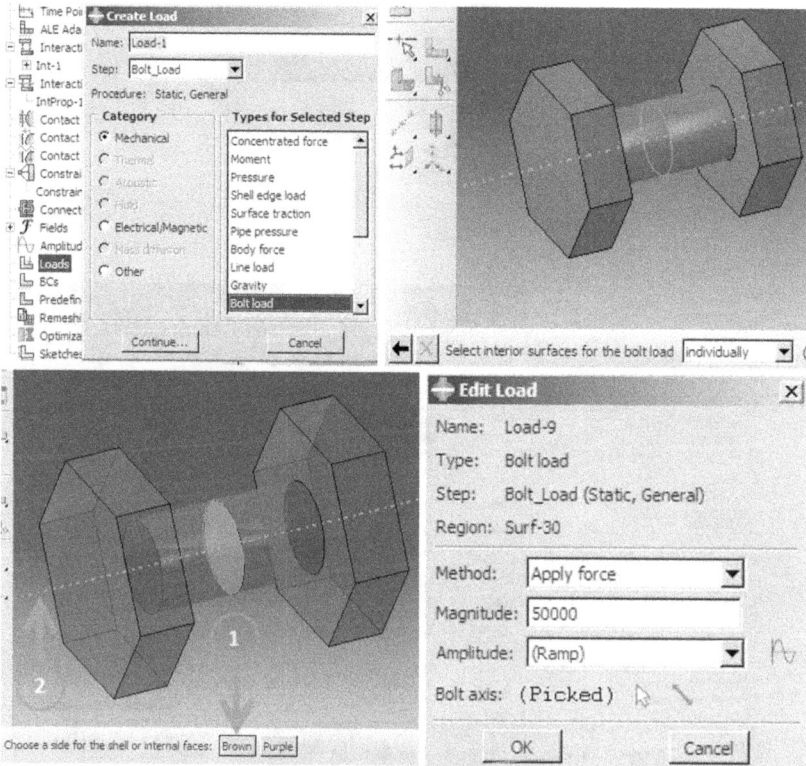

FIGURE 5.59 Defining the bolt loads.

Double click on "BCs" in the "Model tree" and select "Symmetry/Antisymmetry/
Encastre" "Initial" as "Step", then click "Continue". Next, choose upper faces and
lower faces of columns, as shown in Figure 5.61 and click the middle mouse button
to open the "Edit boundary condition". To clamp both sides, select "ENCASTRE",
then click "Ok" to apply and close the dialog box (Figure 5.61).

Next, enforced displacement should be defined as another boundary condition.

Double click on "BCs" in the "Model tree" and choose the step "Cyclic" and
"Displacement/Rotation", and then click "Continue". Select the reference point
and click the middle mouse button to open the "Edit boundary condition". As
the boundary condition should act as an actuator, rotations regarding y and z and
displacement along x should be restrained and displacement along the z direction and should be free. On the other hand, as the enforced displacement should
be defined along y, restrain the direction and consider it to be large enough for
displacement. At this point, consider 1.5 cm for the displacement. To define cyclic
amplitude, click "Create Amplitude" and in the corresponding dialogue box,
choose "Tabular" from "Type" and enter time and amplitude. Finally, click "Ok"
to define the amplitude and choose "Amp-1" from the "Amplitude" and click
"Ok" to define the boundary condition as shown in Figure 5.62.

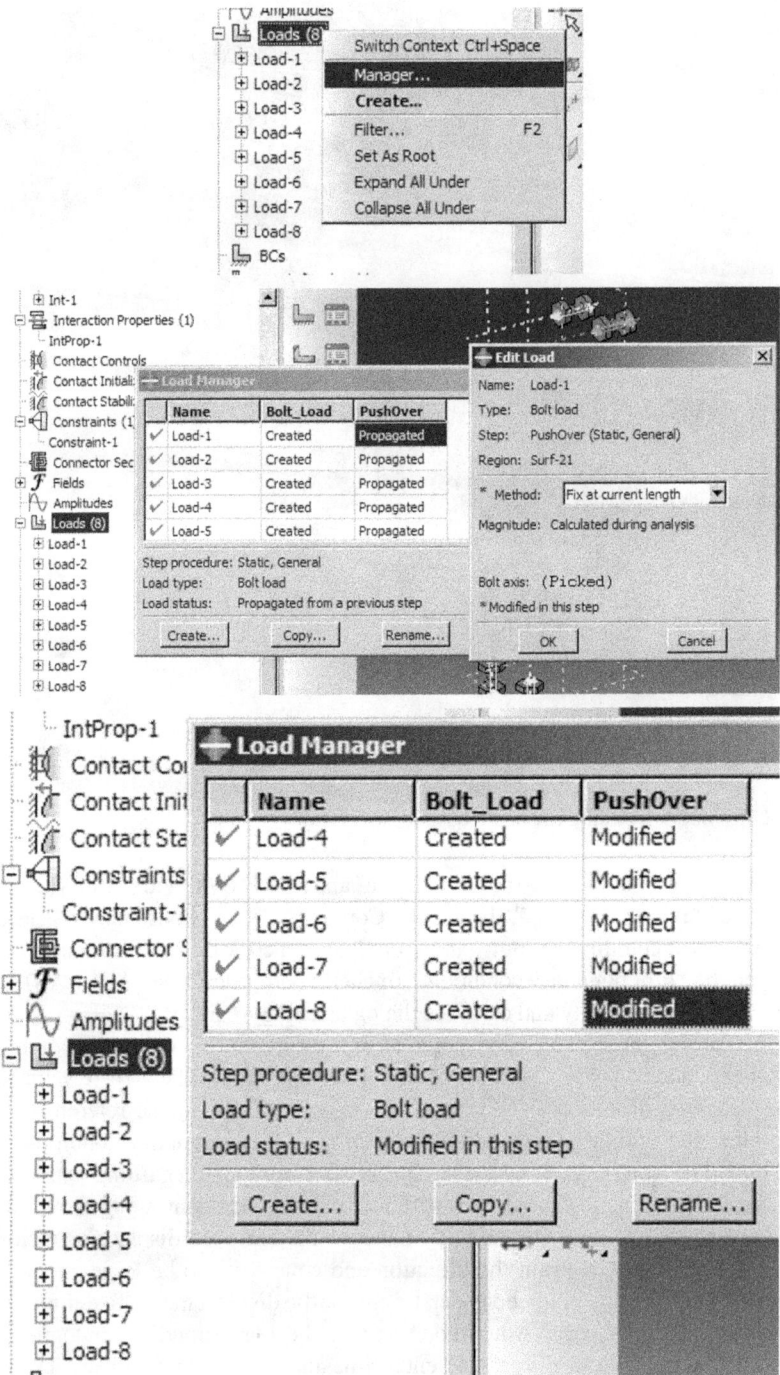

FIGURE 5.60 Modifying bolt loads in the Cyclic step.

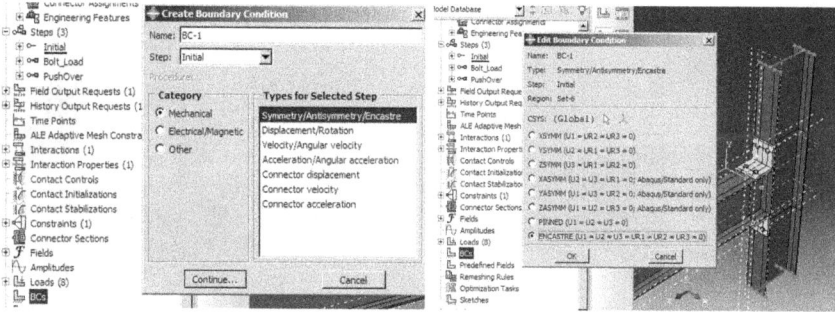

FIGURE 5.61 Defining the fixed boundary condition on the column.

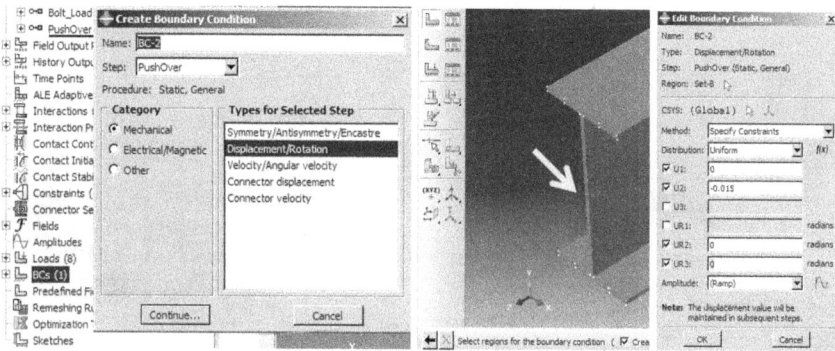

FIGURE 5.62 Defining the boundary condition for the reference point of the beam.

5.5 ANALYSIS: JOB MODULE

Finally, a job should be defined to solve the problem.

Double click on "Jobs" in the "Model tree" and choose a name in the "Create Job" dialog box. Then, click "Continue" to open "Edit Job" and leave the box unchanged by clicking "Ok". Then, right click on the job and click "Submit" to begin analysis (Figure 5.63).

5.6 RESULTS

The results can next be checked, which will be undertaken entirely in the "Visualization" module. However, the solution process is quite time-consuming. After the analysis is completed, the results are displayed and ready to review.

Right click on completed job and select "Results", to open the "Visualization" module (Figure 5.64).

The first purpose is to examine plasticity and the yielded areas of the model.

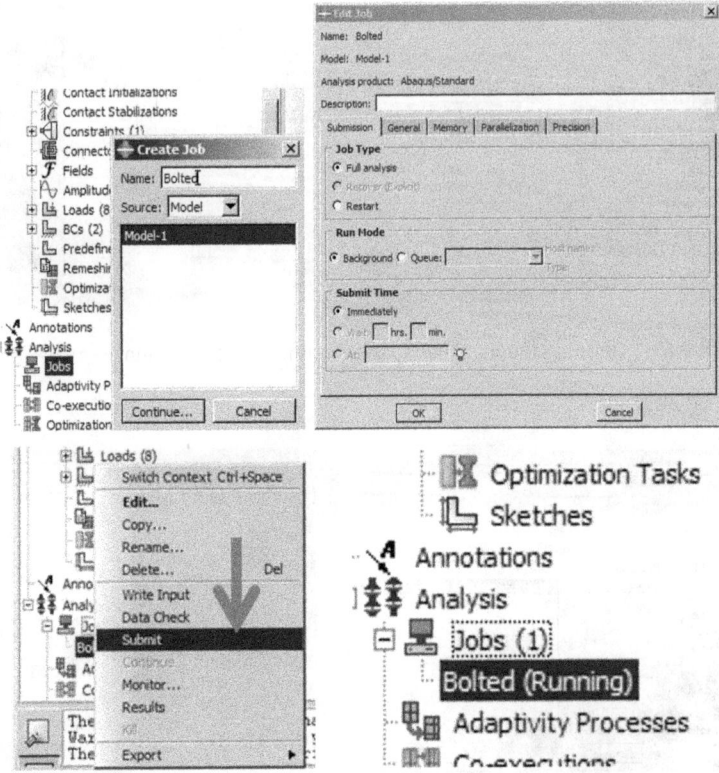

FIGURE 5.63 Defining and submitting the job.

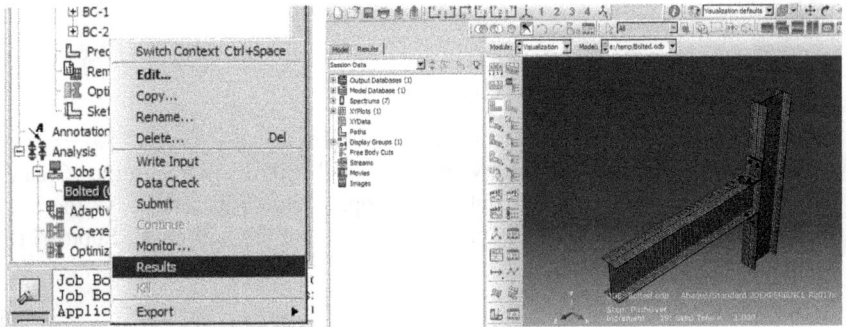

FIGURE 5.64 Opening the visualization module.

Note that the deformed shape is not real; in fact, it is scaled. Select "Options → Common" and change the "Deformation scale factor" to "Uniform" and 1 for the model and click "Ok" to apply and close the box. Select "PEEQ", as the "Primary" variable from the top. It shows that the maximum plastic strain is 1.73e-2. (Figure 5.65).

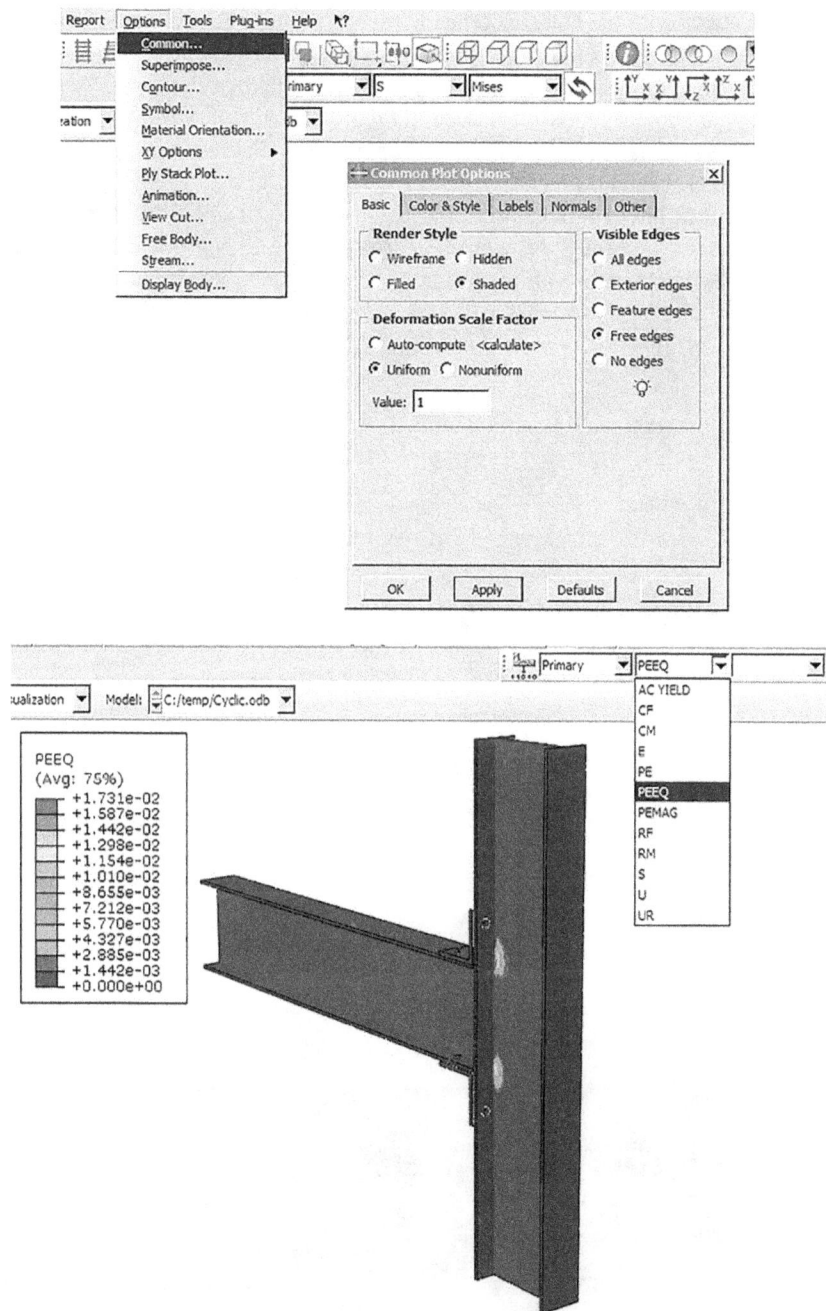

FIGURE 5.65 PEEQ contour plot for the whole model.

Right click on all the bolts underneath instances in the "Result tree" and select "Replace" for more obvious plasticity (Figure 5.66).

By clicking "Replace all" from the top, replot all and then choose "U" to show displacement contour plot as illustrated in Figure 5.67.

The next goal is to achieve a force–displacement diagram that presents the value of the problem nonlinearity.

Double click on "XY Data" from the "Result tree" and select "ODB field output" and click "Continue" to open the corresponding dialog box. Then, select "U2", displacement in the y-direction and "FR2", reaction force in the y-direction. Here, they must be extracted for the assembly constraint reference point from

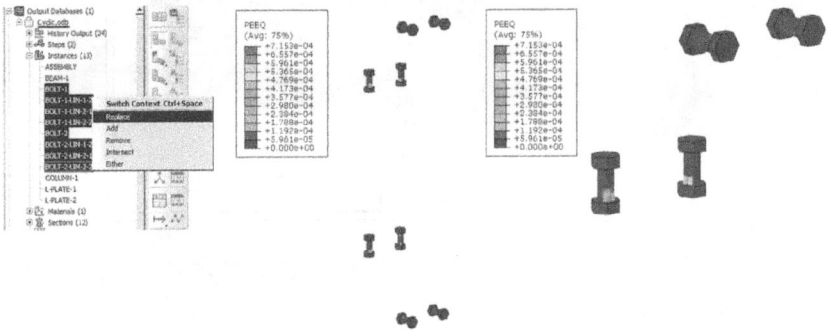

FIGURE 5.66   PEEQ contour plot for the bolts.

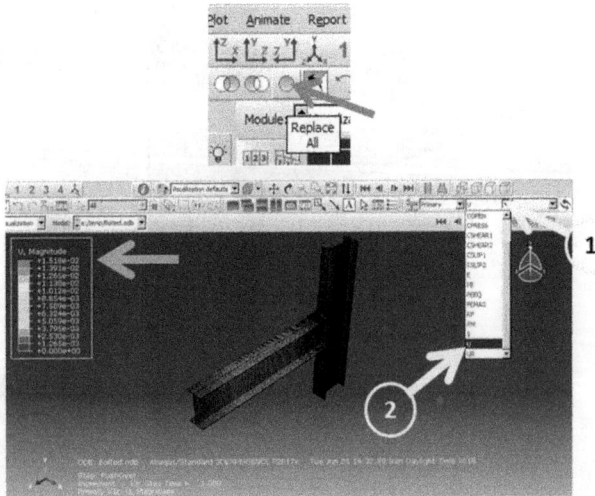

FIGURE 5.67   Displacement contour for the model.

the "Elements/Nodes" tab page. Then click "Save" and "Ok" to save the plots. Finally, close all dialog boxes. Two graphs have been generated in the "Results tree" (Figure 5.68).

Double click on "XY Data" and select "Operate on XY Data" and click "Continue" to open the dialog box. In the dialog box, from the right, choose "Combine" operator. Double click on "U: U2" diagram and double click on "RF: RF2" diagram, then put a "-" before both of them and click "Plot Expression" to plot and create the force–displacement diagram, as shown in Figure 5.69.

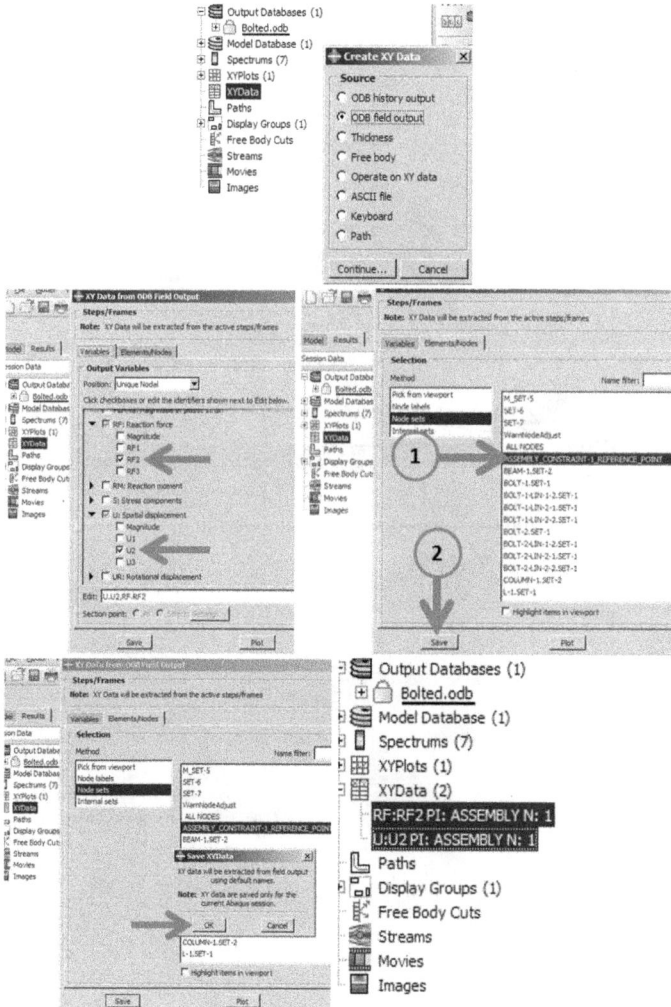

FIGURE 5.68 Extracting force and displacement diagrams.

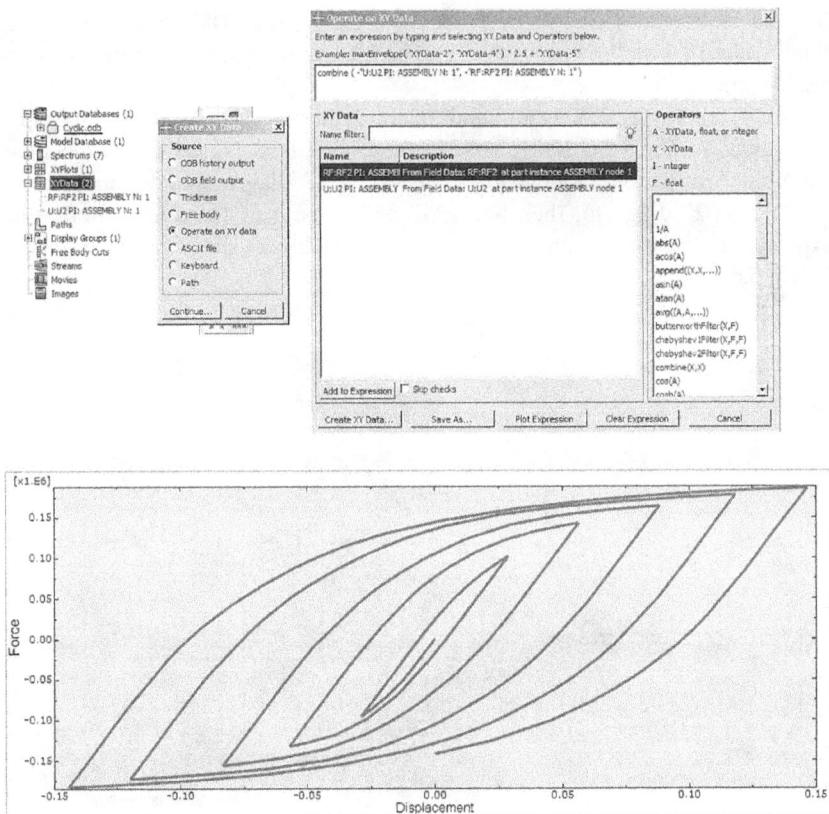

FIGURE 5.69   Extracting the force–displacement diagram.

6 Beam–Column Connection Retrofitted with CFRP Sheets Subjected to Pushover Loading

6.1 INTRODUCTION

Beam–column joints are recognized as the weak points of reinforcement concrete frames. The ductility of reinforced concrete (RC) frames during severe earthquakes can be measured through the dissipation of large energy in the beam–column joints. However, the experience of recent earthquakes proved that the conventional beam–column joints are not exhibiting enough capacity to resist against the lateral seismic load, and retrofitting and rehabilitating of structures through proper methods, such as carbon fiber-reinforced polymer (CFRP), are required to prevent casualties resulted from the collapse of structures. Therefore, in this example, the simulation process of retrofitting the beam–column connection using CFRP sheets and finite element method is demonstrated.

Researcher: Arash Rahimipour (arash.rahimipour19870@gmail.com)

6.2 PROBLEM DESCRIPTION

The main aim of the present example is to evaluate the retrofitting method for a normal ductile beam–column joint using CFRP sheets subjected to pushover loads. Thus, the finite element model of normal ductility RC beam and column is developed. The CFRP sheet is also implemented in the beam–column connection to retrofit the joint against the applied lateral load. In order to evaluate the efficiency of implementing CFRP on the joint capacity, in this study, three beam–column joints are considered that are beam–column joint with partial CFRP wrapping, beam–column joint with full CFRP wrapping, and beam–column joint with normal ductility. The two cases with partial and full CFRP wrapping in the beam–column joints are used to determine the effect of retrofitting with confined CFRP wrapping sheets on the behavior of the beam–column joint. All the models are subjected to pushover loading and the analysis results are investigated.

Figure 6.1 shows three interior beam–column connections and partial wrapping CFRPs fitted on the connection.

DOI: 10.1201/9781003219491-6

FIGURE 6.1   3D model of concrete column reinforced by steel bars and CFRP.

6.3 OBJECTIVES

1. To simulate the beam–column joint with normal ductility retrofitted by CFRP sheets.
2. To investigate the behavior of a beam–column joint partially and fully strengthened by CFRP sheets.
3. To investigate the capacity and energy dissipation of the above-mentioned beam–column connections.

6.4 MODELING

6.4.1 PART MODULE

In this module, the geometry of the considered problem is defined as demonstrated here.

6.4.1.1 Create a New Model Database

Start ABAQUS/CAE software from programs in the "Start menu". Select "Create model database" from the "Start session" dialog box that appears.

Click on "With the Standard/Explicit Model". This step allows the user to begin modeling where the user can create a new file and save it under any name in a new folder (see Figure 6.2).

The Part module toolbox is located on the left side of the software main window. Each module displays its own set of tools in the module toolbox.

6.4.1.2 Create a New Model Database and a New Part

From the main menu bar, select "Part → Create" to create a new part.

The "Create Part" dialog box appears. Use the "Create Part" dialog box to name the part; to choose its modeling space, type, and base feature; and to set the approximate size. Name the part "Column", choose "3D", "Deformable", and "Solid" from the base feature.

In the "Approximate size" text field, type 2000. The value entered in the approximate size text field at the bottom of the dialog box sets the approximate size of the new part (see Figure 6.3).

Click "Continue" to exit the "Create Part" dialog box.

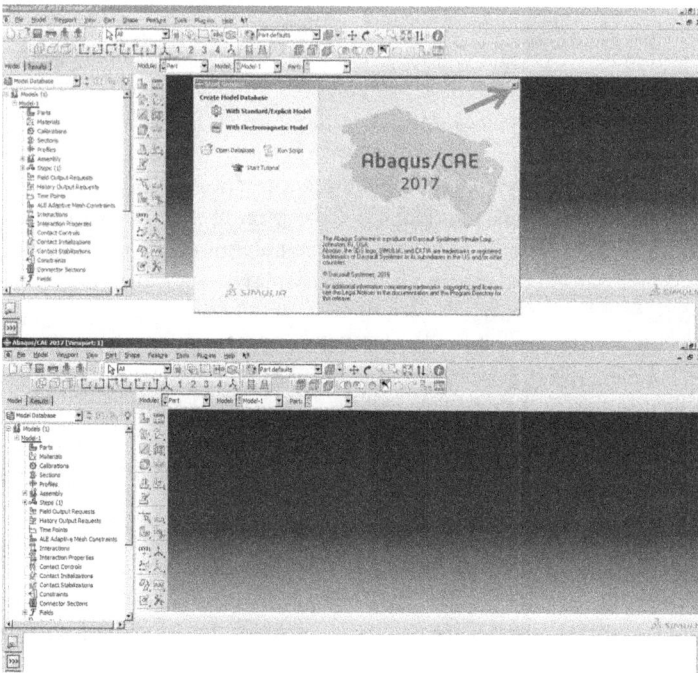

FIGURE 6.2 Getting started.

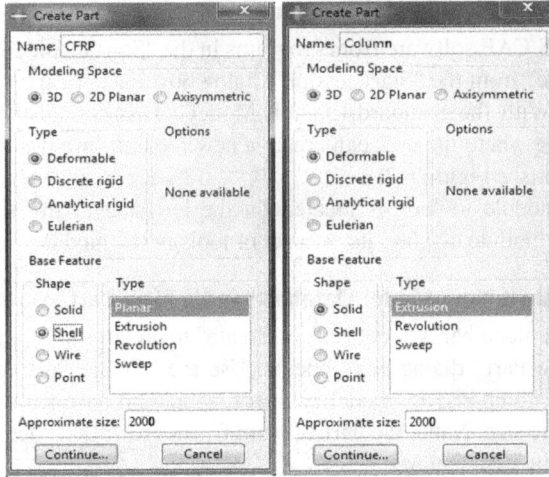

FIGURE 6.3   Create a new part.

6.4.1.3   Define a Rectangle with Dimensions

Use the "Create Lines: Rectangle (4 lines)" tool located in the upper-left corner of the sketcher toolbox to begin sketching the geometry of the plate. The user can choose a starting corner for the rectangle at the viewport or enter the x- and y-coordinates. Create a rectangle in the middle of the sketcher without any dimension, as shown in Figure 6.4. Define the dimensions of the geometry for the rectangle by using the add dimension tool located in the lower corner of the sketcher.

Once sketching is completed, right click and click on "Cancel" Procedure to exit the sketcher. Click on "Done" in the prompt area to exit the sketcher, and define the depth for the extrusion of the column as 2700. Once the depth of the column has been defined, it will be displayed and turn out, as shown in Figure 6.5.

Repeat the above procedure for "CFRP" as 3D, Deformable, and Shell, and then draw a planar sketch as shown in Figure 6.6.

Create three new 3D wireframe parts and draw their sketch as shown in Figure 6.7.

6.4.1.4   Extrude Solid to Create a Beam

Double click on "Column" from the current part. Use the "Shape → Solid → Extrude" tool to create a beam in the middle of the column.

The right side of the column is first chosen as the plane for solid extrusion, and an edge on the surface is chosen to activate the sketcher, as shown in Figure 6.8.

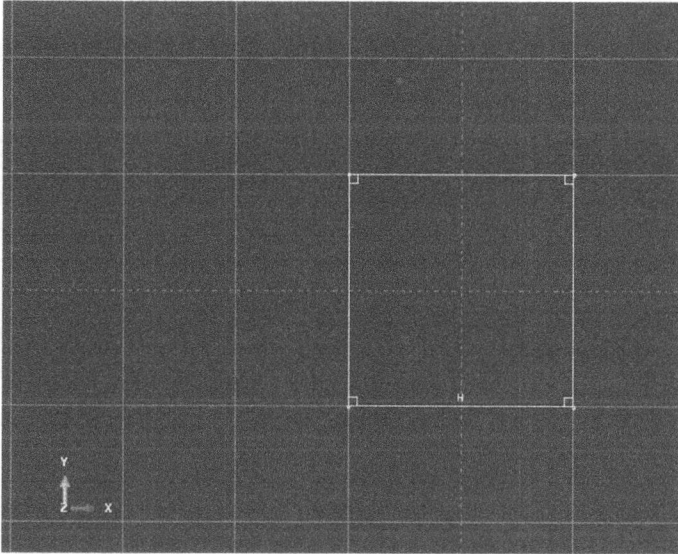

FIGURE 6.4   Draw a rectangle as the column section sketch.

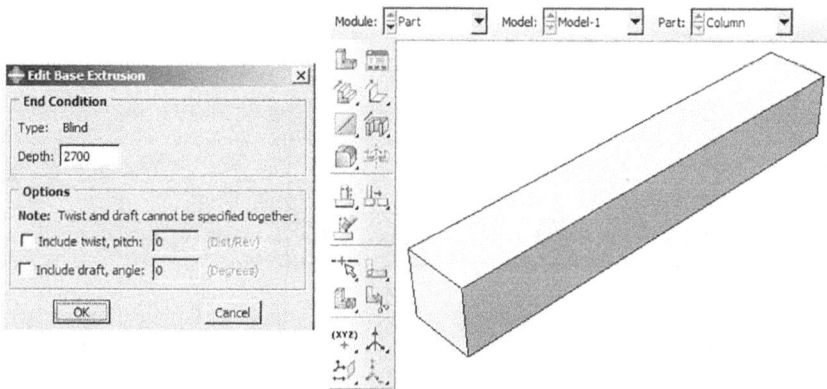

FIGURE 6.5   Depth of the column.

Click on the "Create Lines: Rectangle (4 lines)" tool located in the upper-left corner of the sketcher toolbox to begin drawing the geometry of the solid section for the extruded solid.

The extruded solid beam has a dimension of 300×300 mm. It is sketched at the mid-span length, as shown in Figure 6.9.

Once sketching the section for the extruded solid is completed, click on "Done".

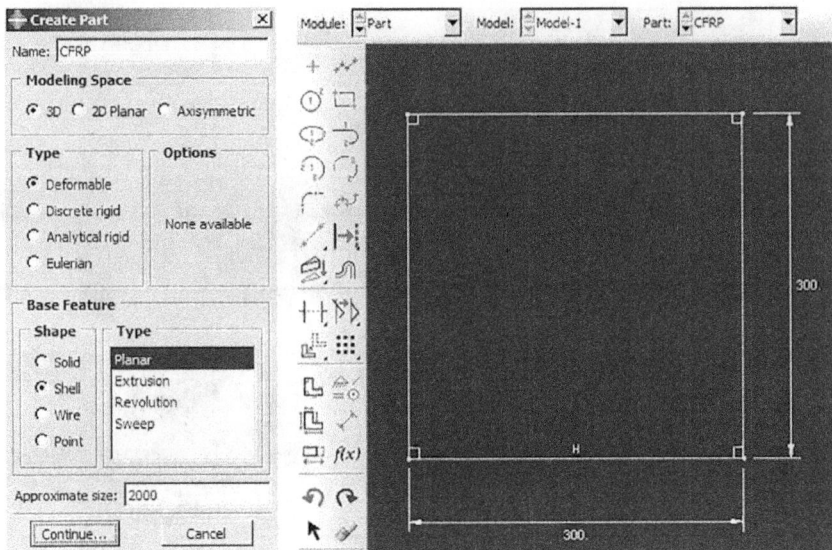

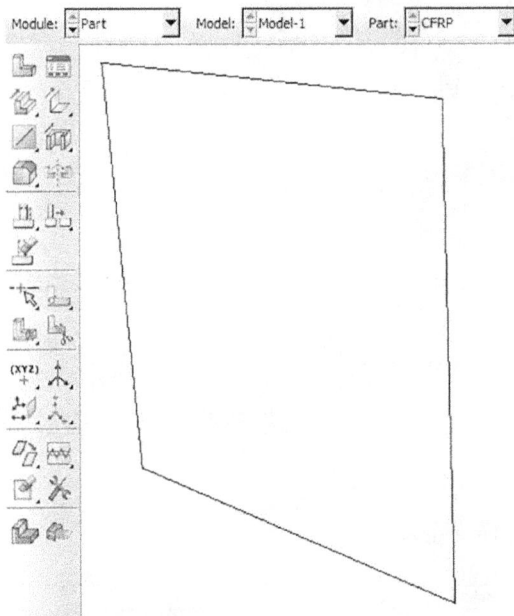

FIGURE 6.6 Define the CFRP.

Then, the "Edit extrusion" dialog box appears. The beam is extruded to a depth of 1800 mm from the surface of the column, as shown in Figure 6.10.

Change the extrude direction to reverse the extrusion direction, if necessary. Click "OK" to exit the "Edit Extrusion" dialog box.

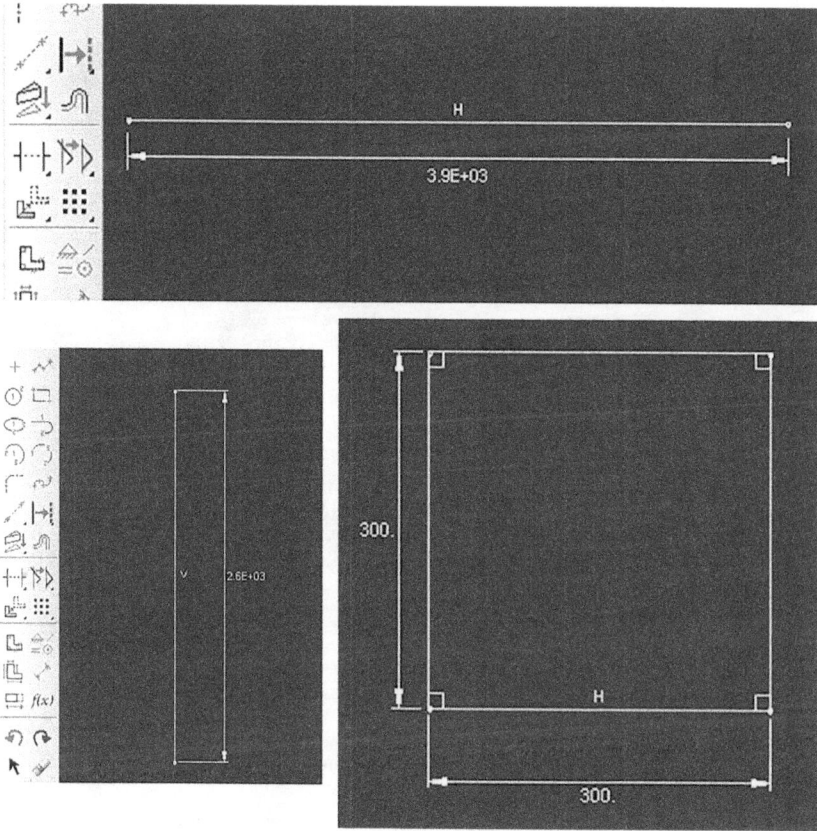

FIGURE 6.7 Define steel parts.

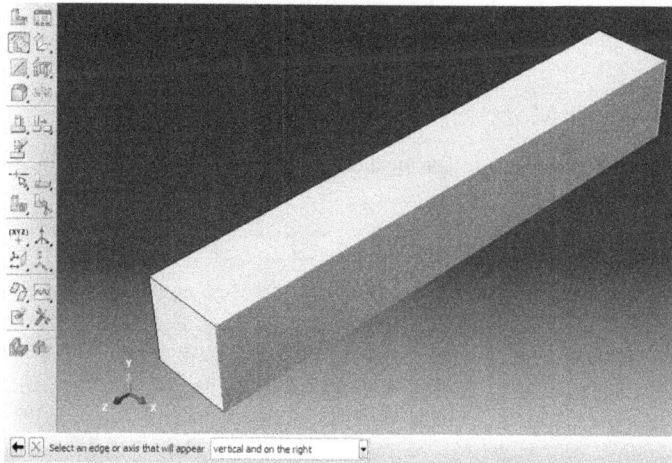

FIGURE 6.8 Solid extrude on the column surface.

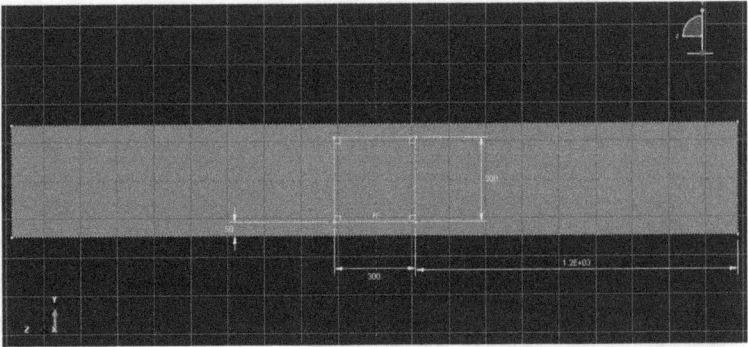

FIGURE 6.9   Sketch the geometry of a beam section.

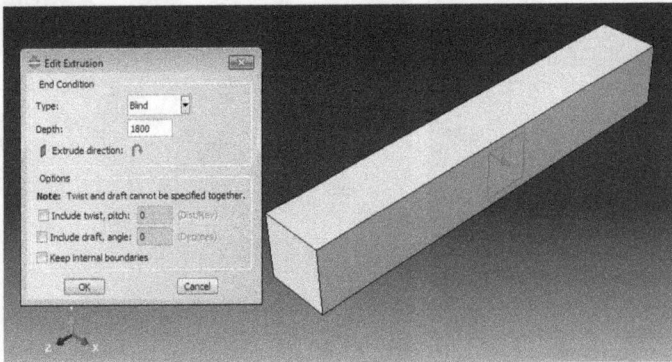

FIGURE 6.10   Define the beam depth.

Repeat the same procedure for solid extrusion on the left side of the column. Figure 6.11 displays the completed extrusion of the beam in the middle of the column.

6.4.2   PROPERTY MODULE

In this module, the material properties should be defined and assigned to the respected parts.

6.4.2.1   Material Properties

The "Property" module is used to define a material and its properties. In this problem, three materials are used, which are concrete, steel, and CFRP. CFRP is defined as an elastic material, but both concrete and steel are defined to have elastic and plastic properties. Thus, nonlinear plastic materials are created for both the steel and the concrete, and a single linear elastic material is created for CFRP.

Double click on "Materials" in the "Model tree". The "Edit Material" dialog box appears.

Name the material as "Concrete". Click "General → Density" and enter the value for mass density.

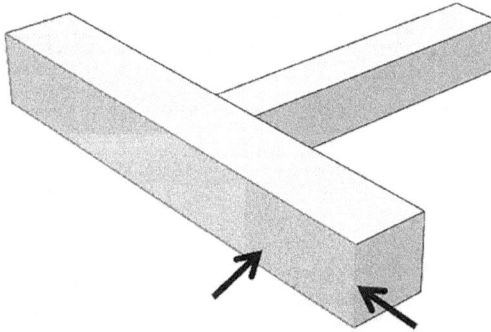

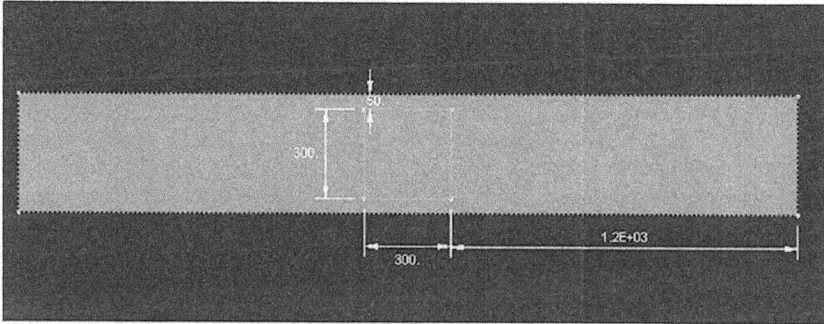

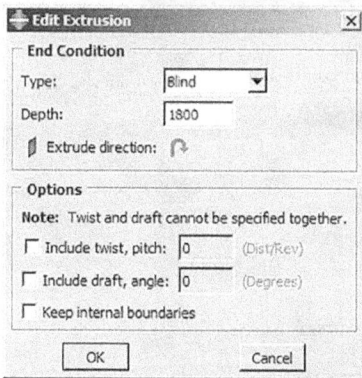

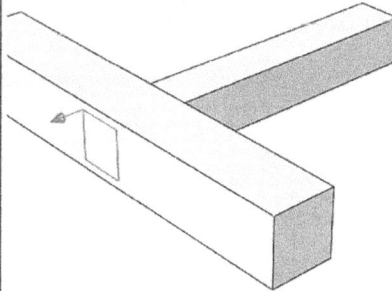

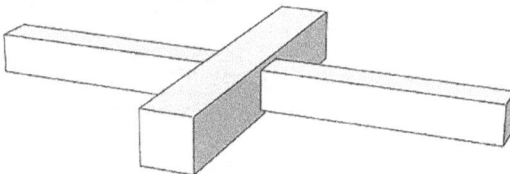

FIGURE 6.11 Complete the column–beam definition.

Then Click on "Mechanical → Elasticity → Elastic". The software displays the elastic data form.

Enter the value of 41,000 MPa for Young's modulus and 0.2 for Poisson's ratio in the respective cells.

In the next step, click on "Mechanical → Plasticity → Concrete Damaged Plasticity" and enter the values shown in Figure 6.12. Click "OK" to exit the material editor.

Repeat the same procedure for the "CFRP" and "Steel" (Figure 6.13).

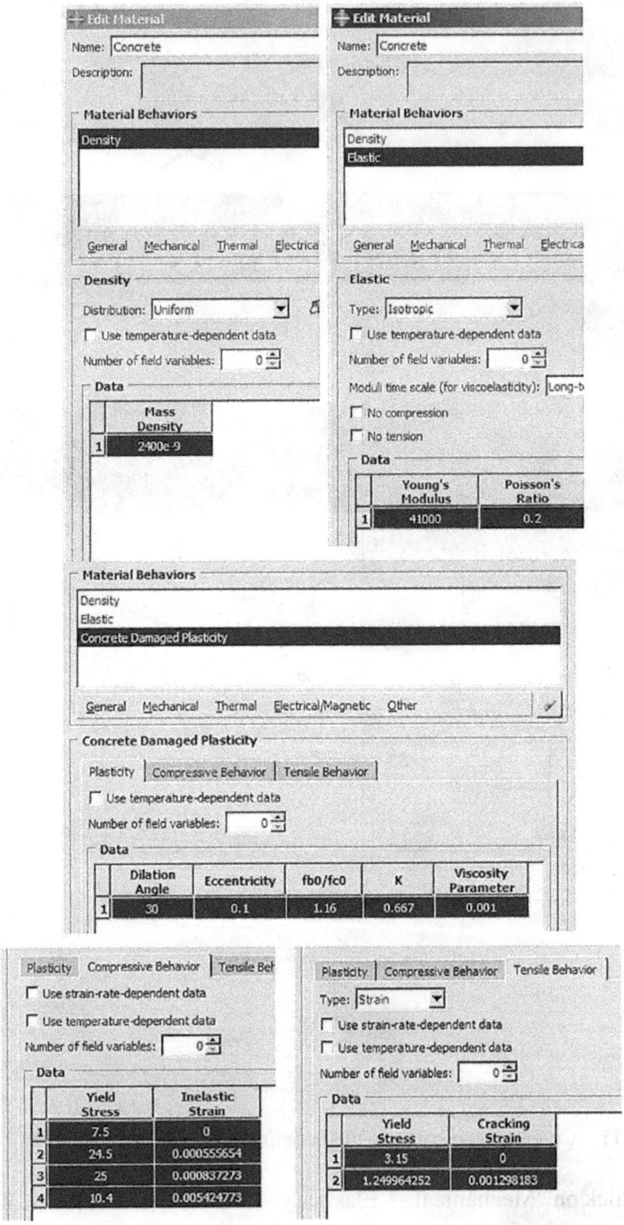

FIGURE 6.12 Identify the material of concrete.

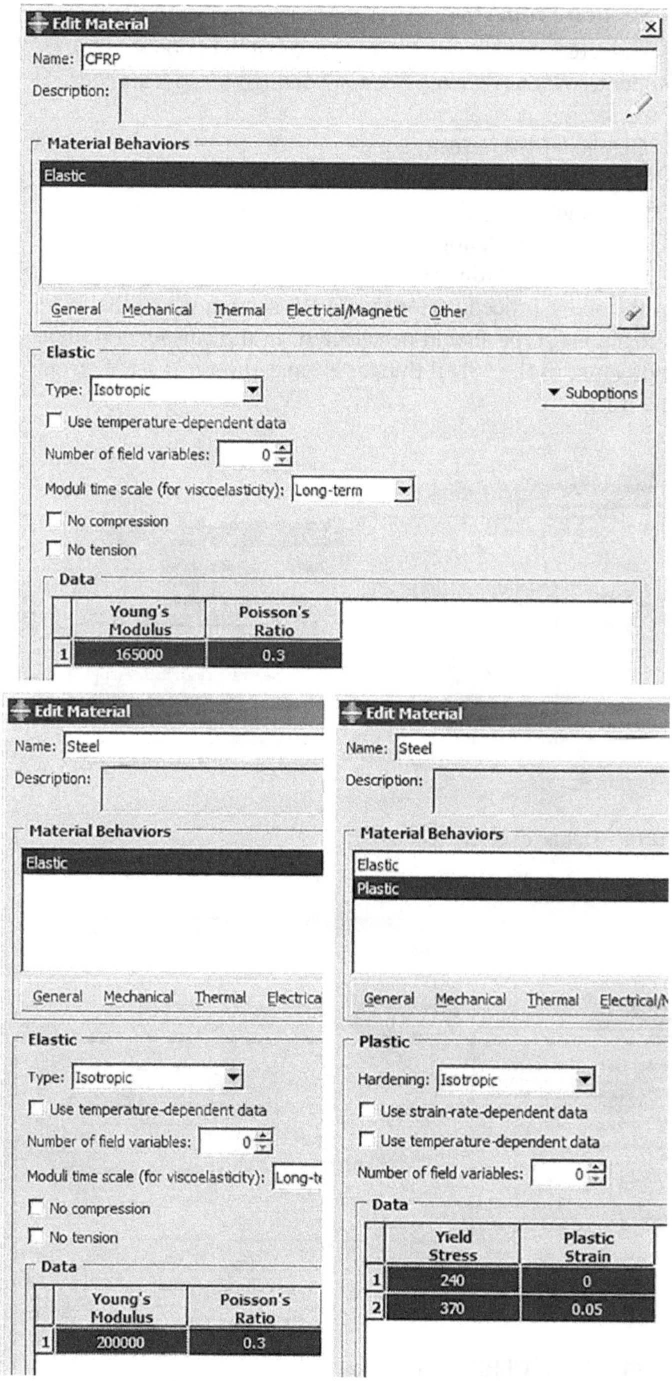

FIGURE 6.13 Identify the material of CFRP and steel.

6.4.2.2   Section Properties

To define a concrete section, from the main menu bar, double click on "Sections" in the "Model tree". The "Create Section" dialog box appears.

Name the section as "Concrete".
In the "Category" list, select "Solid".
In the "Type" list, select "Homogeneous".
Click "Continue". The "Edit Section" dialog box appears.
In the "Edit Section" dialog box:
Select "Concrete" as "Material" and click "OK" (see Figure 6.14).

Repeat the above procedure for the CFRP section where the "Shell" category and "Homogenous" type should be selected. In the edit section dialog box, type the value of 7 mm as the "Shell thickness" and choose "CFRP" from "Material" (see Figure 6.15).

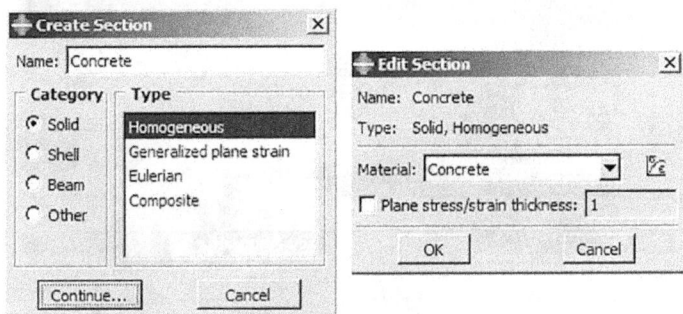

FIGURE 6.14   Define concrete section.

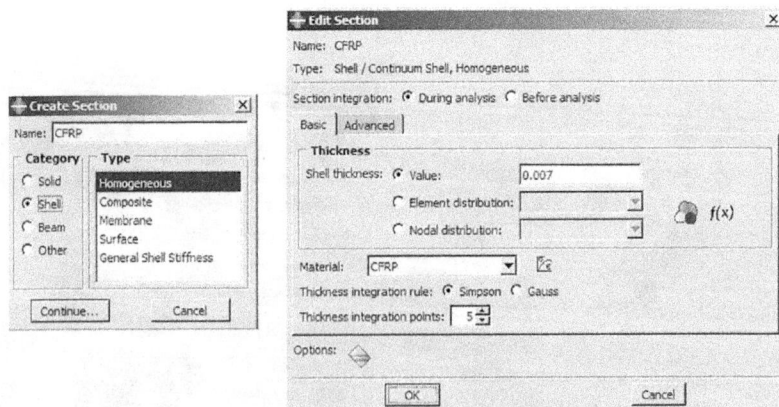

FIGURE 6.15   Define CFRP section.

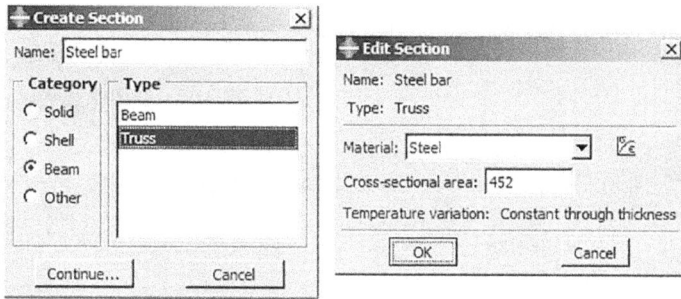

FIGURE 6.16 Define steel bar section.

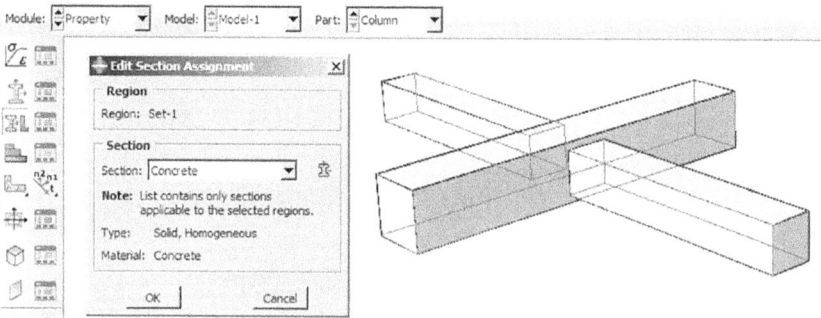

FIGURE 6.17 Concrete section assignment.

Create "Beam/Truss" section for steel reinforcement and select "Steel" from "Material" and enter 452 for "Cross-sectional area" as shown in Figure 6.16.

6.4.2.3 Assign Sections to Parts

As the next step, assign the defined section to the corresponding part.

Double click on "Section assignments" underneath "Column" in the "Model tree" and select the concrete part. Then click the middle mouse button to open the "Edit section assignment" dialog box. Chose "Concrete" from "Section" and click "Ok" to close the box as shown in Figure 6.17.

Perform the same for CFRP as shown in Figure 6.18 and steel parts as illustrated in Figure 6.19.

6.4.3 MESH MODULE

The mesh module is used to generate the finite element mesh. Double click on "Mesh (Empty)" underneath "Column" to activate the "Mesh" module.

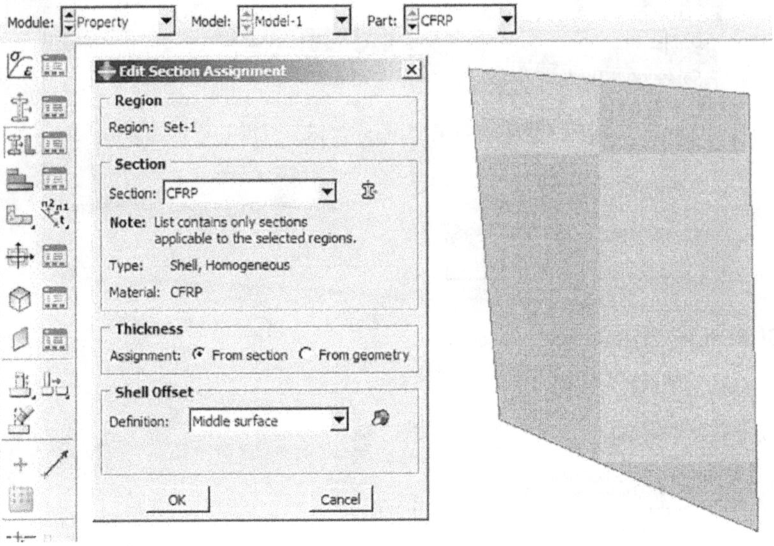

FIGURE 6.18   CFRP section assignment.

From the main menu bar, select "Mesh → Element type".

In the viewport, select the entire column as the region to be assigned. In the prompt area, click "Done". The "Element type" dialog box appears.

In the dialog box, select

- "Standard" from the "Element library" selection (the default).
- "Linear" from the "Geometric order" (the default).
- "3D stress" from the "Family" of elements.

In the lower portion of the dialog box, examine the element options. C3D8R element was selected as default. A brief description of the default element selection is available at the bottom of each tabbed page.

Click "OK" to assign the element type and to close the dialog box.

Repeat the above procedure for the CFRP and select S4R element as a Shell element and for steel parts select T3D2 as a Truss element as shown in Figure 6.20.

For simplicity in "Column" meshing, it should be partitioned into three sections by "Define cutting plane" and "Extrude/Sweep edges" as "Method" (see Figure 6.21).

In the next step, seeding should be defined. Select Seed →Part to open the "Global seeds" dialog box. Consider 50 as "Approximate global size" and click "Ok" to apply and close the box. Perform the same for all parts as shown in Figure 6.22.

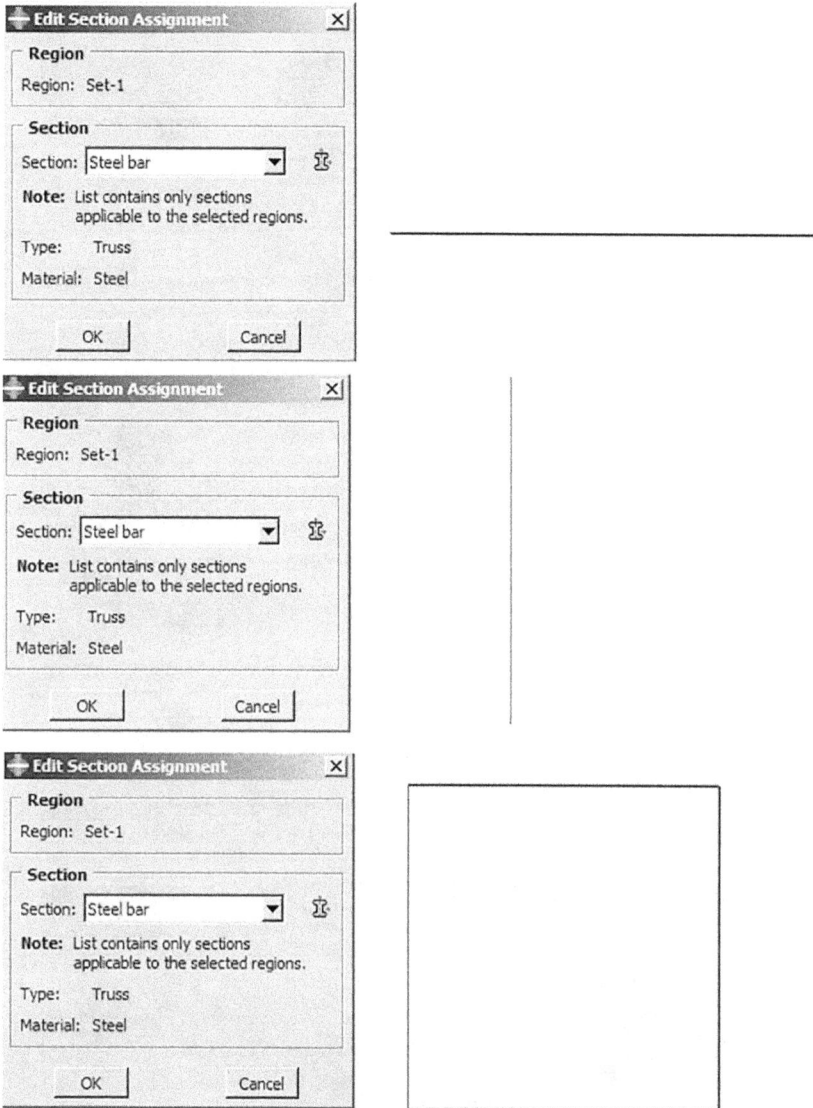

FIGURE 6.19 Steel bars section assignment.

Finally, all parts should be meshed. From the main menu bar, select "Mesh → Part" to mesh the part. Then click "Yes" in the prompt area to confirm the mesh of the part. Once meshed, the color of the plate changes to blue.

Repeat the meshing procedure for other parts as well. The meshed geometry is shown in Figure 6.23

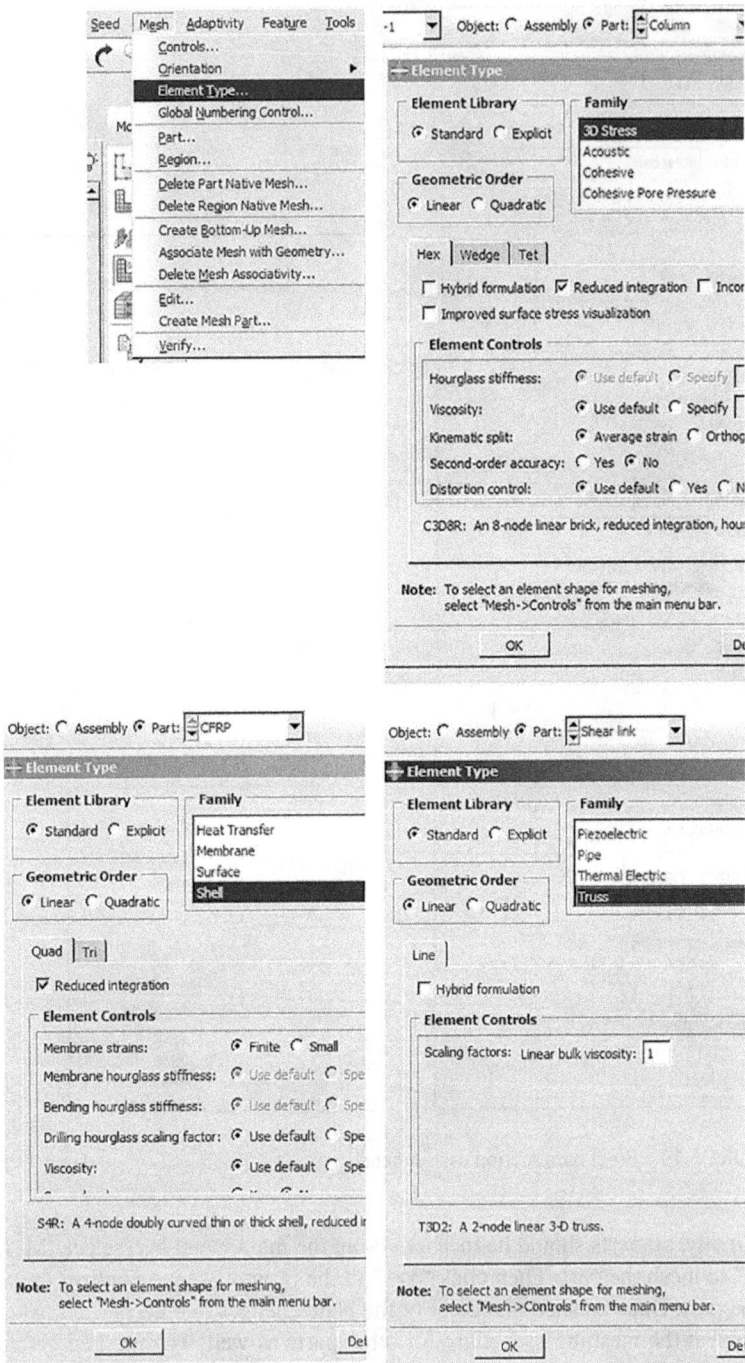

FIGURE 6.20   Define the element type.

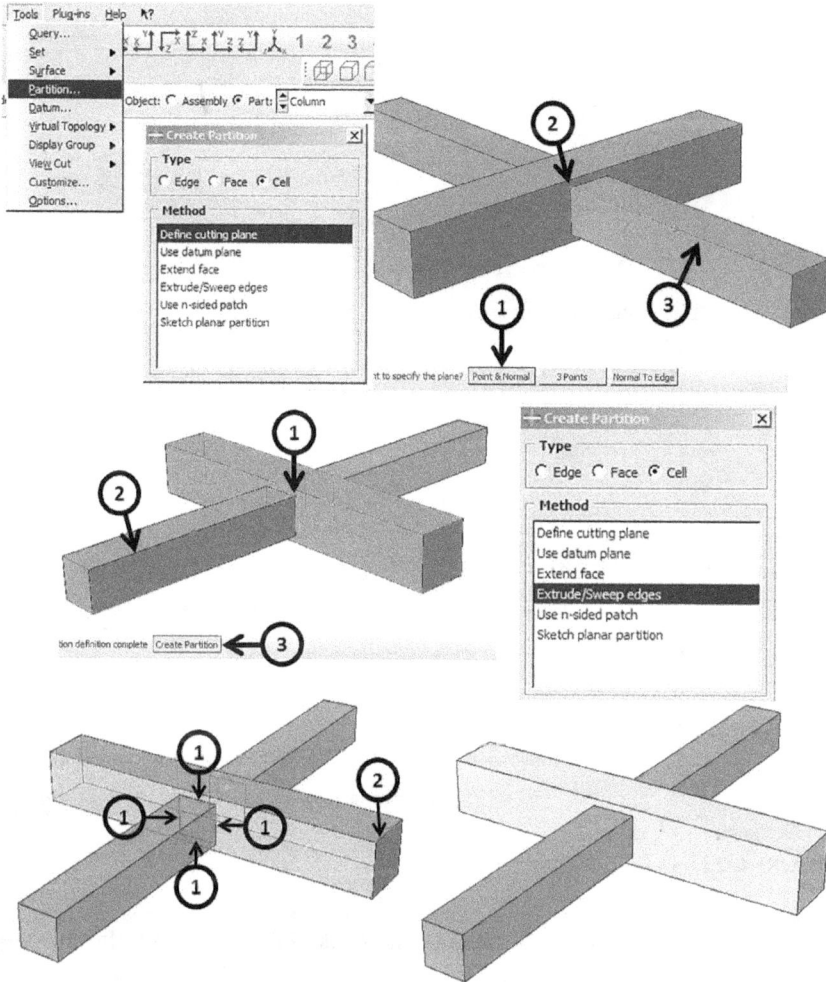

FIGURE 6.21 Partitioning the column.

6.4.4 ASSEMBLY MODULE

In this module, all the parts that are created earlier can be combined (assembly) to obtain the required model.

The parts (column, steel reinforcement bars, and CFRP) are combined to form the required structure. In the "Module" list located under the toolbar, click "Assembly" to activate the "Assembly" module.

From the main menu bar, select "Instance → Create". The "Create Instance" dialog box appears.

In the opened window, under the "Instance type" box, choose "Dependent (mesh on the part)".

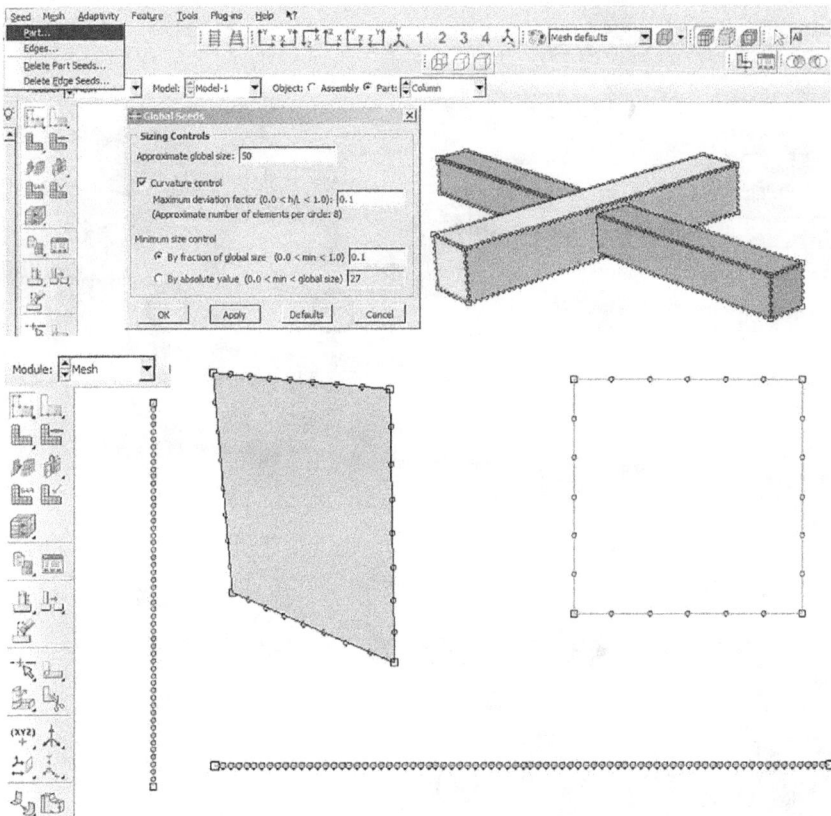

FIGURE 6.22 Define seeds for the parts.

In the dialog box, select "Column" and click "Ok". Then use "Instance →
Rotate" and rotate the instance as illustrated in Figure 6.24 by 90 degrees.

Select "Instance → Create", then choose "CFRP" and click "OK". Use "Rotate"
to rotate it by 90 degrees and "Translate" to translate to the proper position. Repeat
the procedure to define and reposition all CFRP instances (see Figure 6.25).

Add all steel instances to assembly and use Rotate, Translate, and Linear pattern to define the steel reinforcement as shown in Figure 6.26.

Finally, translate the steel reinforcement into the column constraint by the
CFRP sheets (see Figure 6.27)

6.4.5 STEP MODULE

In the next step, the "Step" module can be set up in order to configure the required
analysis.

In the Module list located under the toolbar, click "Step" to activate the "Step"
module.

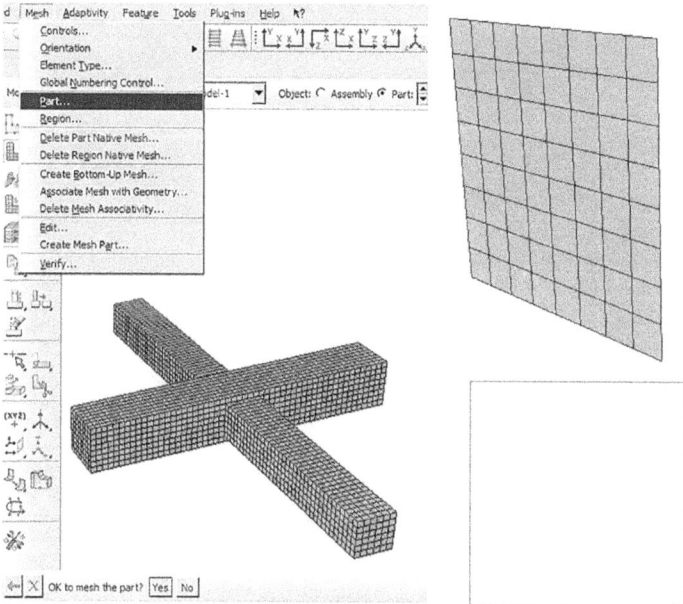

FIGURE 6.23 Meshing the parts.

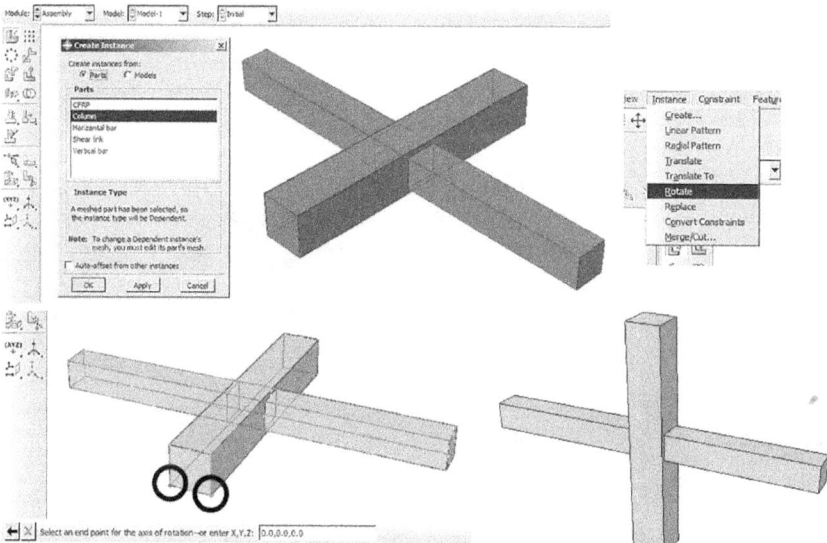

FIGURE 6.24 Define a column instance and rotate it.

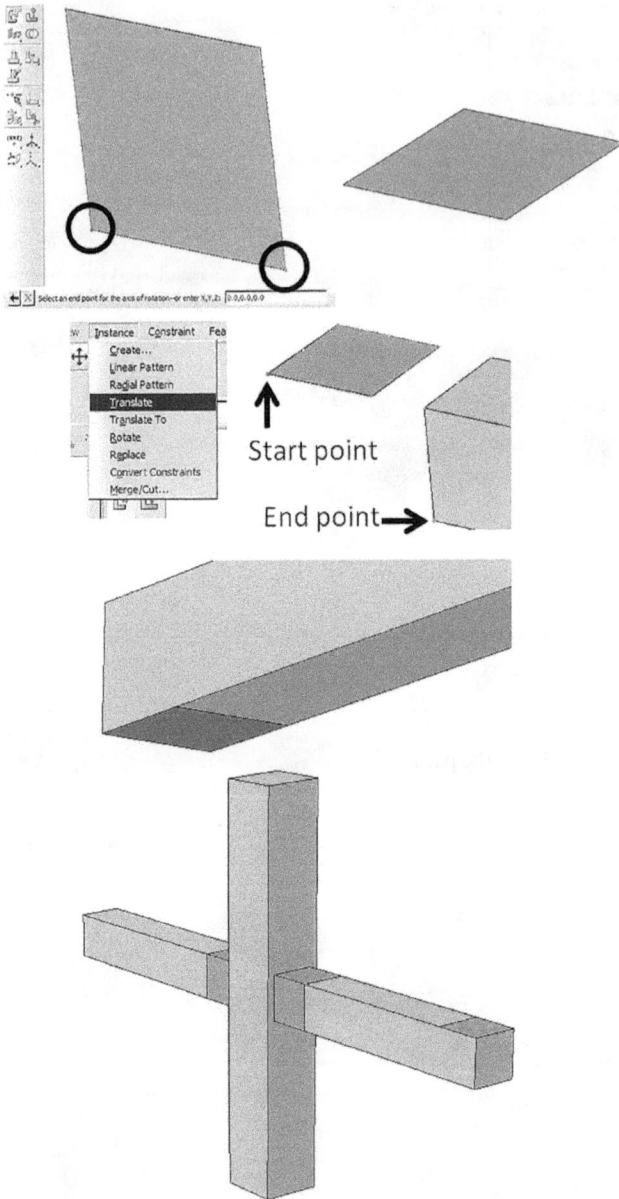

FIGURE 6.25 Define CFRP instances, rotate and translate it to the proper position.

From the main menu bar, select "Step → Create" to create a step. The "Create Step" dialog box appears with a list of general procedures and a default step named "Step-1".

Change the step name to "Static".

Select "General" from the "Procedure type".

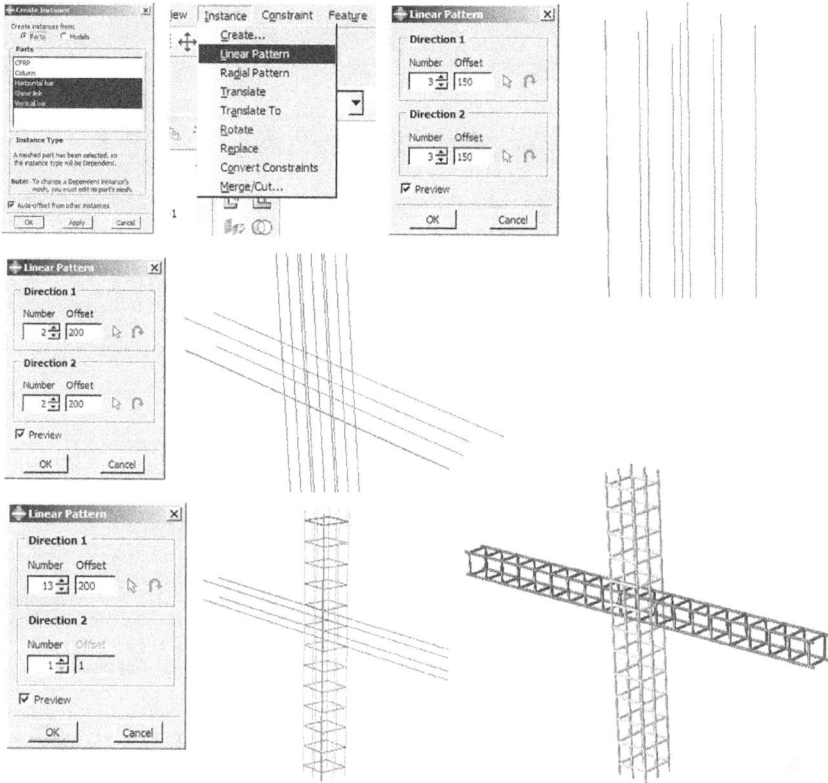

FIGURE 6.26 Define steel part instances and create their assembly by a linear pattern.

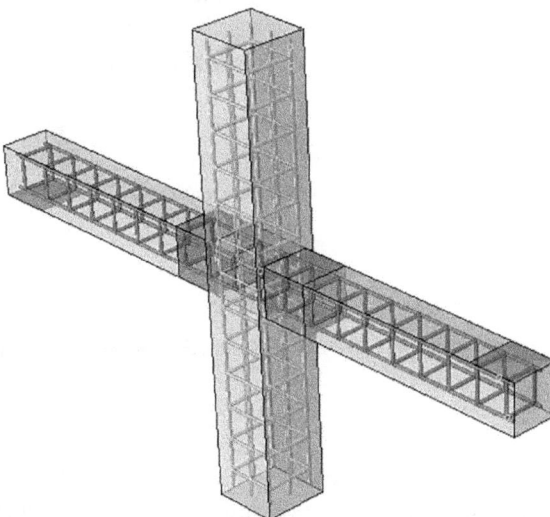

FIGURE 6.27 Final assembly.

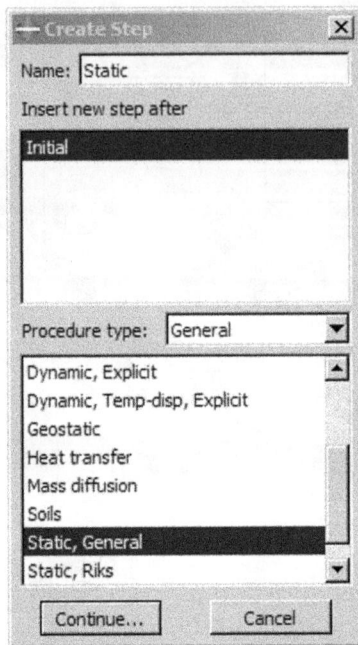

FIGURE 6.28   Create an analysis step.

Scroll through the available list, select "Static, General", and click on "Continue" (see Figure 6.28).

Then, the "Edit Step" dialog box appears.

In the "Basic" tab, consider nonlinear geometries for analysis by toggling "Nlgeom" on.

Click the "Incrementation" tab and change the increment size as shown in Figure 6.29 and then click "OK" to create the step and to exit the "Edit step" dialog box.

Double click on "F-output requests" to open "Edit field output requests". The user can review which variables are requested by default (see Figure 6.30). Leave the dialog box unchanged and click "Ok" to close it.

6.4.6   INTERACTION MODULE

This module is used to define various interactions within the model or interactions between regions of the model and surrounding parts. The interactions used in this example are "Tie" and "Embedded region".

To activate the "Interaction" module, select "Interaction" from the module list.

To define "Tie" constraint between the column and the CFRP sheets, click "Find contact pair" in the toolbox to open its dialog box and then click "Find contact pairs". The software tries to find all near faces and lists them in the dialog box. Click on column "Type", then "Edit" to open the "Edit multiple cells" dialog

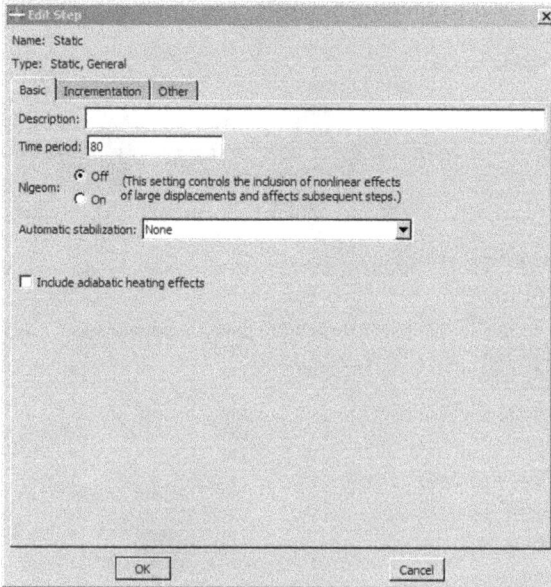

FIGURE 6.29   Analysis step configurations.

box. In the box, select "Tie" constraint from type and click "Ok" to accept and close it. Finally, click "Ok" to close the "Find contact pairs" dialog box.

To define "Embedded region" constraint between steel reinforcement and the column, double click on "Constraints" in the "Model tree" and select "Embedded region" and click "Continue". In the "Model tree" hide the column and CFRPs, then select all steel parts as the embedded regions and click on the middle mouse button. Then choose "Select Region" in prompt area. Show back the column and select the part instance as the "Host region" and click the middle mouse button to open the "Edit constraint" dialog box as shown in Figure 6.31.

6.4.7   LOAD MODULE

The prescribed conditions, such as the loads and boundary conditions, are step-dependent, which means that the user needs to specify the steps they are considered. As the steps in the "Analysis" module have been defined, then the "Load" module can be used to define the prescribed conditions. In this example, the column base is fixed that cannot rotate and translate.

To define the boundary condition, double click on "BCs" in the "Model tree" to open its dialog box. Select "Initial" from "Step" and "Displacement/Rotation" from "Types for selected step" and click "Continue". Then rotate the model to display the bottom face of the column and select the face and click the middle mouse button to open the "Edit boundary condition" dialog box. In the box, check all items to constraint translates and rotates of the face nodes and click "Ok" to define the boundary condition as shown in Figure 6.32.

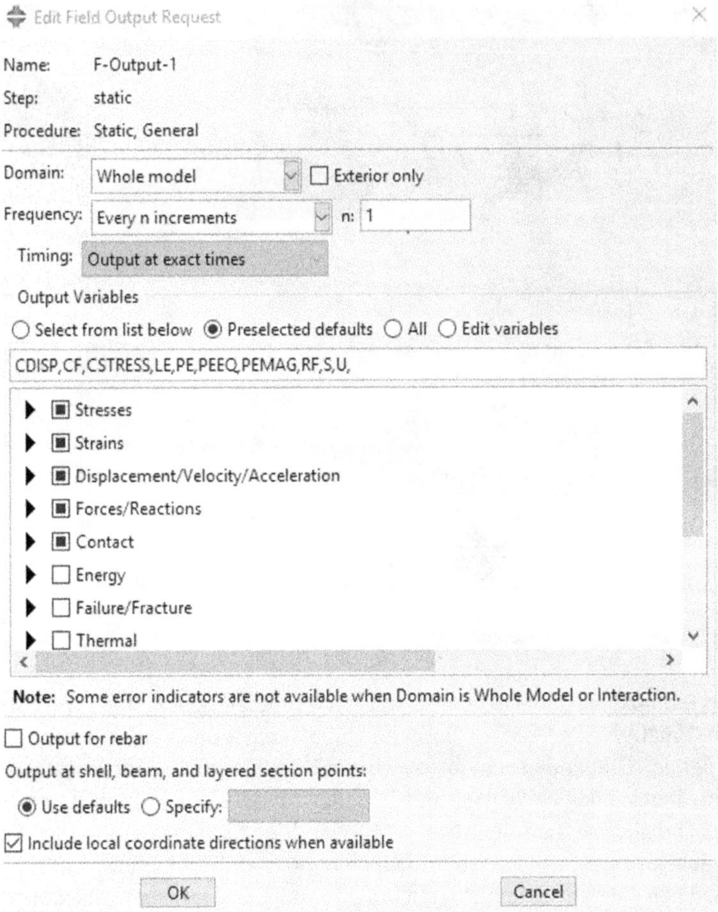

FIGURE 6.30   Edit field output request.

Since the base of the column is constrained, then loading can be applied to the column and beams. In this example, a pressure force of 4 N/mm^2 is applied in the negative direction. The static load is applied as a general step, which was created in the "Step" module.

In the "Model tree" double click on "Loads" to open the "Create load" dialog box. From the list of steps, select "Static" from "Step" that the load will be exerted.

In the "Category" list, accept "Mechanical" from the default category selection. In the "Types for selected step" list, select "Pressure". Click on "Continue". The user is asked to select a region to which the load will be applied.

In the viewport, select the top face of the column, top CFRP on the right beam, and bottom CFRP under the left beam as the elements which load will be applied.

Once the process is finished, click "Done" in the prompt area to complete selecting the regions. The "Edit load" dialog box appears.

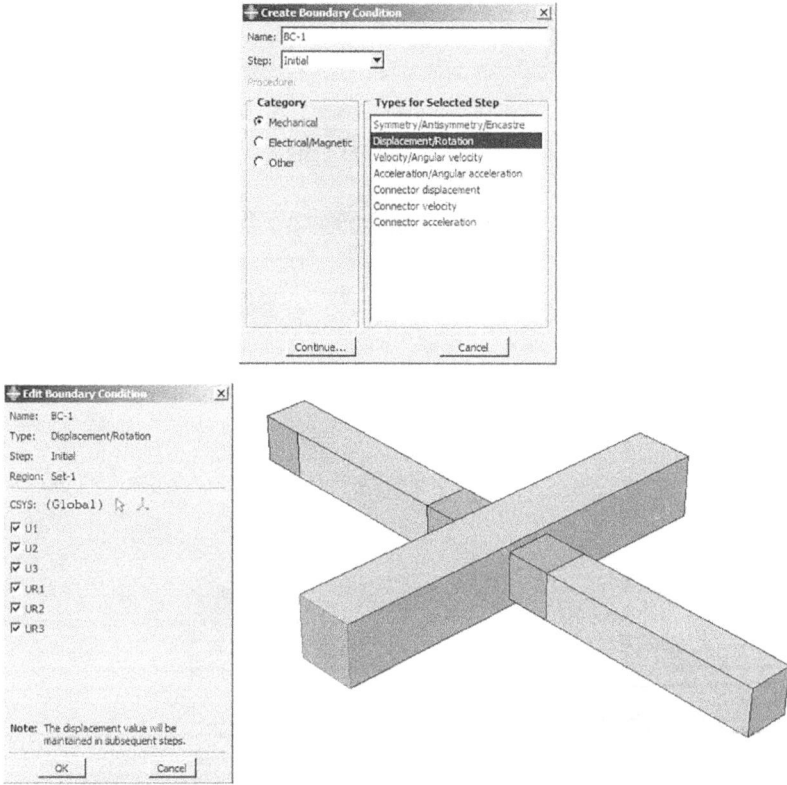

FIGURE 6.31 Tie constraints between CFRPs and Column and Embedded constraint between steel reinforcement and column.

FIGURE 6.32 The boundary condition applied at the base of the column.

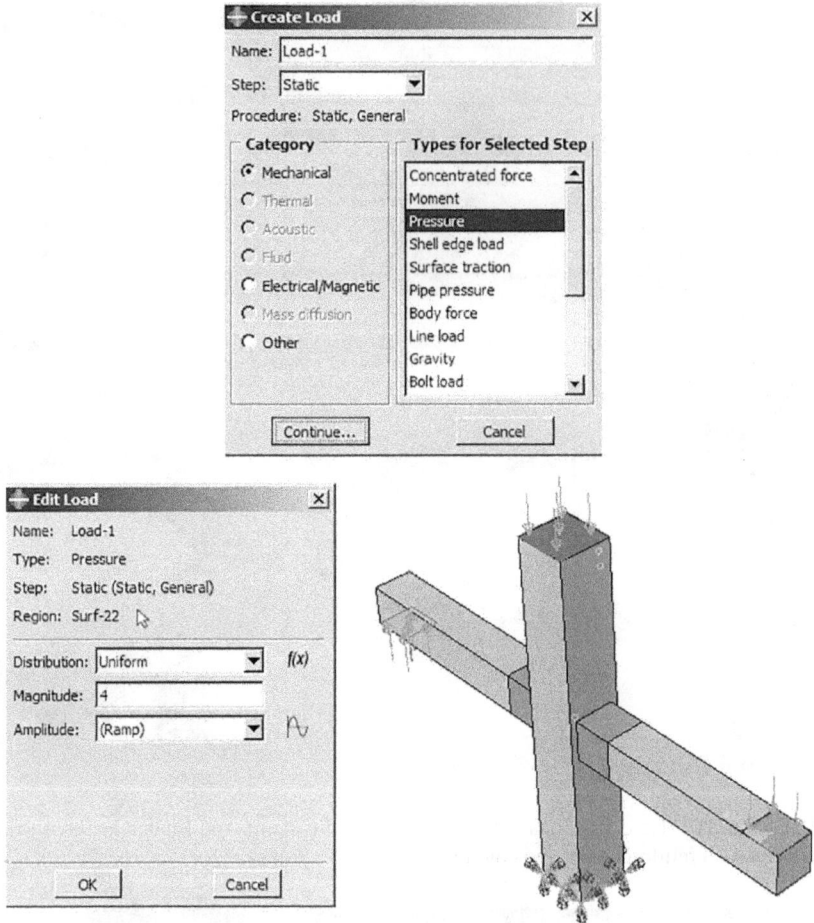

FIGURE 6.33 The load applied to CFRPs.

Enter a value of 4 and click "OK" to create the load and to close the dialog box as shown in Figure 6.33.

6.5 ANALYSIS: JOB MODULE

In the next step, a job should be defined to execute the analysis.

In the Model tree, double click on "Jobs" to open its box (see Figure 6.34).

Accept the default job name as "Job-1" and click "Continue". The "Edit job" dialog box appears.

Accept all defaults and click "Ok" to complete the job definition (see Figure 6.35).

Expand the tree under "Jobs", right click on "Job-1". Then, click on "Submit" to begin the analysis (see Figure 6.36).

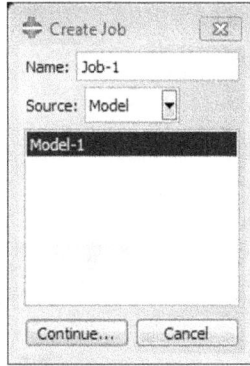

FIGURE 6.34 Create job.

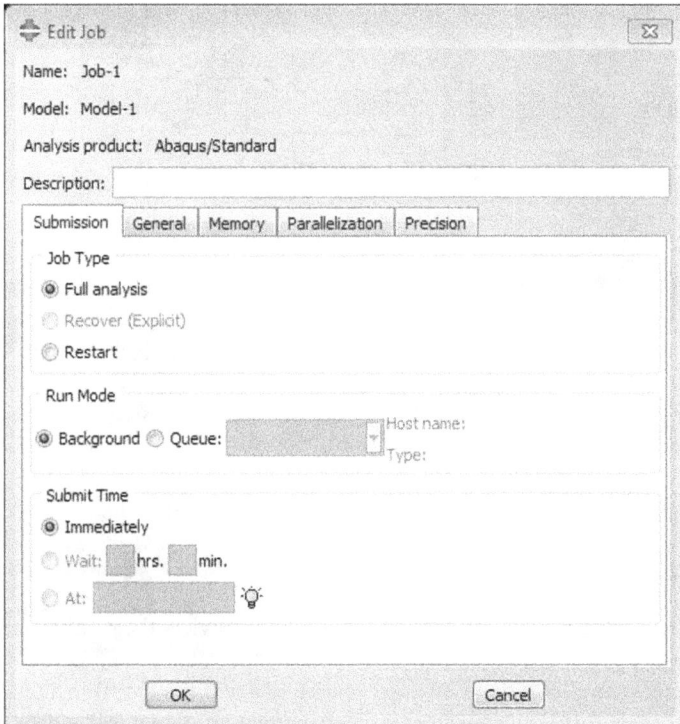

FIGURE 6.35 Job definitions.

Once the running simulation has been completed, the output database of results can be opened.

To open the output database, right click on the completed job and select "Results" to activate the "Visualization" module (see Figure 6.37).

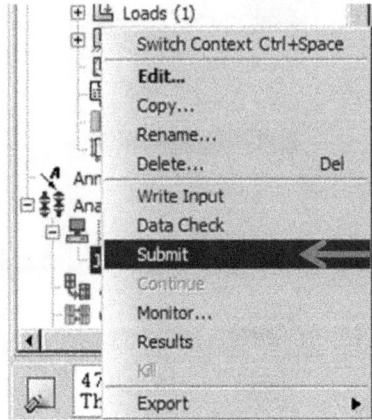

FIGURE 6.36   Submit the job for analysis.

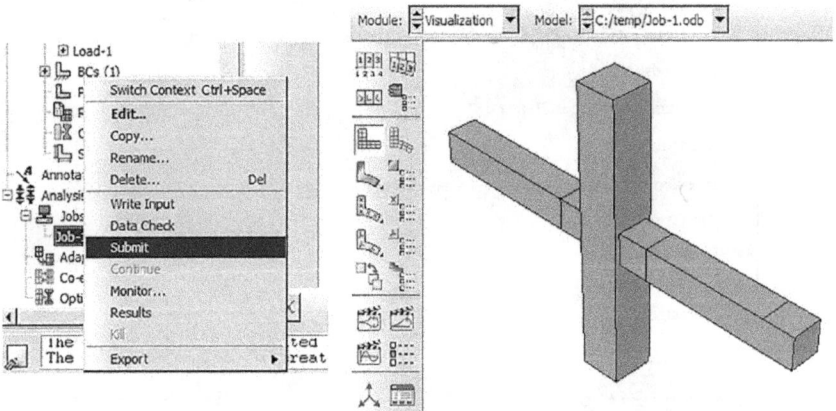

FIGURE 6.37   Activating the Visualization module.

6.6 VISUALIZATION MODULE

As mentioned before, the software allows the user to view the results graphically using a variety of methods.

The software allows the user to write data to a text file (*.rpt) in a tabular format. This feature is very convenient in writing tabular output to the data file. It is also very useful, especially in writing a report. In this problem, the user should generate a report containing the element stress (Von Mises).

To generate the field output report for "Von Mises stress", perform the following steps:

From the main menu bar, select "Report → Field Output". In the "Report field output" dialog box, choose the "Variable" tab and check "Mises" under the expanded list of "S: Stress components".

In the "Setup" tabbed page, name the report as "job-1.rpt." This file can also be saved in the user's favorite directory by clicking on the "Select" button. In the "Data" region at the bottom of the page, toggle off "Column totals" and click on "Apply".

Find the saved report file and open it in "Notepad" program. The stress values (Von Mises) are appended to the report file. The stress contour can be observed as displayed in Figure 6.38.

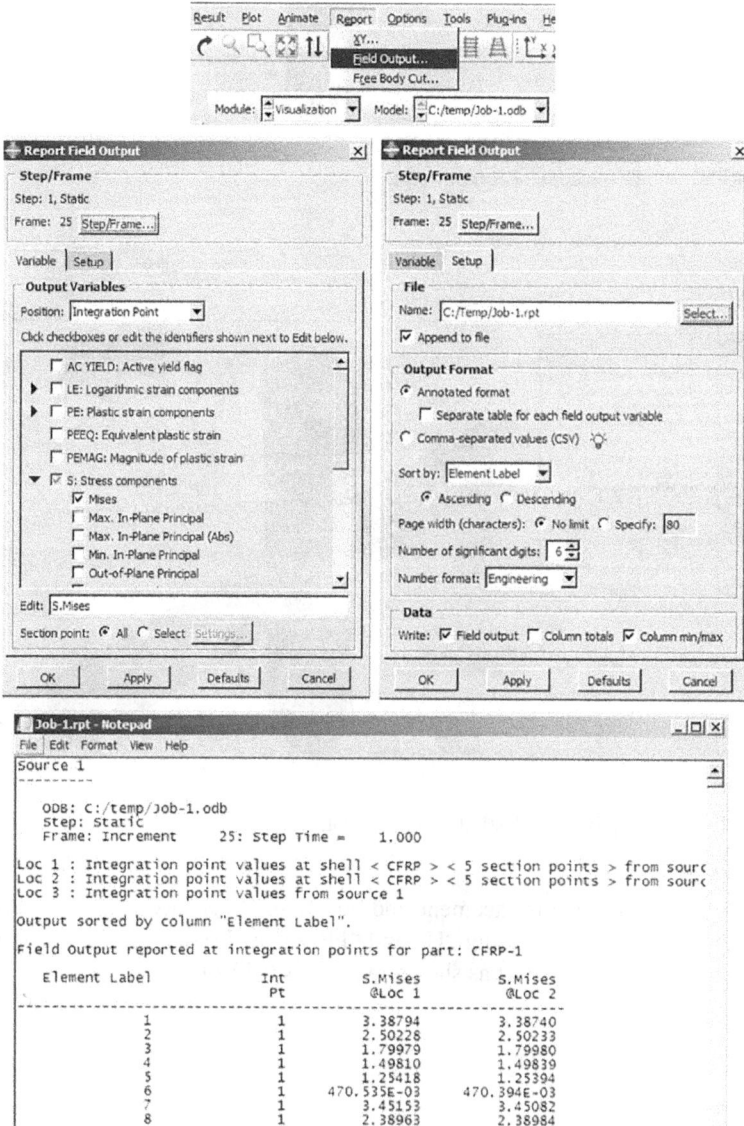

FIGURE 6.38   Von Mises stress report.

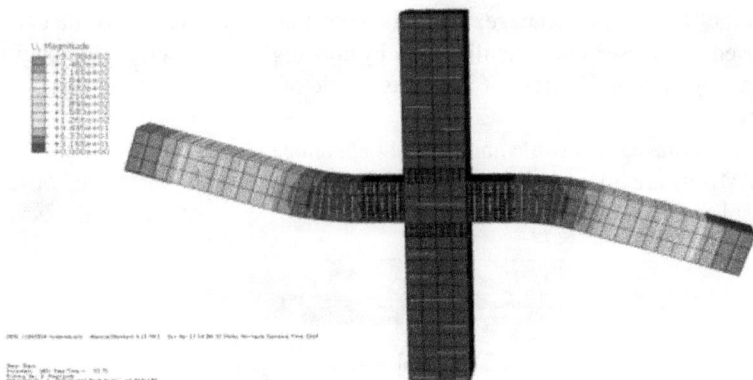

FIGURE 6.39   Displacement contour plot.

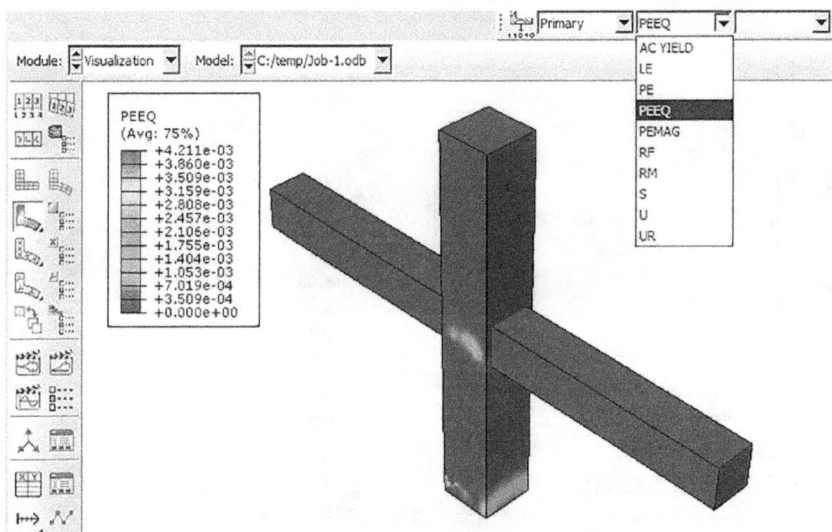

FIGURE 6.40   Equivalent plasticity contour plot.

Also, the plots for displacement and equivalent plasticity contour could be shown in viewport by choosing "U" and "PEEQ" as Primary variables, respectively, on top of the viewport as shown in Figures 6.39 and 6.40.

7 Hollow Circular Ultra-High-Performance Concrete (UHPFRC) Section under Lateral Cyclic Load

7.1 INTRODUCTION

Nowadays, by growing the population, the need for advanced communication systems incredibly increased. The communication towers play an important role in the main infrastructure to enhance the coverage and quality of various communication systems. Recently, there are many attractions for the structural designer for the application of newly developed concretes such as Ultra-High-Performance Fiber-Reinforced Concrete (UHPFRC) in the construction of towers due to their high strength and durability. Therefore, by considering the high loading capacity of UHPFRC material, it was capable to implement in segmental precast communication towers, which makes the possibility for easy transport of the segments of the tower and quick assembly process by using bolts and nuts. For this reason, in this example, a hollow cylindrical tube with UHPFRC materials as a scaled dimension of a segment for the precast communication tower is considered and the modeling and analysis process of the considered part (scaled segment) of the communication tower using the ABAQUS finite element software is demonstrated in detail.

Researcher: Cheyath Ali Abdulameer (gayathchalabi@gmail.com)

7.2 PROBLEM DESCRIPTION

In this example, a hollow cylindrical tube with UHPFRC material as a segment of the precast communication tower is considered and its behavior under cyclic load is investigated using the finite element method. The model consists of the hollow circular section, transverse and longitudinal reinforcement, and a steel sleeve to fully fix the bottom of the section. The hollow tower segments and circular steel tubes (Figure 7.1) were modeled.

DOI: 10.1201/9781003219491-7

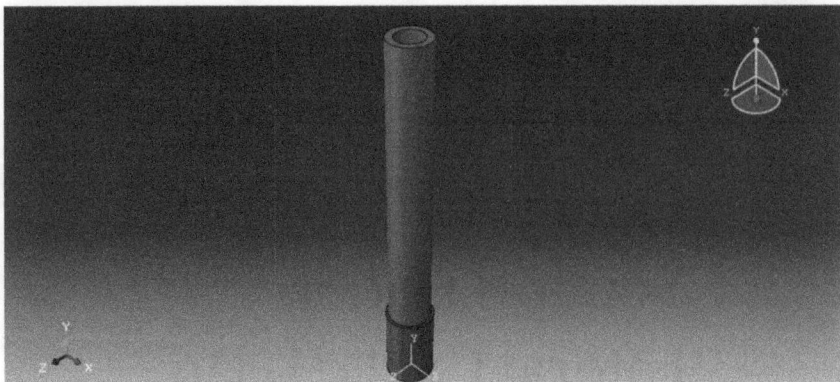

FIGURE 7.1   The complete 3D model.

Rebars in Figure 7.2 were modeled as 3D solid to define the more accurate interaction between rebars and hollow circular sections.

7.2.1 GEOMETRIC PROPERTIES

The numerical program was planned on a hollow circular tower segment in a reduced scale. The test specimen has a wall thickness of 50 mm, a height equal to 2500 mm, and a diameter of 300 mm. Figures 7.3 and 7.4 show the segment height and dimensions of the hollow circular tower segment.

Half meter steel sleeve was used to fix the bottom of the section, and this was carried out to capture what occurred during the lab experiment. Figure 7.5 shows the dimensions of the steel sleeve.

Ten mm diameter bars were used for the longitudinal reinforcement, and 6 mm diameter plain round bars were used for the transverse reinforcement. Figure 7.6 shows the reinforcement details.

7.3   OBJECTIVES

1. To develop the finite element model of the hollow circular section with UHPFRC material.
2. To define the concrete damage plasticity (CDP) model for UHPFRC material in ABAQUS software.
3. To evaluate the failure mode of the hollow cylindrical tube under cyclic load in terms of cracking and crushing damage.

7.4   MODELING

7.4.1 PART MODULE

This module is used to create various parts of the model. In this case, the model is divided into three parts: hollow circular section, sleeve, and steel reinforcements.

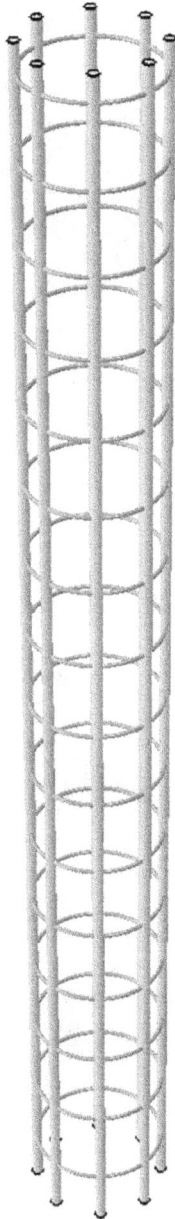

FIGURE 7.2 Rebars and stirrups.

Start the software from programs in the "Start menu".

From the Start Session dialog box that appears, click on "With Standard/Explicit Model". This step allows the user to start modeling where the user can create a new file and save it under any name in a new folder.

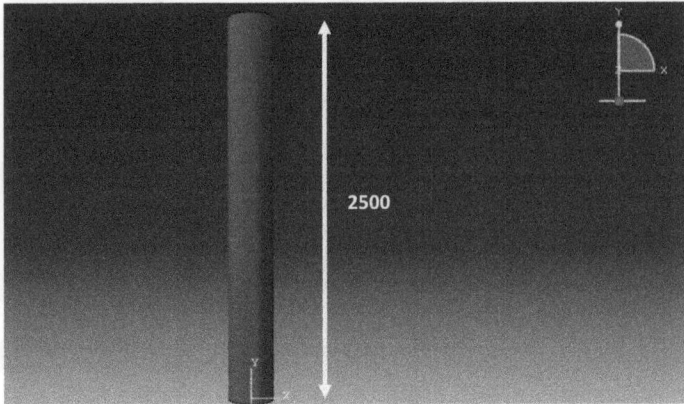

FIGURE 7.3   Segment height.

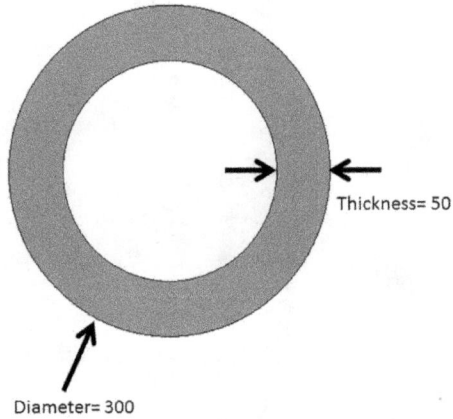

FIGURE 7.4   Segment diameter and thickness.

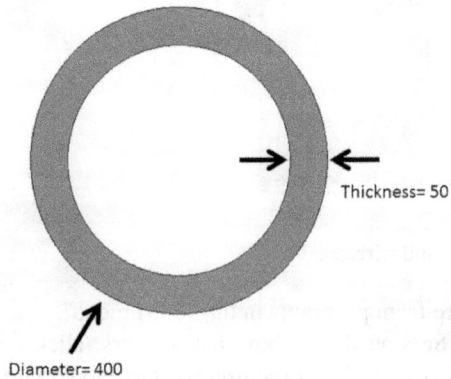

FIGURE 7.5   Steel sleeve.

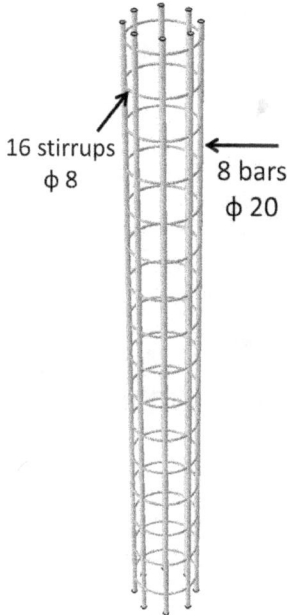

16 stirrups
φ 8

8 bars
φ 20

FIGURE 7.6 Reinforcement details.

From the menu bar, select "Part → Create" to create a new part and the "Create Part" dialog box appears. Use the "Create Part" dialog box to name the part and define related details.

Hollow circular section:

Name the part "Hollow circular section". For this section, choose "3D" and "Deformable", "Solid" and "Extrusion" from the base feature (see Figure 7.7).

Enter an "Approximate size" as 2000. The value entered in the approximate size text field at the bottom of the dialog box sets the approximate size of the new part. Click "Continue" to exit the "Create Part" dialog box.

Use the "Create circle: (Center and Perimeter)" tool located in the upper-left corner of the sketcher toolbox to begin drawing the geometry of the hollow circular section. Two circles will be drawn according to the inner and outer diameters of the section, as shown in Figure 7.8. The user can select a center point for the circle at the viewport or enter the x- and y-coordinates. Then, select a perimeter point for the circle at the viewport or enter the x- and y-coordinates. Instead of using known coordinates, the user can also define the dimension of the geometry by clicking on the add dimension tool.

After completing the sketching, right click and click on "Cancel procedure" to exit the sketcher. Click on "Done" in the prompt area, and it will turn out, as shown in Figure 7.9, where the user needs to define the depth for the extrusion of the hollow circular section.

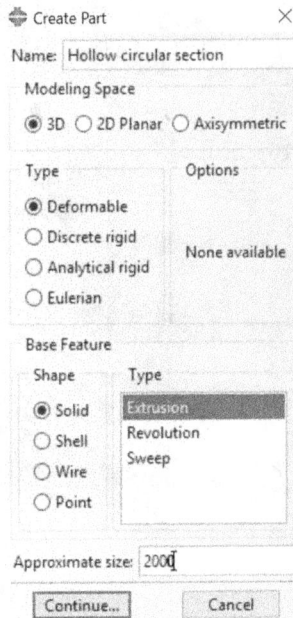

FIGURE 7.7   Create a hollow circular section.

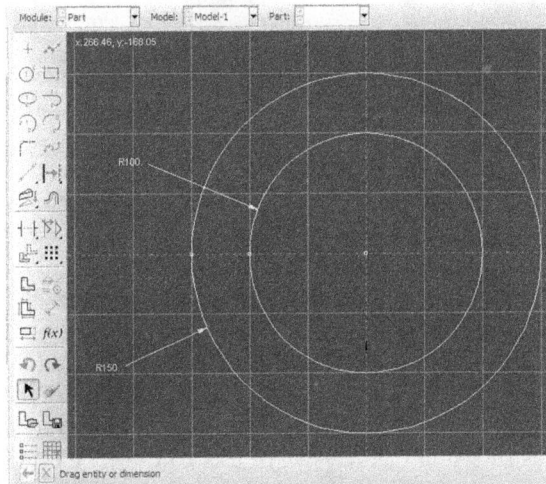

FIGURE 7.8   Draw the circular sketch of the column.

Once the depth of the hollow circular section has been defined, it will turn out, as shown in Figure 7.10.

By following the previous procedure, the "Steel sleeve" is drawn according to its dimensions. Figure 7.11 shows the steel sleeve.

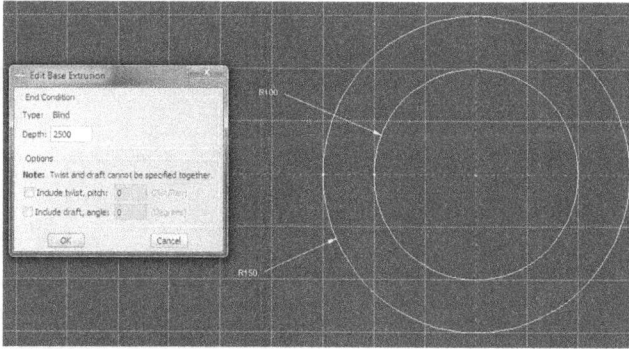

FIGURE 7.9   Edit base extrusion.

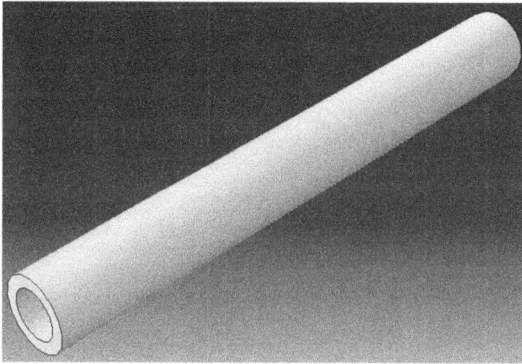

FIGURE 7.10   Hollow circular column.

Longitudinal and transverse reinforcement:

A total of eight (8) longitudinal reinforcements are embedded in the column wall. To perform this, name the part as "Longitudinal Bars". Then, in the sketcher, draw a reference circle with a diameter equal to the outer plus inner diameter of the hollow circular section divided by two. After that, we draw a circle at x equal to zero and y equal to the radius of the reference circle, as shown in Figure 7.12 (the circle represents a rebar section).

In this example, a total of eight rebars are drawn. Use the "Radial pattern" tool located in the lower right corner of the sketcher toolbox to duplicate the part instance. First, the user needs to select the rebar circle as the entity to the pattern. After clicking on "Done", the "Radial pattern" box appears. Enter 8 for "Number" and 360° for the "Total angle" and click "Ok" (see Figure 7.13)

Once finished patterning the rebar circles, it is necessary to erase the reference circle by using the "Delete" tool from the sketcher toolbox. Click on "Done" in the prompt area, and it will turn out, where the user needs to define the depth for the extrusion of rebars.

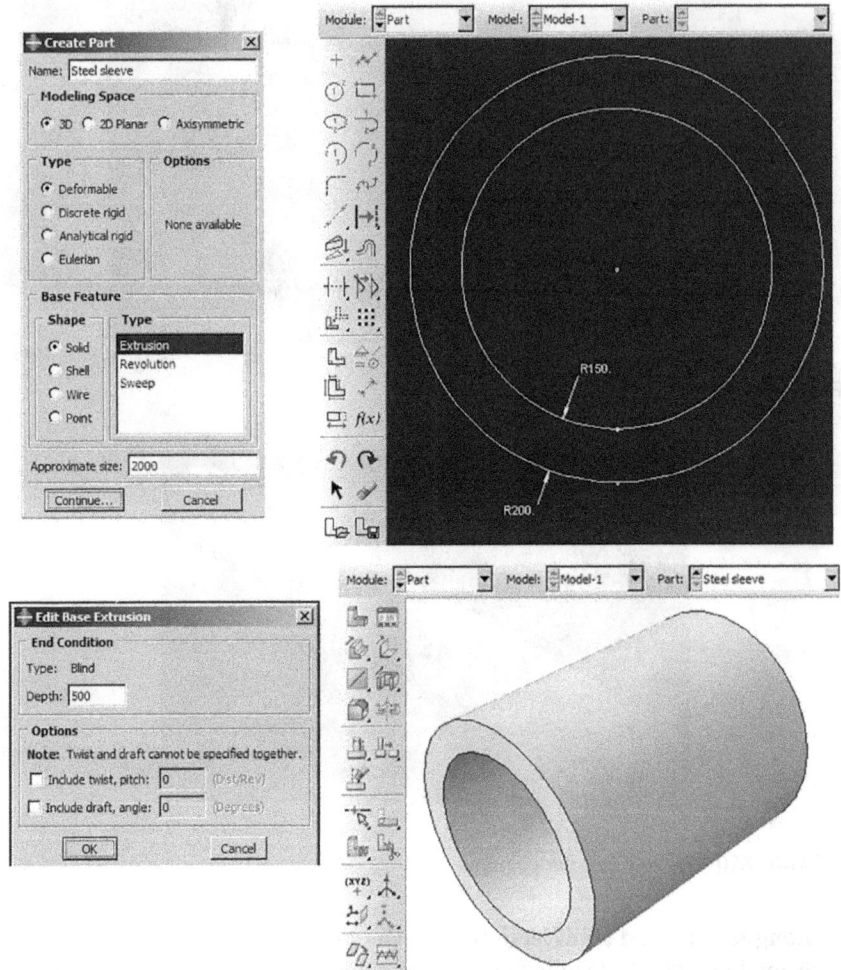

FIGURE 7.11   Steel sleeve definition.

The final geometry is shown in Figure 7.14.

Finally, create "Stirrup" as a swept solid. Consider 200 as the path approximate size and try to draw a circle with 100 as radius in the center of the plane. Then exit the sketcher by clicking "Done" and consider 20 as the scale of section sketch and draw a circle with a radius of 4 and click done to exit the sketcher and define the stirrup geometry as shown in Figure 7.15.

7.4.2   PROPERTY MODULE

In this module, the material properties are defined and assigned to the sections as demonstrated here.

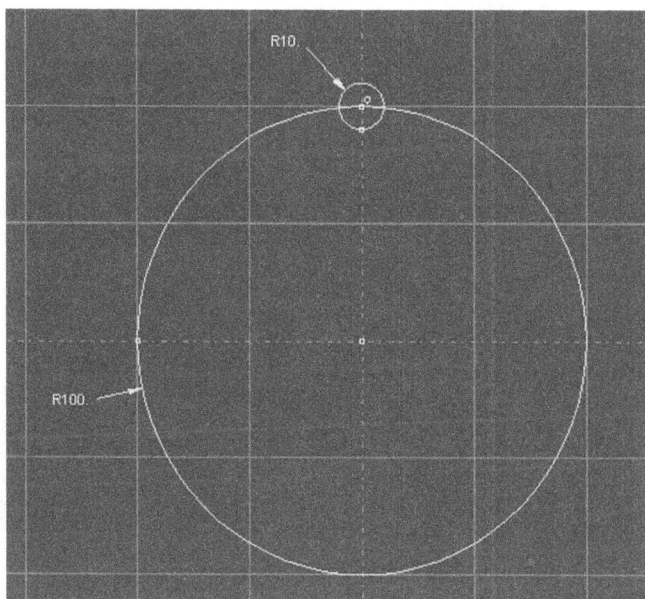

FIGURE 7.12 Longitudinal bars sketch.

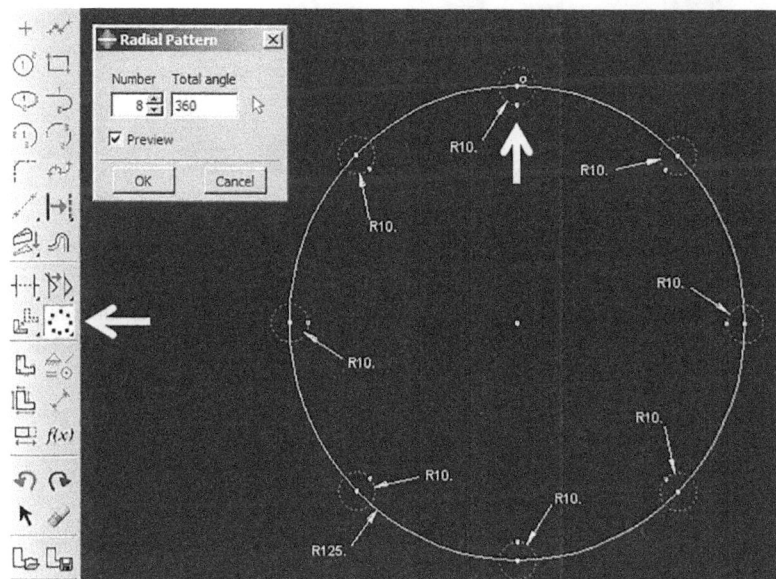

FIGURE 7.13 Radial pattern.

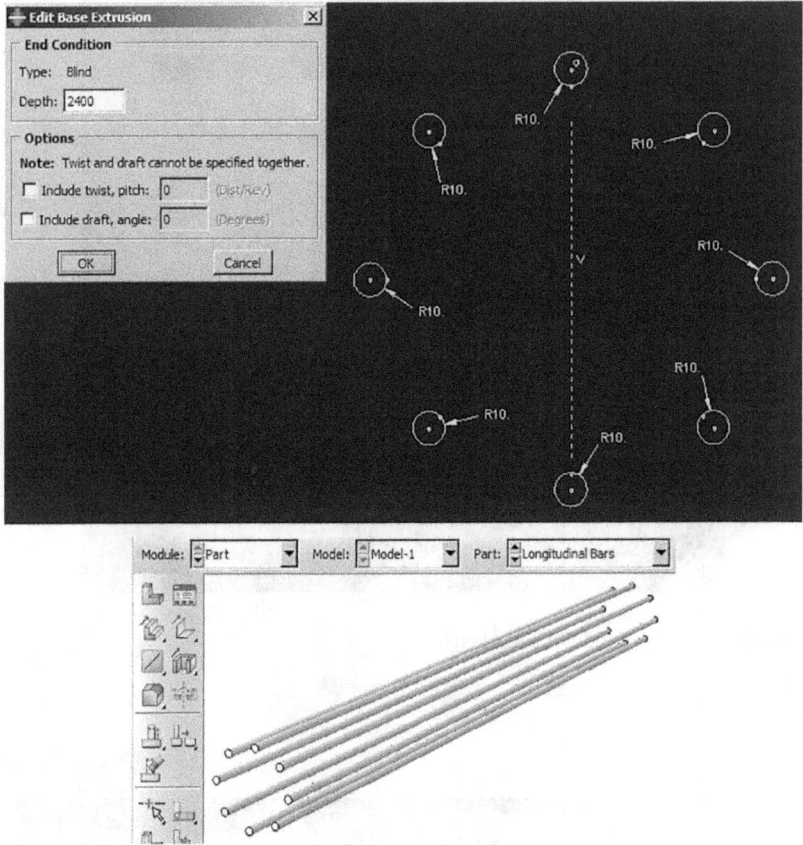

FIGURE 7.14 Final geometry of longitudinal bars.

7.4.2.1 Material Properties

Sleeve, rebars, and stirrups are made of steel and are assumed to be the classic elasto-plastic model. Also, the Concrete damaged plasticity (CDP) material model should be defined for the hollow circular section.

Define the material models, in the module list located under the toolbar, select "Property" to open the "Property" module. The cursor changes to an hourglass while the "Property" module loads.

From the menu bar, select "Material → Create" to create a new material.

The "Edit Material" dialog box appears.

Name the material "Steel".

From the material editor's menu bar, click on "Mechanical → Elasticity → Elastic". The software displays the Elastic data form. Enter the value of 200,000 for "Young's modulus" and 0.3 for "Poisson's ratio" in the respective cells.

Click on "General → Density". Then, enter the value of 7850E-009.

In the next step, click on "Mechanical → Plasticity → Plastic".

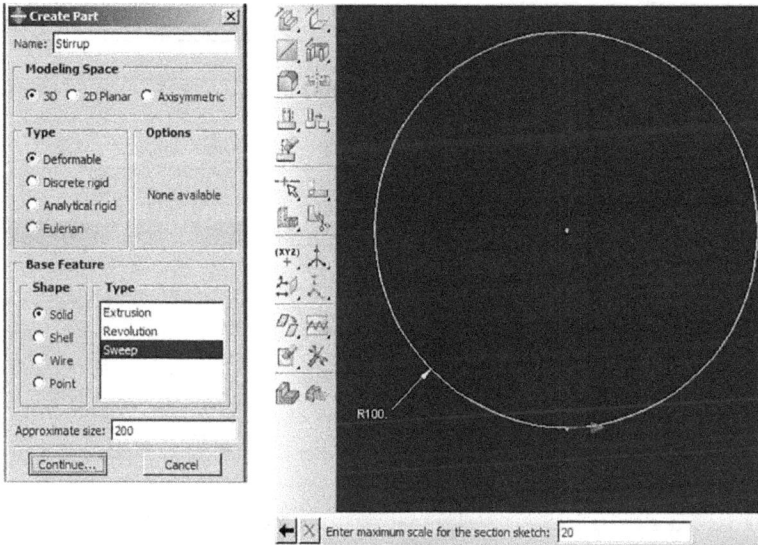

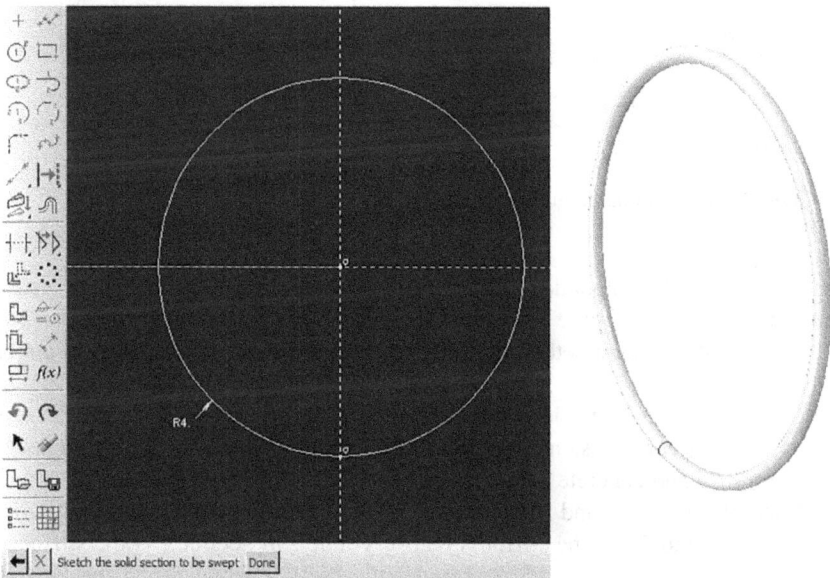

FIGURE 7.15 Define stirrup.

Enter the values for yield stress and plastic strain, as shown in Figure 7.16. Click OK to exit the material editor.

The damage plasticity model (CDP) was used for concrete modeling, which needs to input the tension–compression constitutive model of concrete and also concrete plasticity index. Tension and compression damage factor are introduced

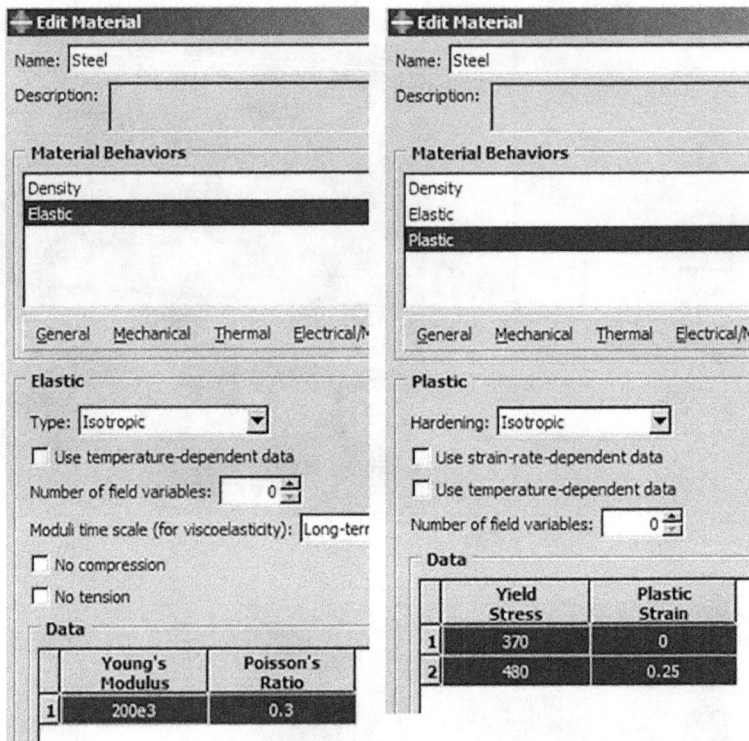

FIGURE 7.16 Material definition for steel.

and calculated. The input data of the CDP model are determined using the tested material properties. Figure 7.17 shows the plastic flow constants for the concrete damage plasticity parameters; all values are unitless except the dilation angle which is in degrees.

Compressive and tensile behavior is derived from a dog-bone tensile test and a cylinder compression test. Figures 7.18 and 7.19 show the concrete tensile behavior data and concrete compressive behavior data, respectively.

Concrete tension and compression damage parameters are shown in Figures 7.20 and 7.21. The stiffness recovery index, ω_c, defined as the stiffness recovery of concrete, is controlled under the cyclic loading, when the load is changed from tension to compression, as long as the crack is closed, then the compressive stiffness would be recovered ($\omega_c = 1$); when the load is changed from compression to tension, as long as the crack appeared, then the tensile stiffness could be recovered ($\omega_c = 0$) and the crack appeared.

7.4.2.2 Section Properties

The section properties of a model are defined by creating sections in the property module.

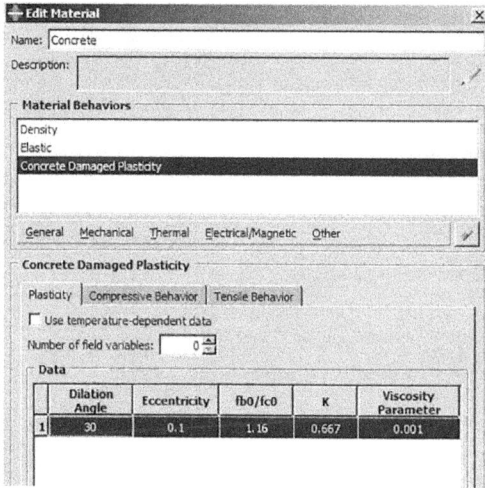

FIGURE 7.17 Plasticity flow for concrete damage plasticity.

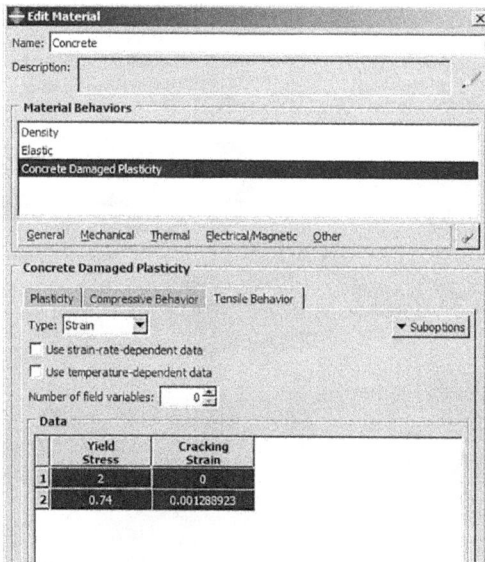

FIGURE 7.18 Concrete tensile behavior.

From the main menu bar, select "Section → Create". The "Create section" dialog box appears.

In the "Create section" dialog box:

Name the section "Steel".

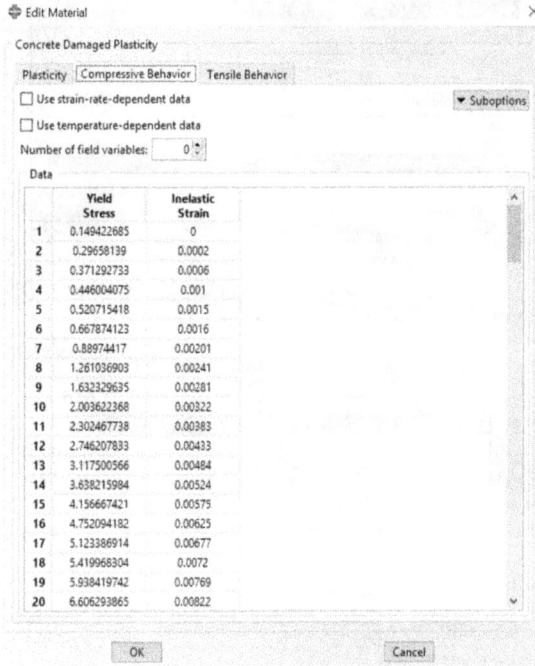

FIGURE 7.19 Concrete compressive behavior.

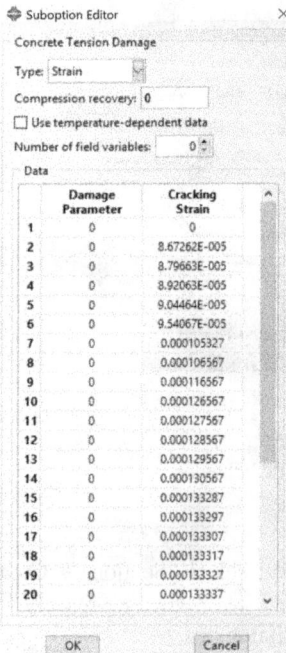

FIGURE 7.20 Tension damage parameters.

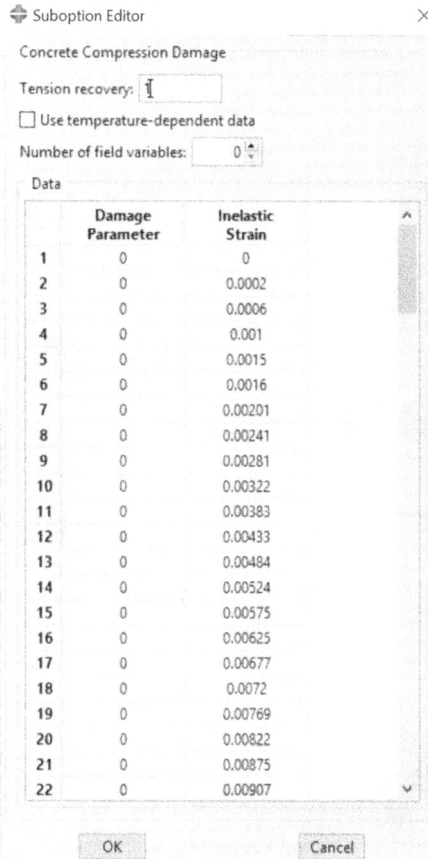

FIGURE 7.21   Compression damage parameters.

FIGURE 7.22   Create the solid section for steel.

In the "Category" list, select "Solid". In the "Type" list, select "Homogeneous" (see Figure 7.22).

Click "Continue". The "Edit section" dialog box appears. In the "Edit section" dialog box, select "Steel" from the "Material" and click "Ok" to define the section as shown in Figure7.23.

Perform the same for the Concrete section as shown in Figure 7.24.

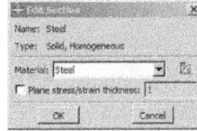

FIGURE 7.23   Steel section definition.

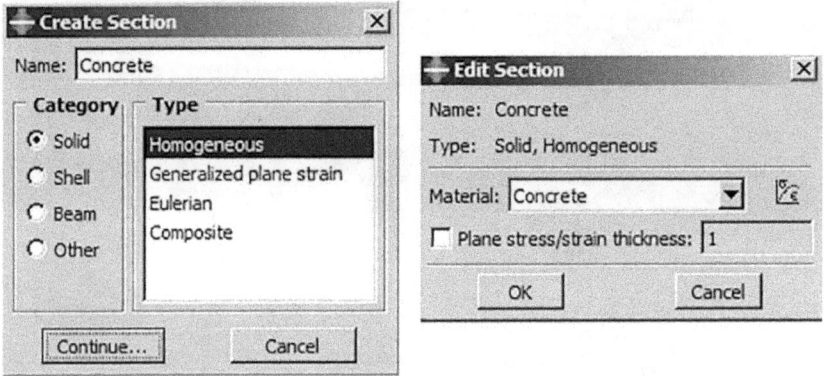

FIGURE 7.24   Concrete section definition.

7.4.2.3   Assigning the Defined Section to the Parts

In the next step, assign the defined sections to the corresponding parts.

Double click on "Section assignments" underneath the "Hollow Circular Section" part in the "Model tree", then choose the hollow circular section as the region to be assigned to a section and click the middle mouse button. In the dialog box, select "Concrete" from the section and click "Ok", as illustrated in Figure 7.25.

Perform the same for other parts (see Figure 7.26).

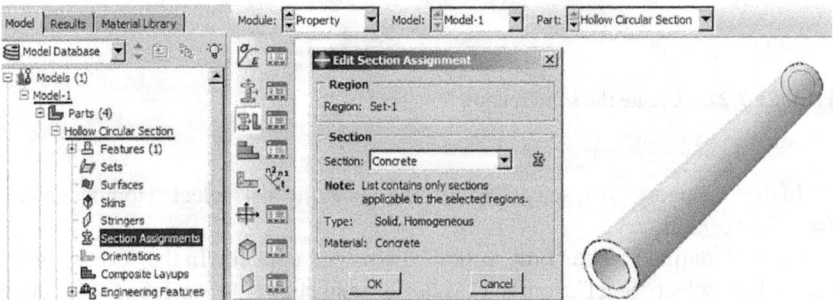

FIGURE 7.25   Section assignment for a concrete hollow cylinder.

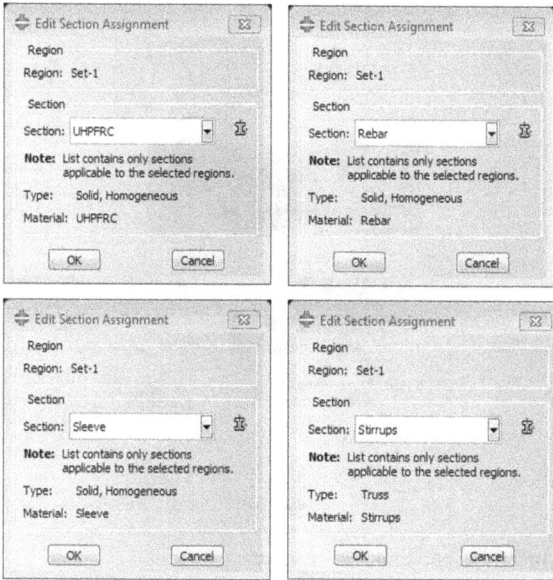

FIGURE 7.26 Section assignments to steel parts.

7.4.3 ASSEMBLY MODULE

In this module, all the parts that were created earlier can be combined (assembly) to achieve the required model. The constraints and loads can be applied to the model once all the individual part instances are assembled. Each part is oriented in its own coordinate system and it is independent of the other parts. Although a model may contain many parts, it can only have one assembly.

Define the geometry of the assembly by creating instances of a part and then positioning the instances relative to each other in a global coordinate system. An instance may be independent or dependent. Independent part instances are meshed individually, while the mesh of a dependent part instance is associated with the mesh of the original part.

In the module list located under the toolbar, click "Assembly" to open the "Assembly" module. From the main menu bar, select "Instance → Create". The "Create Instance" dialog box appears. Selects all the part instances for assembly, as shown in Figure 7.27.

In the opened window, under the "Instance Type" box, choose "Dependent (mesh on the part)".

In the dialog box, select all parts and click "OK".

Use "Instance → Rotate" and rotate instances to be in the same direction. Rotate the hollow circular section, sleeve, longitudinal bars and stirrup by 90 degrees. Then use "Instance → Translate" and move the sleeve and longitudinal bars and stirrup in the right position. Finally, use "Instance → Linear pattern" to define a proper array of the "Stirrup" by 16 stirrups and a vertical offset of 150 mm. Figure 7.28 displays the tools that should be used by the user.

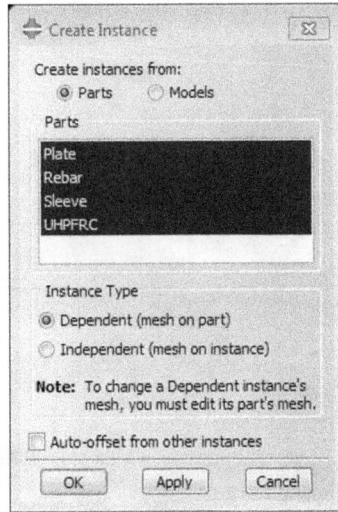

FIGURE 7.27   Insert the parts into the assemble platform.

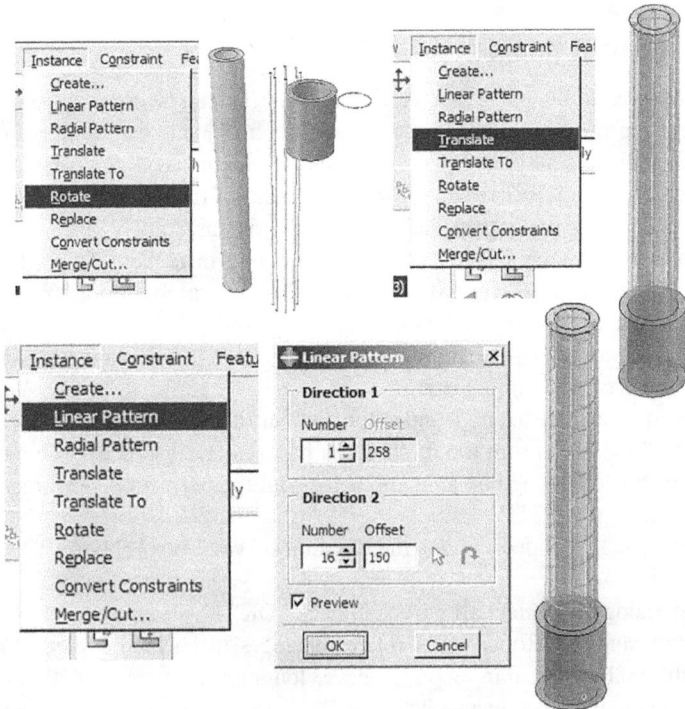

FIGURE 7.28   Using rotate, translate, and linear pattern tools to define the final assembly.

7.4.4   Step Module

Upon finishing the assembly section, the configuration of the analysis should be defined. In this example, the static response of the hollow circular section is considered.

7.4.4.1   Create an Analysis Step: Cyclic

From the main menu bar, select "Step → Create" to create a step. The "Create Step" dialog box appears with a list of all general procedures and a default step named Step-1.

Change the step name to "Cyclic".

Select "General" as the "Procedure type" and select "Static, General", and click on "Continue".

The "Edit step" dialog box appears. In the "Basic" tab, enter 30 for the "Time period" and select "Nlgeom" to include nonlinear effects of geometry in the analysis. Click on the "Incrementation" tab and change the increment size, as shown in Figure 7.29. Click "OK" to create the step and to exit the "Edit Step" dialog box.

7.4.4.2   Create an Amplitude

At the amplitude field, a tabular form of amplitude-time step is defined according to the experimental testing protocols, as shown in Figure 7.30. This option allows arbitrary time variations of load, displacement, and other prescribed variable magnitudes to be given throughout the step.

To create amplitude, double click on the Amplitudes in the "Model tree".

The "Create Amplitude" dialog box appears. Name the amplitude as "Cyclic loading".

Select the "Tabular" type of amplitude and click "Continue". Input the incremental of the amplitude data, as shown in Figure 7.30. Click "OK" to exit from the dialog box.

7.4.4.3   Create Set for Request History Output

To obtain the required outputs (especially for the force–displacement graph) at the visualization module, it is necessary to request the field outputs and history outputs in the "Step" module.

First, a set for displacement and reaction force should be defined.

Select "Tools → Set → Create" and name the set "Displacement" and consider a vertex top of the hollow circular section. Perform the same to define a set, named "Reaction" and consider the base of the model as shown in Figure 7.31.

Double click on "History output requests" in the "Model tree" and select "Set" from "Domain" and "Displacement" from "Set". Then check "U1" underneath "Displacement/Velocity/Acceleration" as the output variable and click "Ok" to define the output. Perform the same to define a history output for the "RF1" for set "Reaction" as illustrated in Figure 7.32.

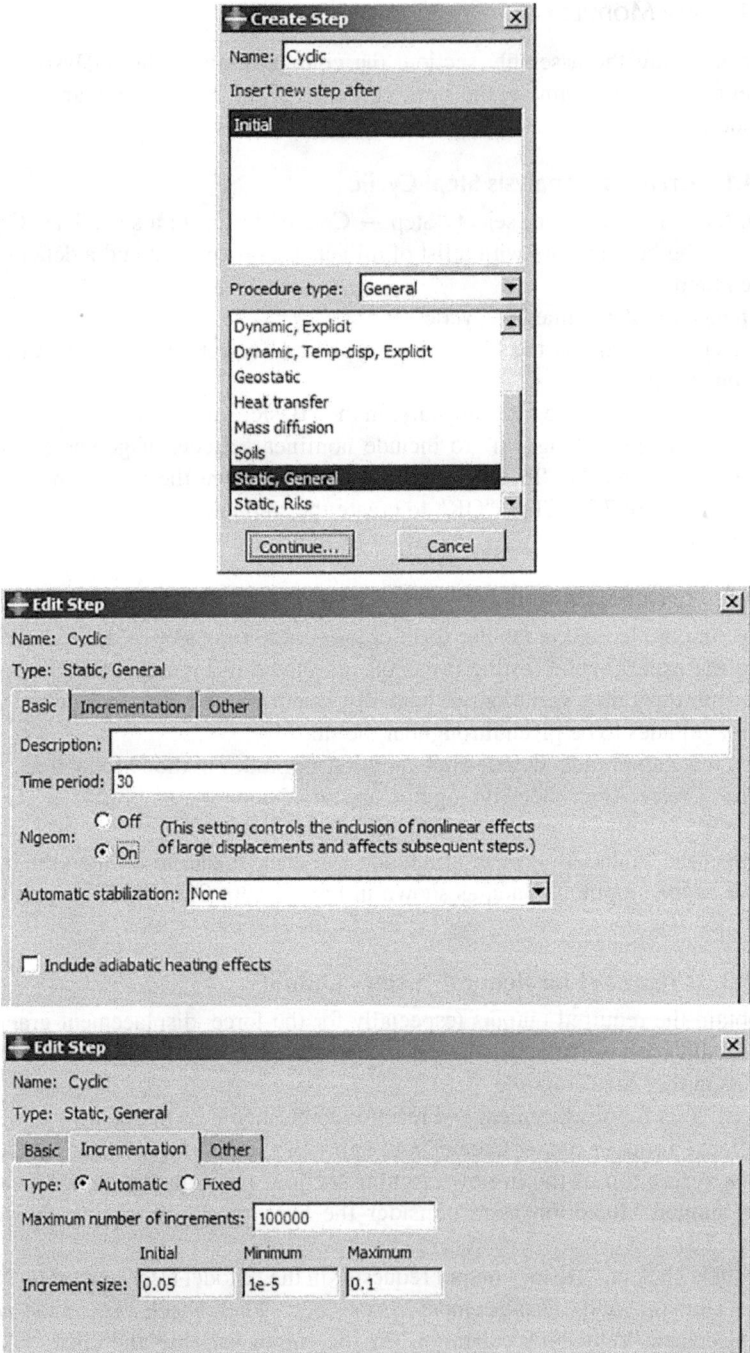

FIGURE 7.29 Static step definition.

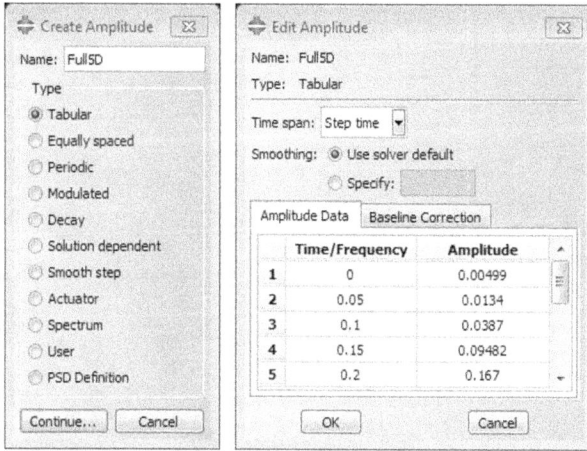

FIGURE 7.30   Create amplitude.

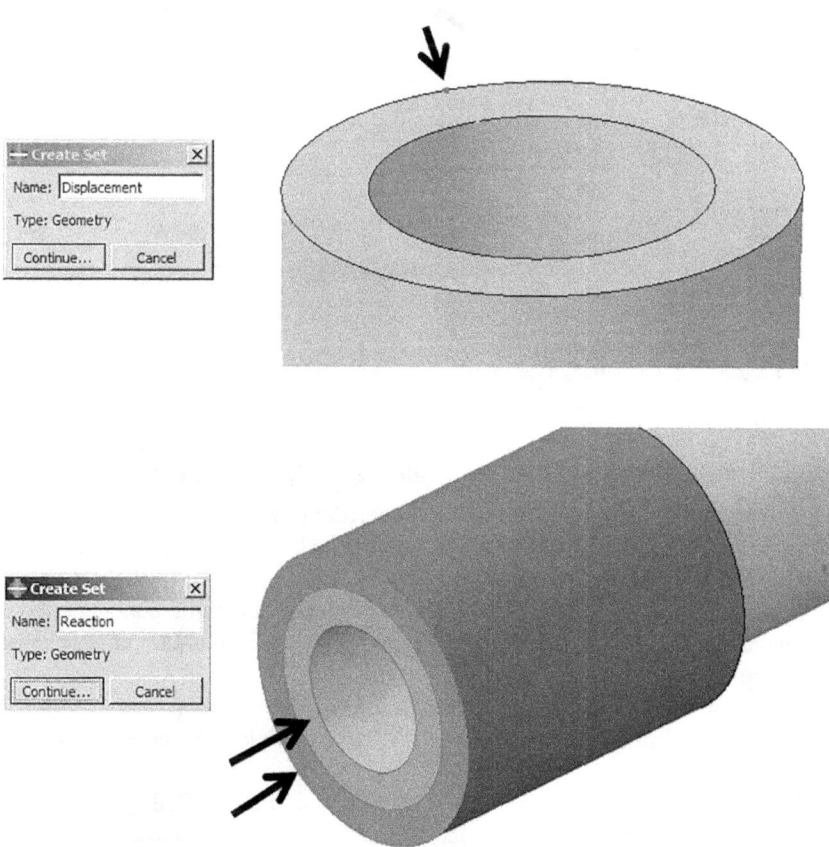

FIGURE 7.31   Define sets for history outputs.

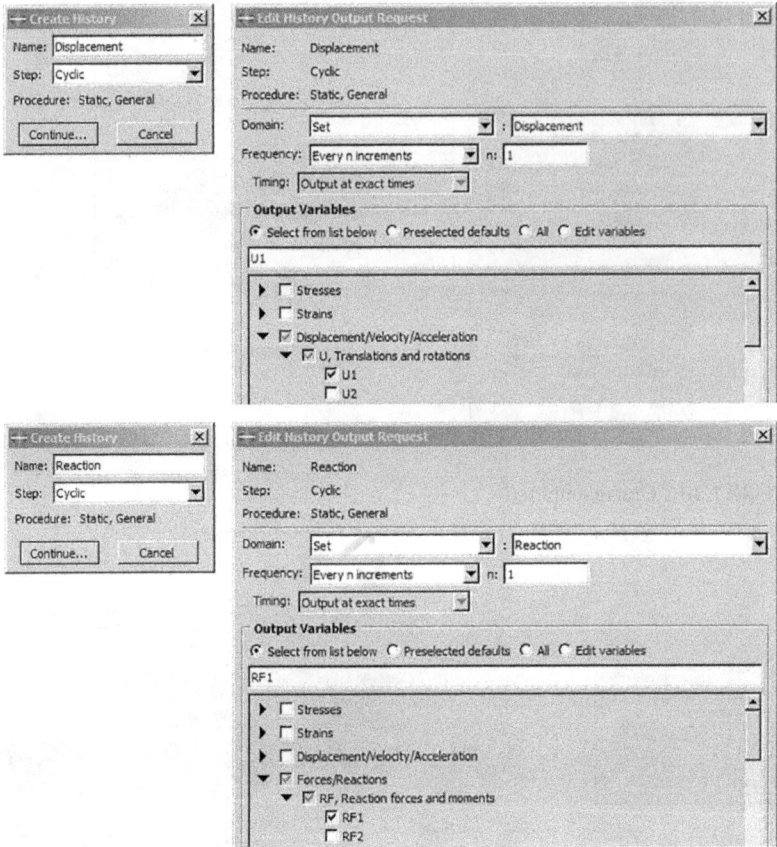

FIGURE 7.32 History output definition.

7.4.4.4 Create Partitions for the Hollow Circular Section

To define load properly, it is necessary to partition the concrete hollow circular section. First, two datum planes should be defined. Then the cylinder could be partitioned by the datum planes.

For this purpose, select "Tools → Datum" to open its dialog box and choose "Plane" from the "Type" and "Offset from plane" from the "Method" as shown in Figure 7.33.

Then chose the top face of the cylinder and add 200 as an offset in the proper direction. Perform the same to define another datum plane by entering 200 for offset, as illustrated in Figure 7.34.

Then, select "Tools → Partition" to open its box. Select "Cell" from "Type" and "Use datum plane" from the "Method". Choose a datum plane and then click "Create partition". Perform the same to partition by other datum planes as shown in Figure 7.35.

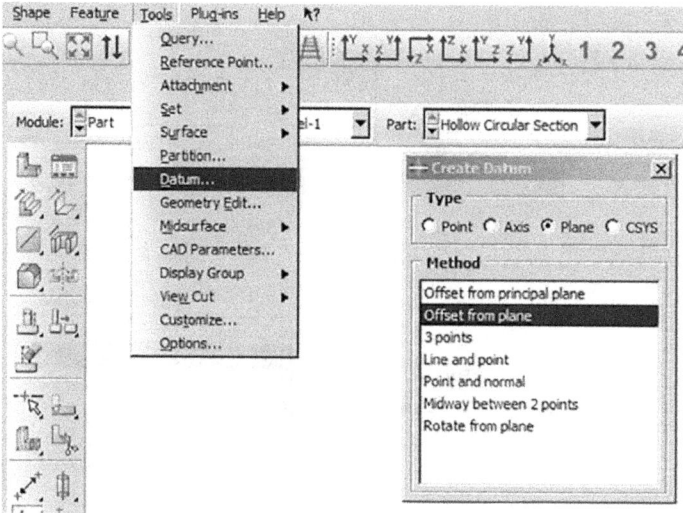

FIGURE 7.33   Create datum planes.

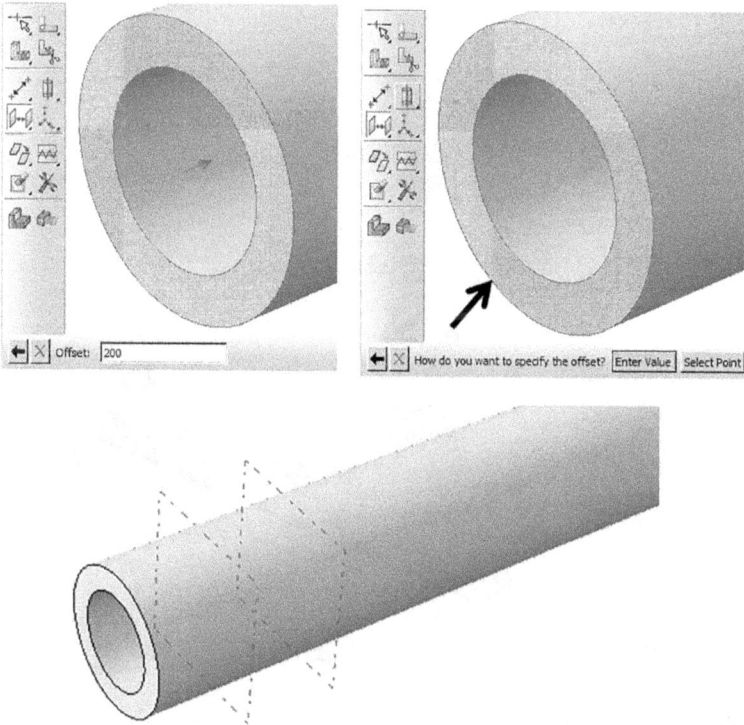

FIGURE 7.34   Datum plane definition.

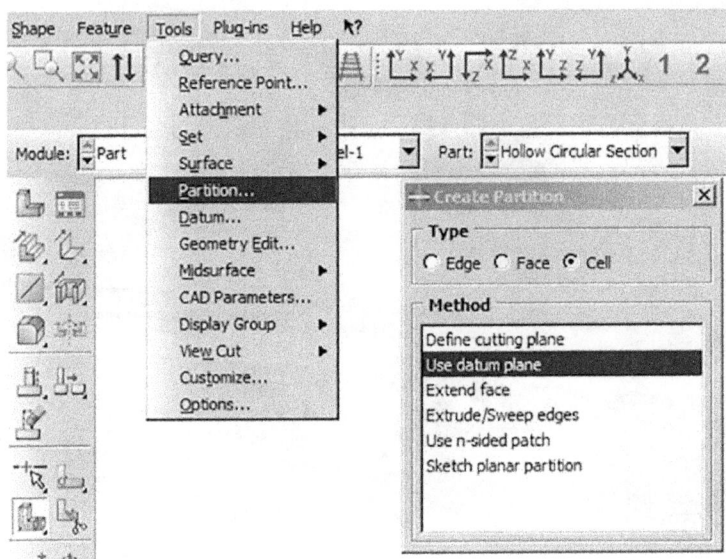

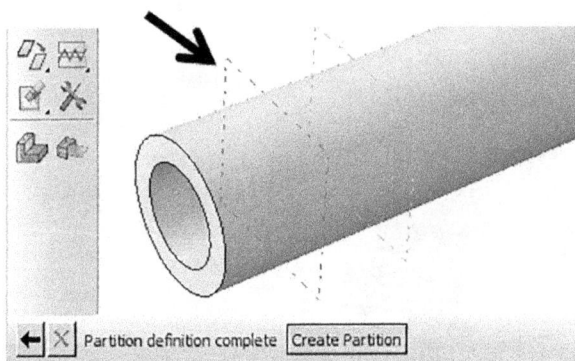

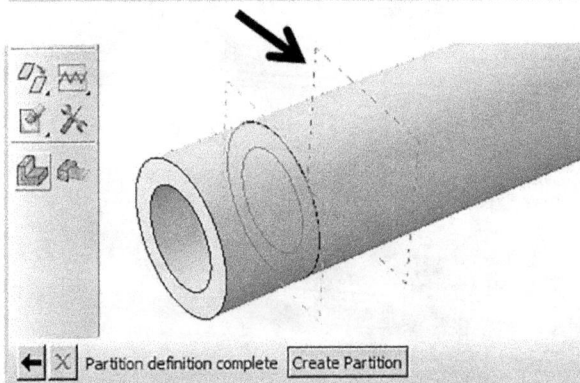

FIGURE 7.35   Partition definition.

7.4.5   Interaction Module

This module is used to define various interactions within the model or interactions between regions of the model and its surrounding parts. The interactions used in this study are "General contact" interaction, where it describes the relationship between the hollow circular section and sleeve. Another type of constraint utilized in this study is the "Embedded region", which has been used to represent the connection between the hollow circular section and steel reinforcements.

7.4.5.1   General Contact Interaction

General contact defines the interactions between all surfaces (by default) in a model. The user can define the interaction in the initial step.

To define general contact and contact properties, double click on Interactions in the "Model tree" to open the "Create interaction" dialog box. Select "Initial" from "Step" and "General contact" from "Types for selected step" and click "Continue". In the subsequent dialog box, click "Create interaction property" to open its dialog box. Select the "Contact" from the "Type" of interaction property and click "Continue". The "Edit Contact Property" dialog box appears.

Click on "Mechanical → Tangential behavior". For "Friction formulation", consider "Penalty" as the method that permits some relative motion of the surfaces (an "elastic slip") when they should be sticking. While the surfaces are sticking (i.e., $\overline{\tau} < \overline{\tau}_{crit}$), the amount of sliding is limited to this elastic slip. Then the software continually adjusts the value of the penalty constraint to enforce this condition. Next, specify 0.4 as the "Friction coefficient".

Keep the default setting in the "Shear Stress" and "Elastic Slip" tabs. Click "OK" to complete the definition of contact property.

In the "Edit interaction" dialog box, choose "Intprop-1" as the "Global property assignment" and click "Ok" to define the interaction as shown in Figure 7.36.

7.4.5.2   Embedded Region Constraint

An embedded region constraint can be used to embed a region of the model within the host region or within the whole model. As mentioned before, an embedded region constraint can be simply created by specifying the embedded region, the host region, a weight factor round-off tolerance, and an absolute exterior tolerance or fractional exterior tolerance.

To create an embedded region constraint, perform the following steps:

From the main menu bar, select "Constraint → Create" to create a constraint to open its box.

Select "Embedded region" from the type of constraint and click on "Continue". From the "Model tree", hide "Hollow circular section-1" and "Steel sleeve-1" as shown in Figure 7.37.

Select all steel bars in the viewport from the "Embedded region" and click "Done". Then choose "Select region" and click on all three parts of the hollow circular section by holding the "Shift" key on the keyboard and click "Done" to

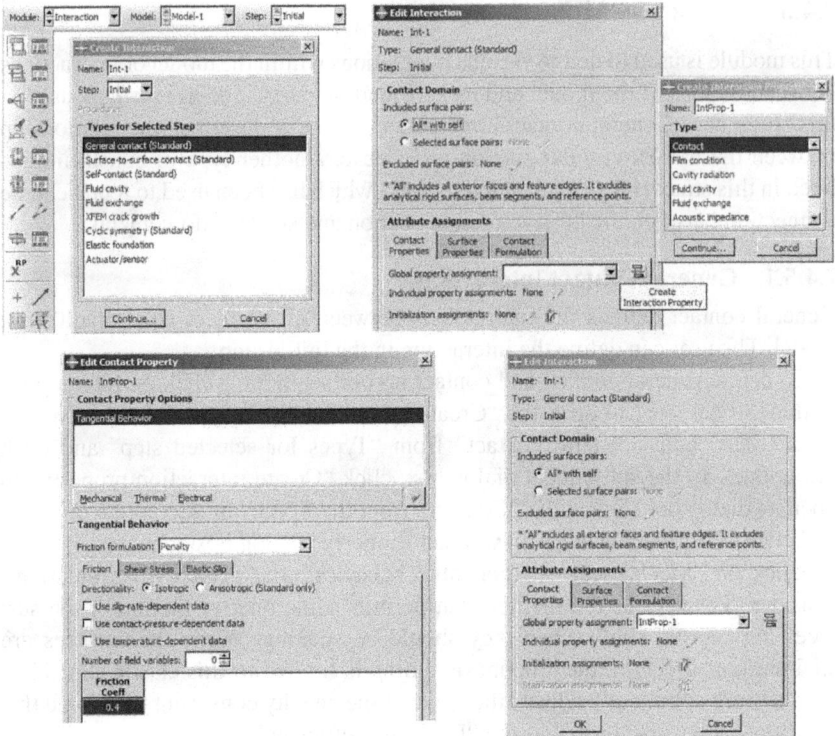

FIGURE 7.36 General contact definition.

open the "Edit constraint" dialog box as shown in Figure 7.38 to complete the constraint definition.

7.4.6 LOAD CONDITION MODULE

Using this command, the user can define the loading that needs to be assigned to the elements or any parts of the structure.

7.4.6.1 Apply the Cyclic Displacement as a Boundary Condition

In the Module list located under the toolbar, click "Load" to open the "Load" module.

From the main menu bar, select "BC → Create". The "Create boundary condition" dialog box appears.

In the "Create Boundary Condition" dialog box, name the boundary condition as "Displacement", and select "Displacement/Rotation" from the "Types for selected step" and click "Continue" (see Figure 7.39).

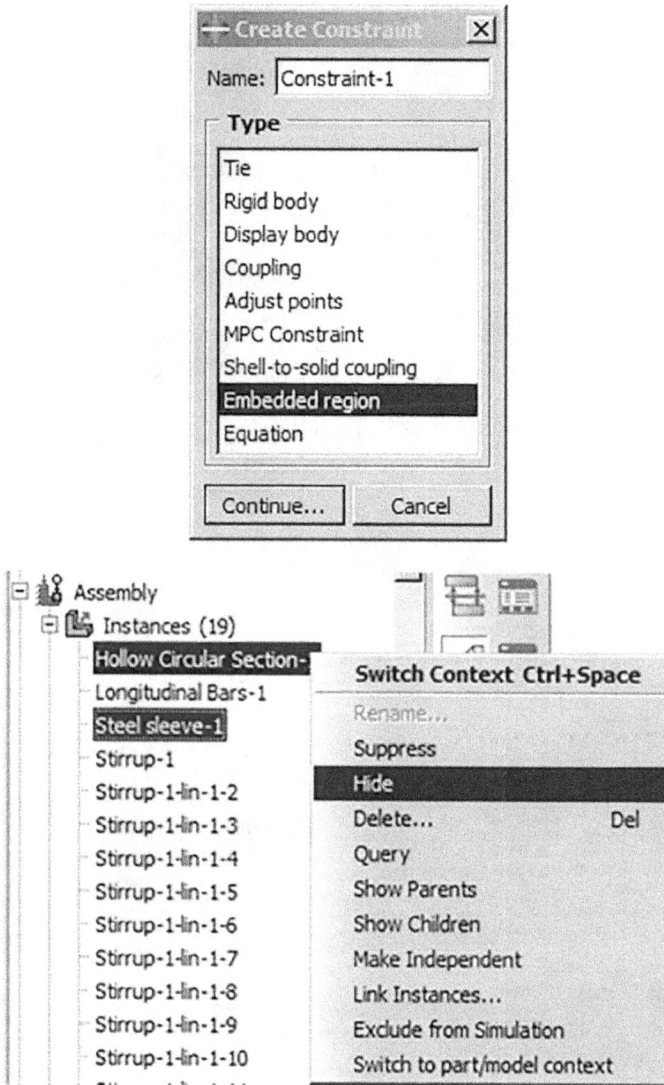

FIGURE 7.37 Create embedded region constraint.

Select the cylindrical face shown in Figure 7.39. Then click the middle mouse button to open the "Edit boundary condition" dialog box. In the box, check "U1" and enter 1 as the multiplication factor and choose "Cyclic Loading" as the "Amplitude" as shown in Figure 7.40.

Click "OK" to save the load and to close the dialog box. The cyclic load boundary condition will appear on the selected column surface (see Figure 7.41).

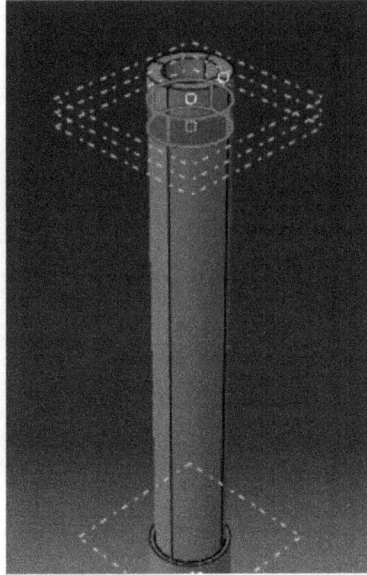

FIGURE 7.38   Embedded region constraint definition.

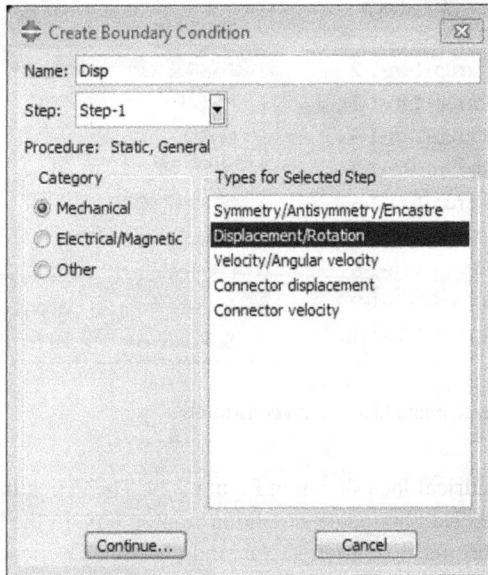

FIGURE 7.39   Create cyclic displacement on the model.

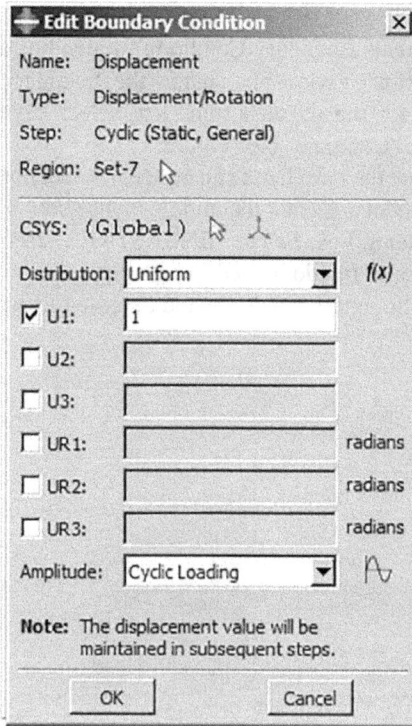

FIGURE 7.40 Input the displacement value and amplitude.

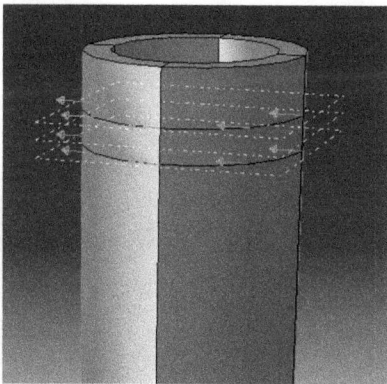

FIGURE 7.41 Applied displacement.

7.4.6.2 Apply Boundary Condition to the Column Base

Select "BC → Create" and the "Create Boundary Condition" dialog box appears. In the "Create Boundary Condition" dialog box, name the boundary condition "Fixed-End". In the "Types for the selected step" list, select "Symmetry/Axisymmetry/Encastre" and click "Continue" (see Figure 7.42).

Rotate the model properly so that the base faces can be seen. Select both circular surfaces at the bottom of the model and click the middle mouse button to open the "Edit boundary condition" dialog box. Select "ENCASTRE", since all the translational and rotational degrees of freedom need to be constrained as a fixed support. Click "OK" to create the boundary condition and to close the dialog box as shown in Figure 7.43.

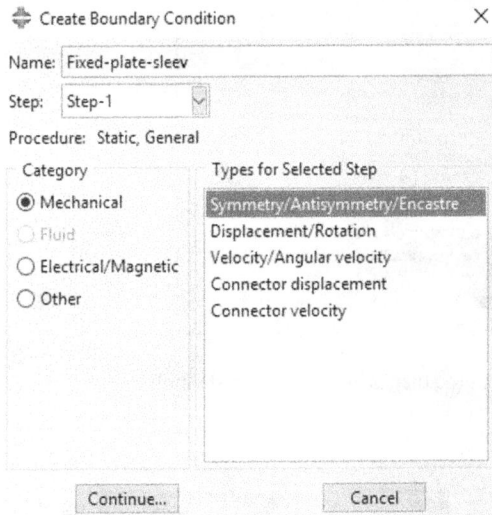

FIGURE 7.42   Create fixed-end boundary conditions.

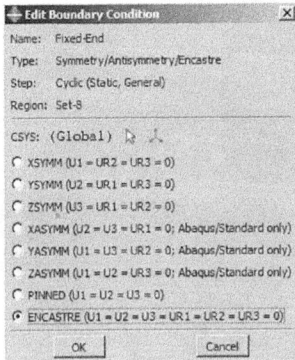

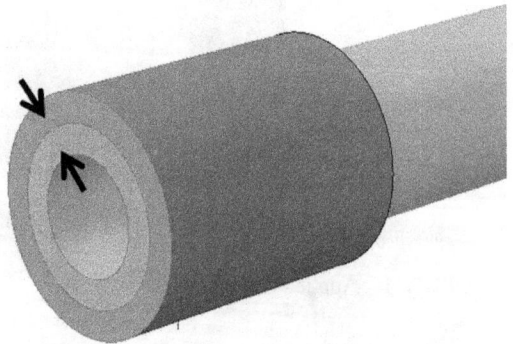

FIGURE 7.43   Edit boundary condition for the fixed end.

FIGURE 7.44 Boundary conditions.

Then, the boundary conditions at the column base will be illustrated, as shown in Figure 7.44.

7.4.7 MESH MODULE

The Mesh module is used to generate the finite element mesh. To assign an ABAQUS element type to the hollow circular section, in the module list located under the toolbar, click on "Mesh" to open the "Mesh" module. At the context bar, click on "Part" and select "Hollow Circular Section".

From the main menu bar, select "Mesh → Element Type". In the viewport, select the entire part as the region to be assigned to an element type. In the prompt area, click "Done". The "Element Type" dialog box appears.

In the box, select "Standard" from the "Element Library" (the default). Also, select "Linear" from the "Geometric Order" (the default) and "3D stress" from the "Family" of elements. In the lower portion of the dialog box, examine the element shape options. A brief description of the default element selection is available at the bottom of each tabbed page. Click "OK" to assign the element type and to close the dialog box (see Figure 7.45).

From the main menu bar, select "Seed → Part" to seed the part instance. Alternatively, select the "Seed Part" in the upper-left corner of the meshing toolbox and the "Global Seeds" dialog box will appear, as shown in Figure 7.46. Type 50 for the approximate global size of the mesh elements and click "OK" to accept the seeding.

Perform the same for seeding other parts as shown in Figure 7.47.

From the main menu bar, select "Mesh → Part" to mesh the part. Select the part to be meshed. Once the selection is done, click "Yes" in the prompt area to confirm the meshing. Once meshed, then the part color changes to

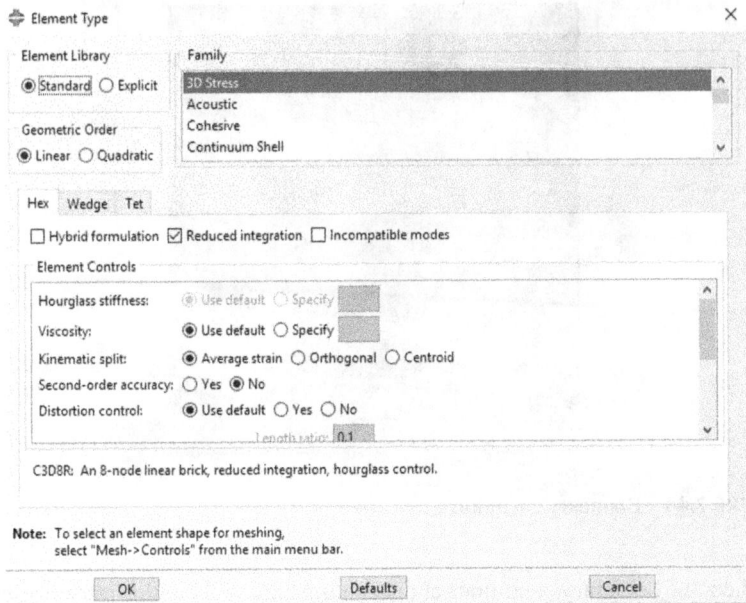

FIGURE 7.45   Select element type for the hollow circular section.

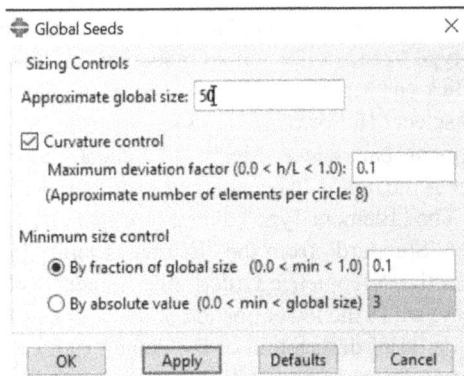

FIGURE 7.46   Assign the approximate global size for the hollow circular section.

blue. Perform the same for all parts and the meshed geometry is shown in Figure 7.48.

7.5 ANALYSIS: JOB MODULE

The "Job" module is used to create and manage analysis jobs and to submit them for analysis.

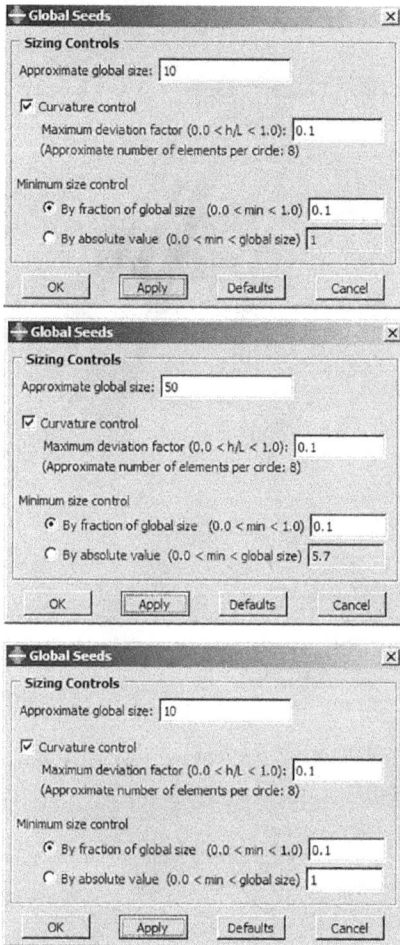

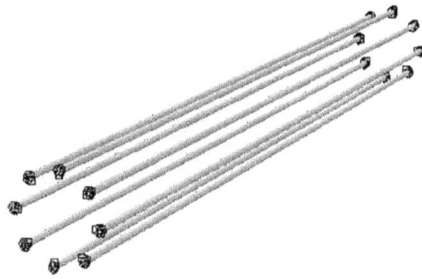

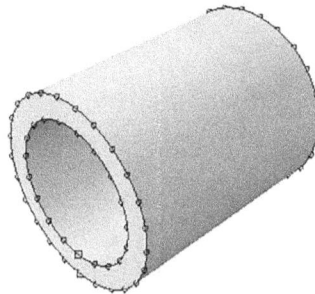

FIGURE 7.47 Assign the approximate global size for steel parts.

Double click on the "Jobs" from the "Model tree" to open its dialog box. Name the job as "Cyclic" and click "Continue". The "Edit job" dialog box appears and leave it unchanged and click "Ok" to accept all other default job settings in the job editor and to close the dialog box. Click on "Submit" to start checking the input file and run the analysis (see Figure 7.49).

7.6 VISUALIZATION MODULE

When the job is completed, the results are ready in the output database. To obtain the outputs of the model, right click on the completed job and click "Results" to open "Cyclic.odb" in the "Visualization" module as shown in Figure 7.50.

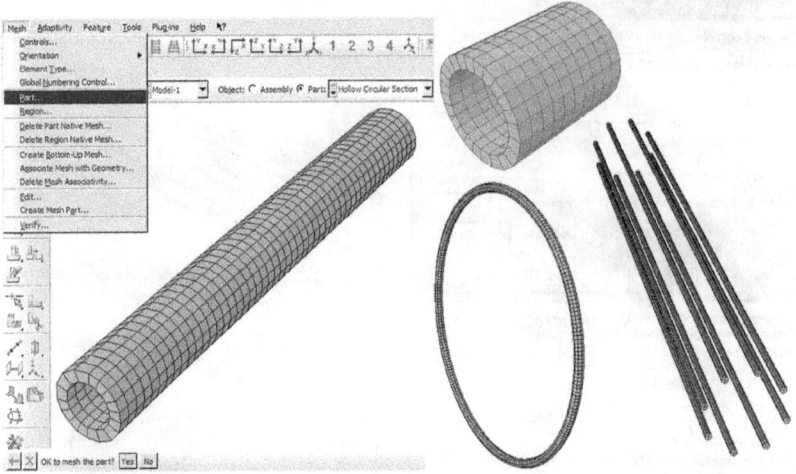

FIGURE 7.48   Meshing parts.

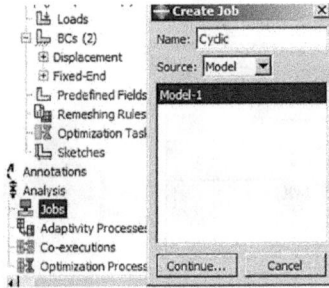

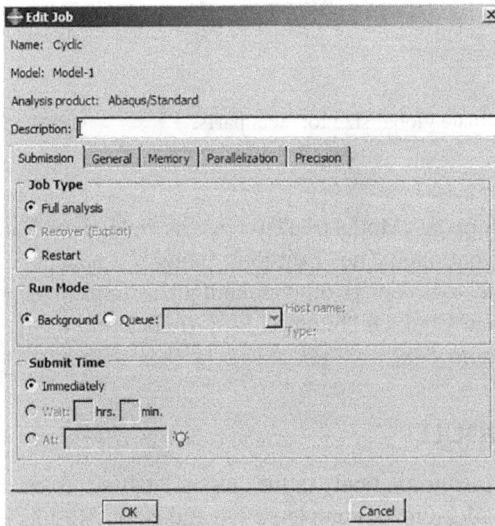

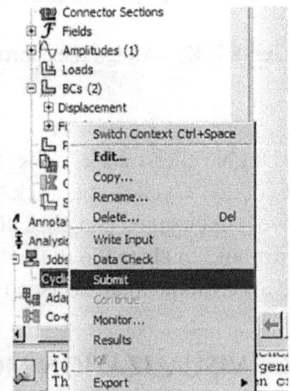

FIGURE 7.49   Job definition and submission.

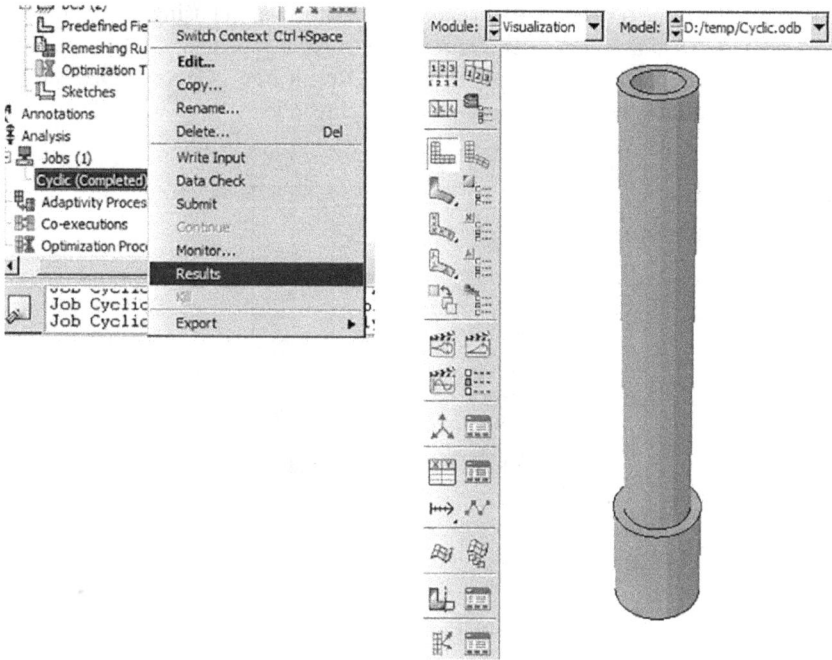

FIGURE 7.50   Visualization module.

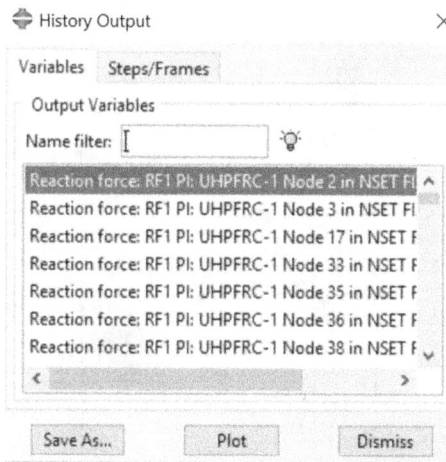

FIGURE 7.51   Save all reaction force graph.

Double click on "XY Data" in the "Result tree" and select "ODB history output" and click "Continue". Then chose all reaction forces (RF1 items) and click "Save as", and then the "Save XY Data As" dialog box appears. Name the data as "Force", choose Sum ((XY,XY,...)) from the "Operation" and click "Ok" to save the graph as shown in Figure 7.51.

Repeat the same procedure to displacement history output using "As is" operation, named "Displacement" (see Figure 7.52).

Double click on the "XY Data" from the "Result tree". Choose "Operate on XY data" and click "Continue" as shown in Figure 7.53.

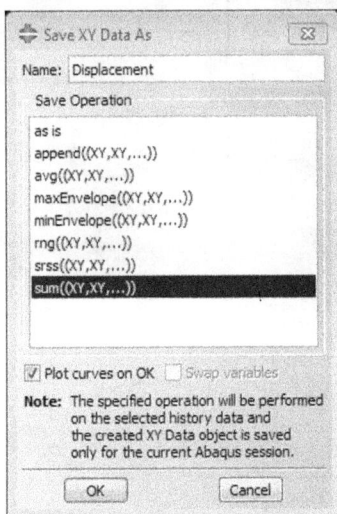

FIGURE 7.52   Save displacement graph.

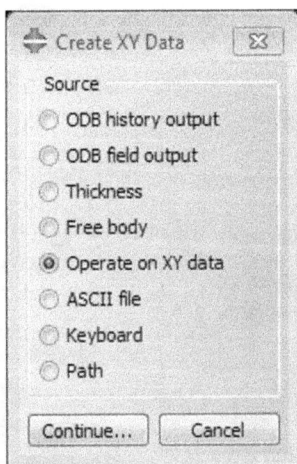

FIGURE 7.53   Operate on XY data.

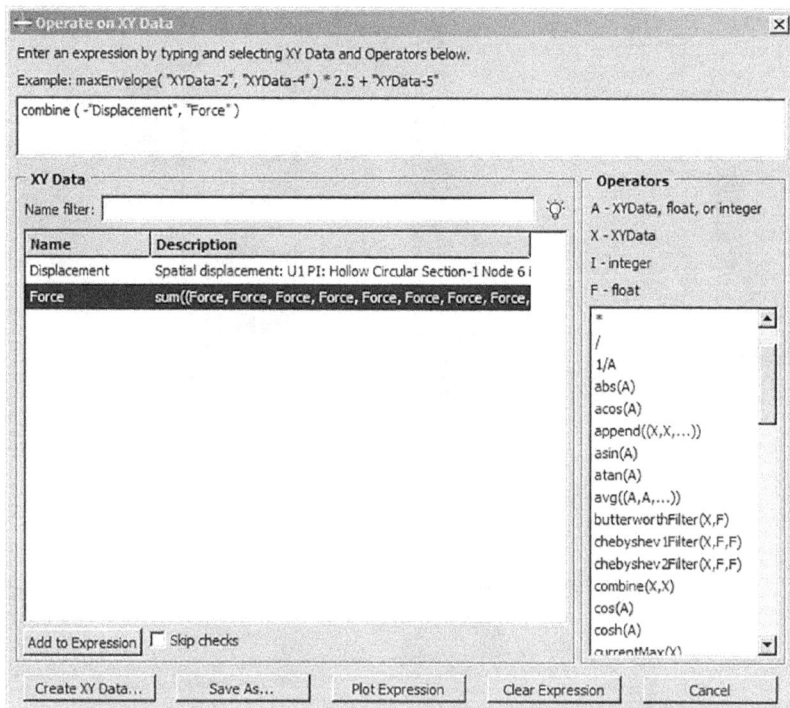

FIGURE 7.54 Combining displacement and force graphs.

Select "combine (X,X)" from the "Operators" list, then double click on "Displacement" and "Force" graph, respectively. Consider minus before "Displacement", click "Plot Expression" to display the graph as shown in Figure 7.54.

A force vs. displacement curve is plotted. All the results can be extracted from this module as further details are demonstrated in the next section.

7.7 ANALYSIS RESULT

Select "PEEQ" as the primary variable to display the equivalent plastic strain contour plot (see Figure 7.55).

Select "DAMAGEC" as the primary variable to displace the compressive damage contour plot (see Figure 7.56).

Select "DAMAGET" as the primary variable to displace the tensile damage contour plot (see Figure 7.57).

FIGURE 7.55   PEEQ contour plot.

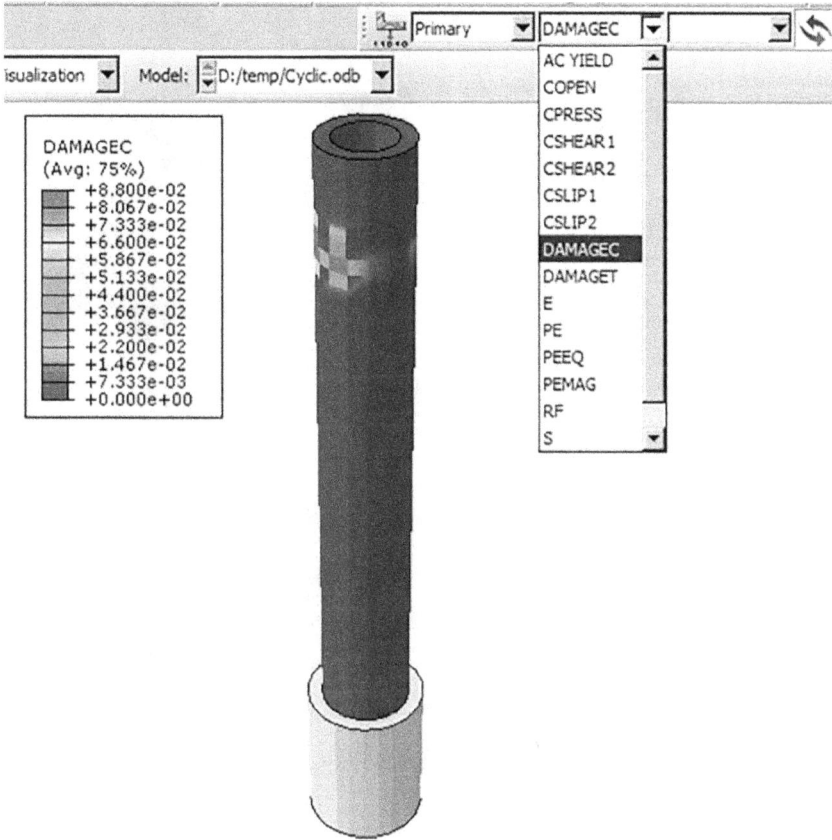

FIGURE 7.56 DAMAGEC contour plot.

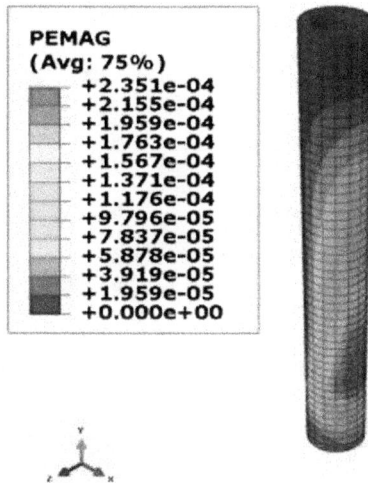

FIGURE 7.57 DAMAGET contour plot.

8 Modal Analysis of a Three-Story Building

8.1 INTRODUCTION

The dynamic characteristics of structures are very essential parameters for the design of buildings under various dynamic loads such as vibration generator machines, vehicles, trains, wind, ground motions, or any other sources of vibration. Therefore, in this example, a simple structure is considered and the process for conducting modal analysis and obtaining the natural frequency of the structure and mode shapes using ABAQUS software is explained step by step.

8.2 PROBLEM DESCRIPTION

In this example, a simple three-story steel structure with a fixed base is considered and an attempt has been made to perform modal analysis and to determine the natural frequency of the structure and mode shapes. The considered three-story structure is shown in Figure 8.1. As a result, natural frequencies and shape modes are extracted to detect the critical regions.

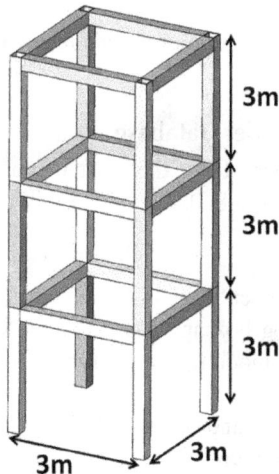

FIGURE 8.1 The three-story building.

DOI: 10.1201/9781003219491-8

Material properties:

Density = 7850 E-09 kg/mm^3
Young's modulus = 210,000 MPa
Poisson's ratio = 0.3
Yield stress = 370 MPa
Ultimate stress = 460 MPa

Dimensions:

Section (300 × 300 mm)
All members are square hollow section steel, 25 mm in thickness
Beam length 3000 mm
Column length 3000 mm

8.3   OBJECTIVES

- To conduct the frequency response analysis of the structure using the ABAQUS finite element software.
- To develop the finite element model of a three-story steel structure.
- To determine and investigate the dynamic characteristics of the structure in terms of the natural frequency and mode shapes.

8.4   MODELING

8.4.1   PART MODULE

This module allows for the creation of the geometry required for the considered problem.

8.4.1.1   Create a New Model Database

Run ABAQUS/CAE software from the start menu and then close the "Start Session" dialog box (see Figure 8.2).

8.4.1.2   Create Part

From the main menu bar, select "Part → Create" to create a new part.

The "Create Part" dialog box appears. Use the "Create Part" dialog box to name the part; to choose its modeling space, type, and base feature; and to set the approximate size.

Name the part "Beam" and choose "3D", "Deformable", "Solid", and "Extrusion" from the base feature.

In the "Approximate size" text field, type 2000 to set the approximate size of the new part (see Figure 8.3).

Click "Continue" to open the sketcher.

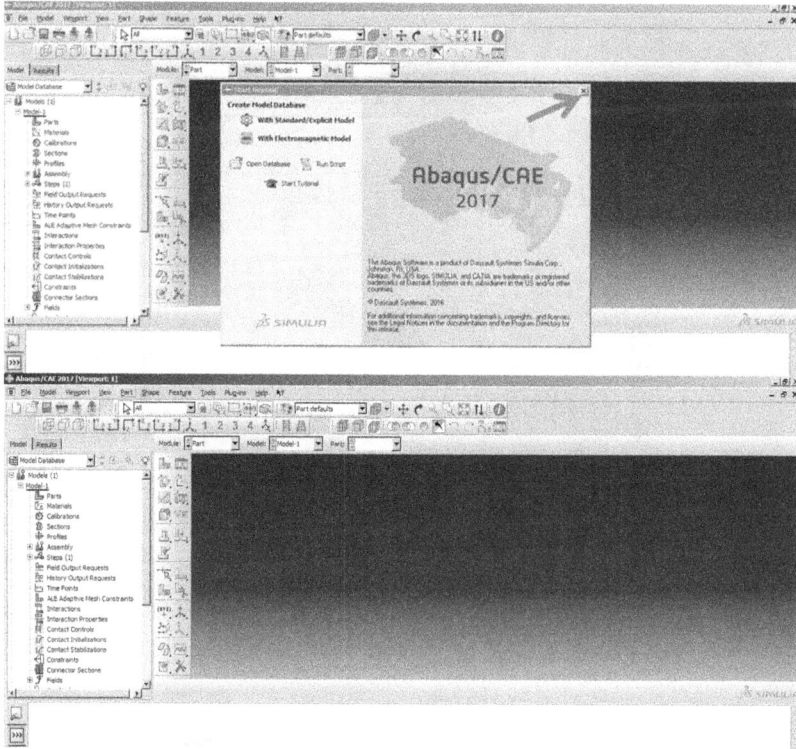

FIGURE 8.2 Create a new model database.

Click on "Create line: Connected" from the draw menu on the left side of the layout. Draw two rectangles and then use "Add dimension" to define the dimensions as shown in Figure 8.5.

Click the middle mouse button to open the "Edit base extrusion" dialog box. Then enter 3000 for "Depth" and click "Ok" to define the part as shown in Figure 8.5.

8.4.2 Property Module

In this module, the material properties for the analysis should be defined and assign those properties to the available parts.

8.4.2.1 Material Properties

In this problem, all parts of the frame are made of steel.

In the module list located under the toolbar, select "Property" to open the "Property" module. The cursor changes to an hourglass while the "Property" module loads.

From the main menu bar, select "Material → Create" to create a new material. The "Edit Material" dialog box appears.

Name the material "Steel".

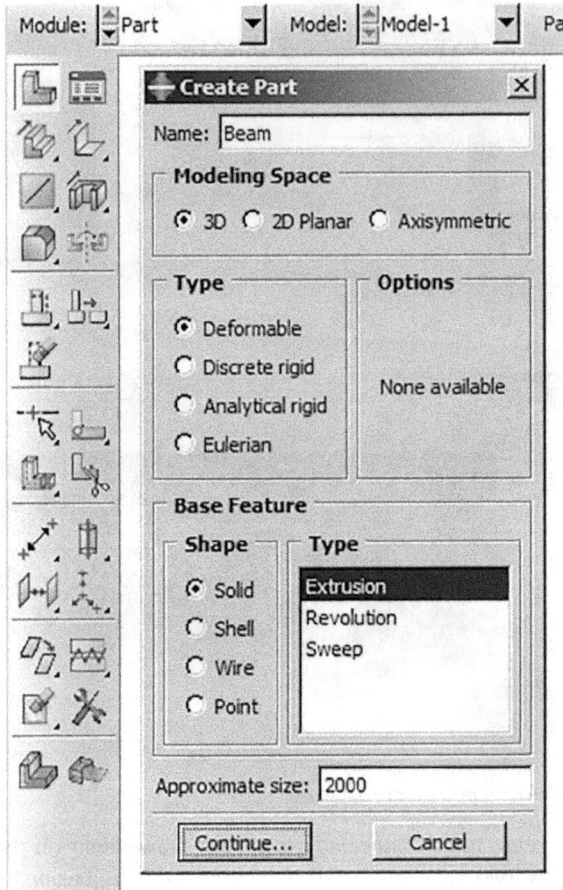

FIGURE 8.3 Create a new part.

Select "General → Density" and enter 7850e-9 as illustrated in Figure 8.6.

To define elasticity, select "Mechanical → Elasticity → Elastic" and enter the value of 20,000 for "Young's modulus" and 0.3 for "Poisson's ratio" in the respective fields (see Figure 8.7).

To define plasticity, select "Mechanical → Plasticity → Plastic" and enter values as shown in Figure 8.8.

Click OK to exit the material editor.

8.4.2.2 Section Properties

The section properties of a model can be defined by creating sections in the property module.

From the main menu bar, select "Section → Create". The "Create Section" dialog box appears. In the "Category" list, select "Solid". In the "Type" list, select "Homogeneous" (see Figure 8.9).

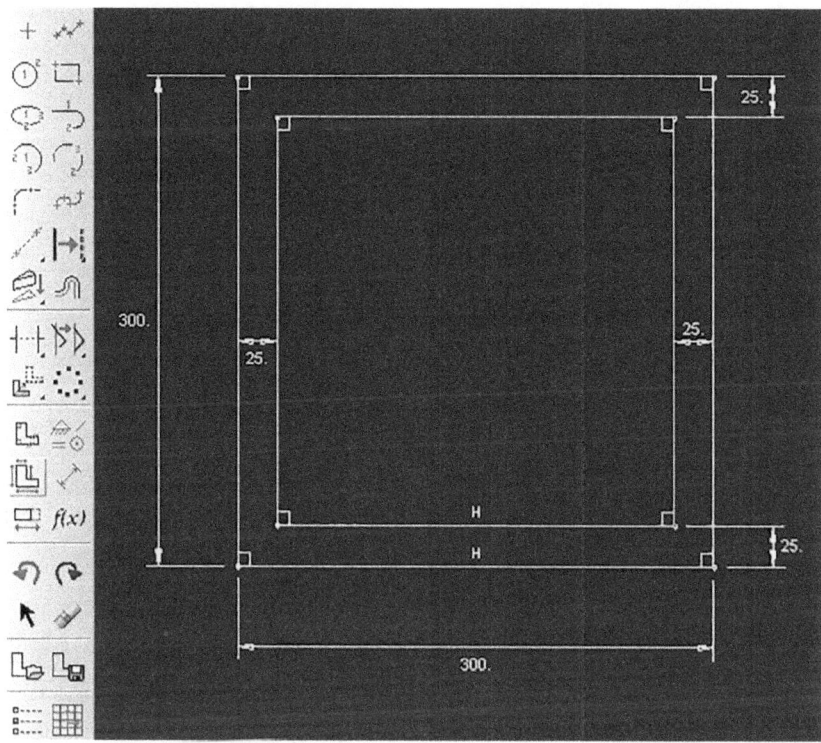

FIGURE 8.4 Draw the section sketch of the model.

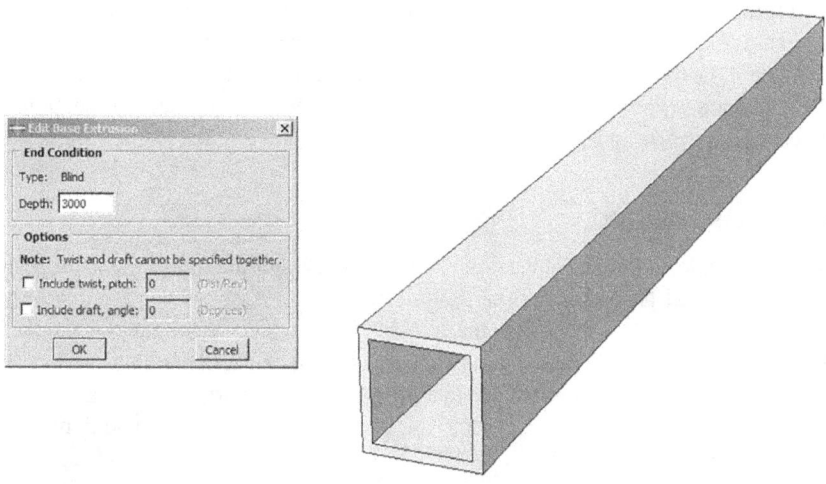

FIGURE 8.5 Define the depth of the beam.

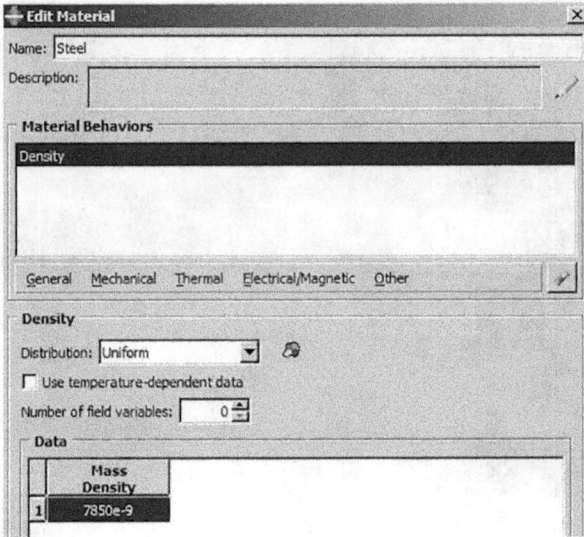

FIGURE 8.6   Identify the steel density.

Click "Continue". The "Edit Section" dialog box appears.

Accept defaults and click "Ok" to complete the section definition as shown in Figure 8.10.

8.4.2.3   Section Assignment

The "Assign" menu is used in the "Property" module to assign the section to the plate. To assign the section to the part, follow the procedure described here:

From the main menu bar, select "Assign → Section". Select the entire part as the region in which the section will be applied and click the middle mouse button to open the "Edit section assignment" dialog box (see Figure 8.11)

Accept the default section which was defined recently and click "Ok" to define the section assignment. The "Part" turns to a green color once the section is assigned, as shown in Figure 8.12.

8.4.3   Assembly Module

In this module, the user can define part instances and assemble them properly to create a final assembly. In the module list, click "Assembly" to activate the "Assembly" module. From the main menu bar, select "Instance → Create". The "Create Instance" dialog box appears. In the opened window, under the "Instance Type" box, choose "Dependent (mesh on the part)" and click "Ok" to define the part instance. Then, select "Instance → Linear pattern" and define the linear pattern as shown in Figure 8.13.

Create another part instance and use "Instance → Rotate" and rotate it by 90 degrees, then use "Instance → Translate" and move it to the instances patterned

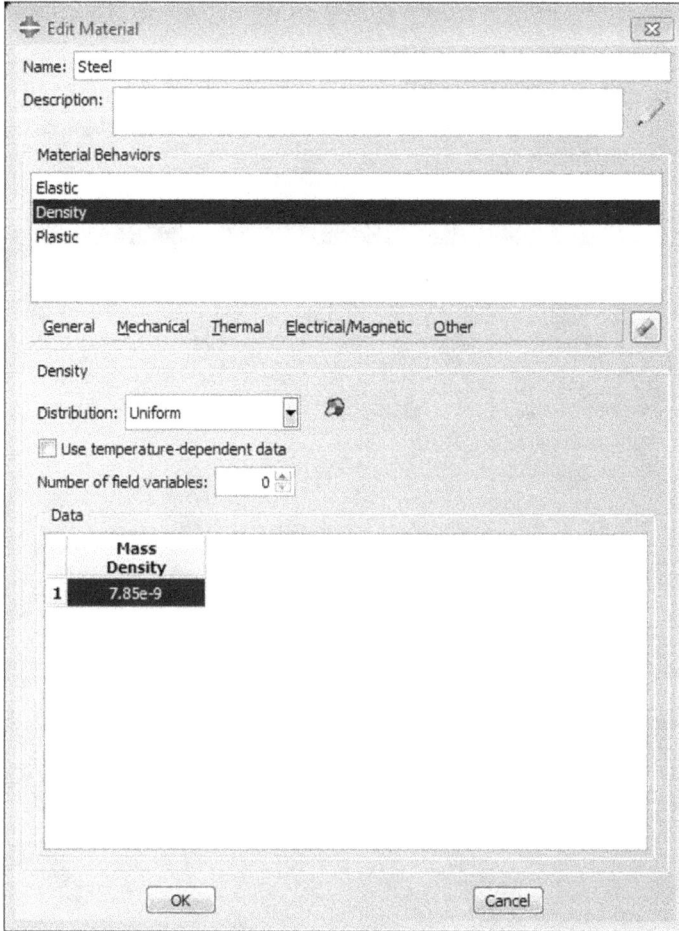

FIGURE 8.7 Identify the steel elasticity.

recently. Then, use "Instance → Linear Pattern" and define the right pattern (see Figure 8.14).

Perform the same and create other part instances, rotate it by 90 degrees, and set it vertically. Then move it to the patterned instances and define the pattern shown in Figure 8.15.

Perform the same and reposition them correctly (see Figure 8.16).

8.4.4 STEP MODULE

After finishing the assembly section, then the configuration of the analysis should be defined. In this simulation, only the extraction of the natural frequency of the structure is considered.

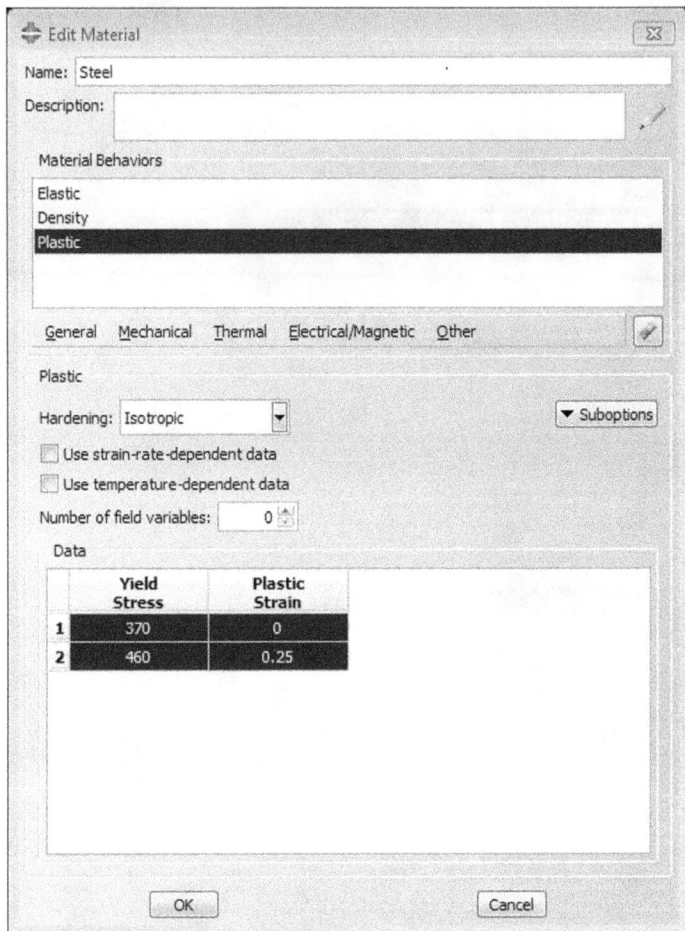

FIGURE 8.8   Identify the material plasticity.

This is a single event, so only a single analysis step is needed for the simulation. Natural frequency extraction is a linear step; therefore, all sources of nonlinear behavior will be neglected.

In the module list, select "Step" to activate the "Step" module as shown in Figure 8.17.

Then, select "Step → Create" to open the "Create step" dialog box. Select "Linear perturbation" as "Procedure type" and "Frequency" (see Figure 8.18).

Then click on "Continue" and the "Edit Step" dialog box appears. Insert a value of 10 as the number of eigenvalues requested in the Basic tab (see Figure 8.19).

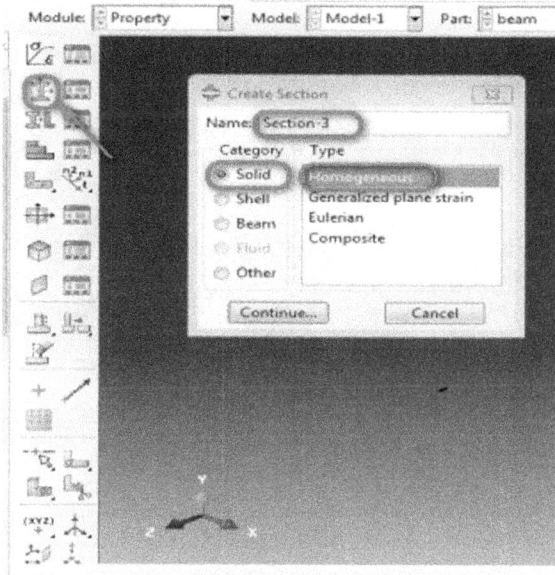

FIGURE 8.9   Create solid section.

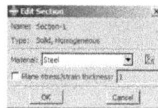

FIGURE 8.10   Solid, homogenous section definition.

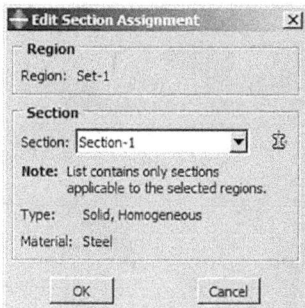

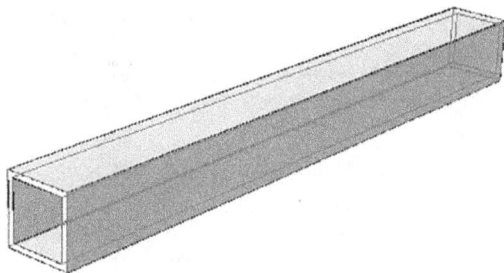

FIGURE 8.11   Section assignment.

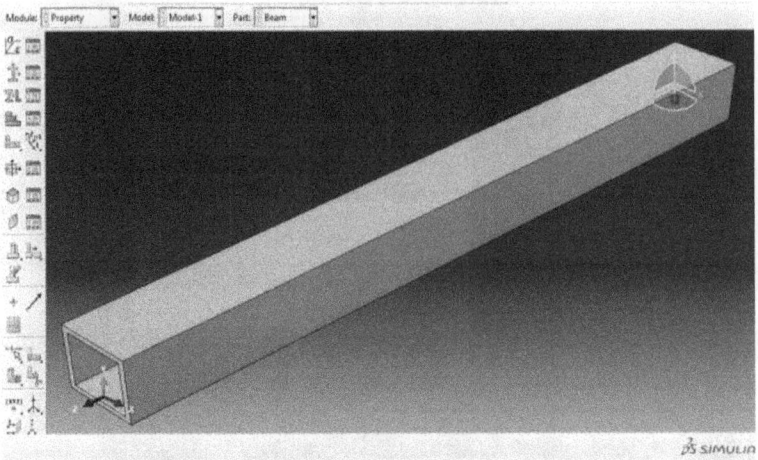

FIGURE 8.12   Beam section assigned.

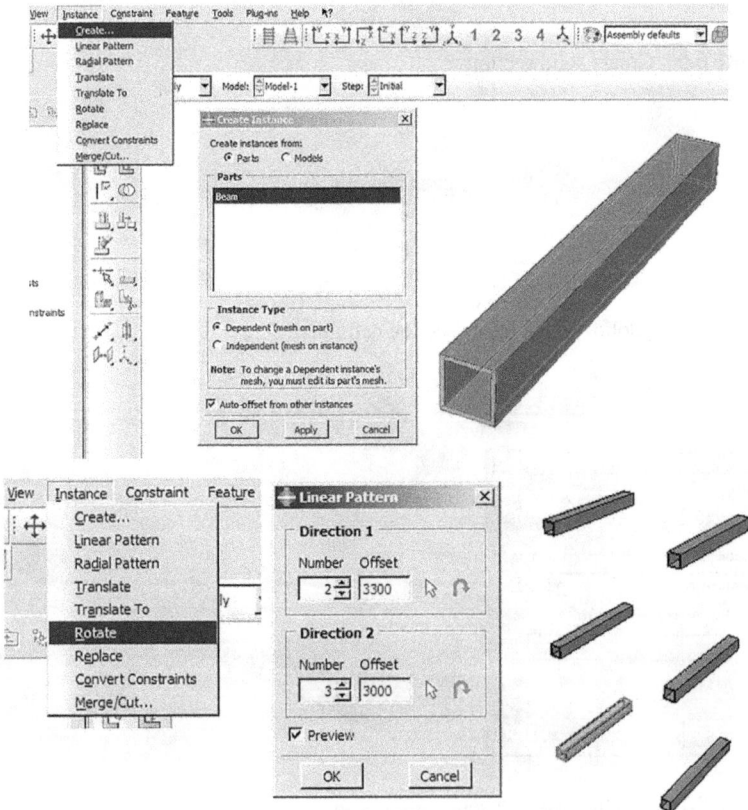

FIGURE 8.13   Create instance and linear pattern.

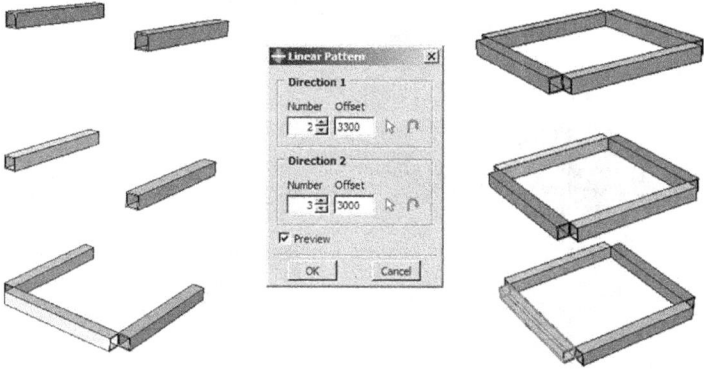

FIGURE 8.14   Create instance and linear pattern as beams.

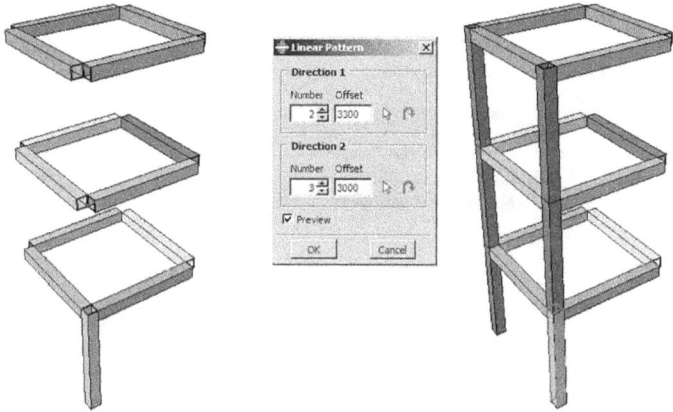

FIGURE 8.15   Create instance and linear pattern as columns.

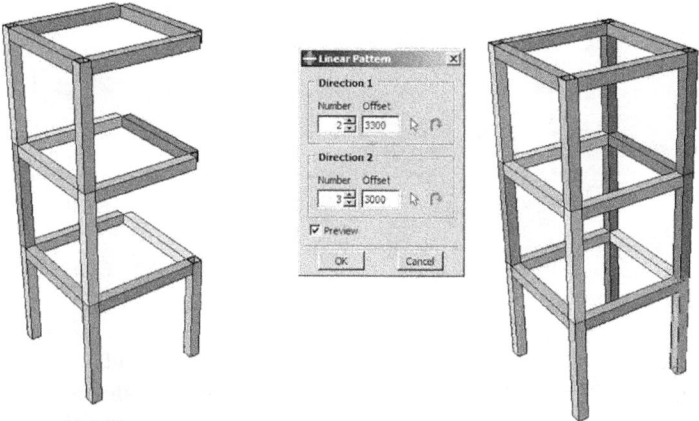

FIGURE 8.16   Finalizing the assembly.

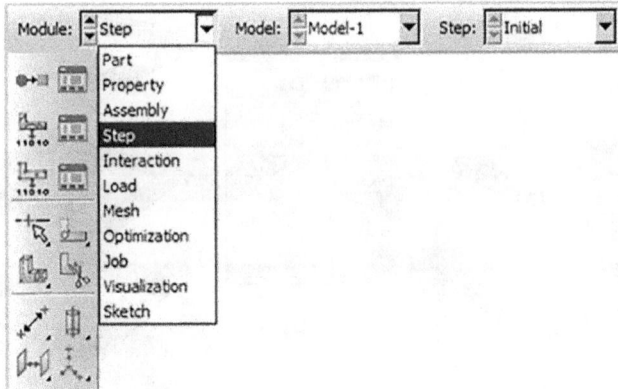

FIGURE 8.17   Step module.

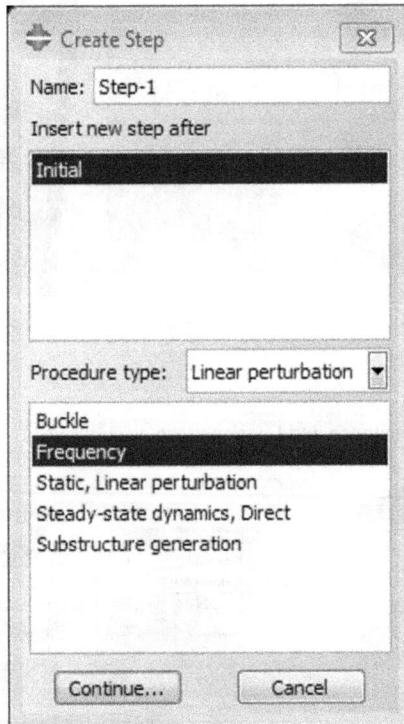

FIGURE 8.18   Define frequency step.

To check the variables which software is required by default, double click on "F-output-1" underneath "Field output requests" from the "Model tree" to open its box. As shown in Figure 8.20, only "U" was considered as the variable to extract. The variable indicated the mode shapes corresponding to 10 natural frequencies extracted by the solver.

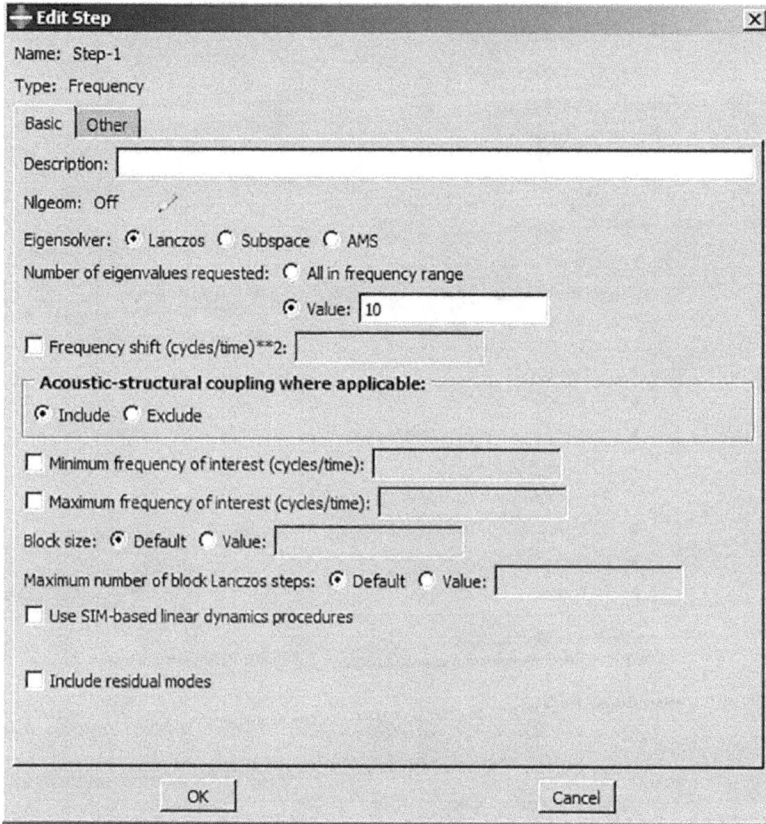

FIGURE 8.19   Edit step for frequency extraction.

8.4.5   INTERACTION MODULE

This module can be used to define the type of connection and constraints between part instances such as the Tie connection, Rigid, embedded region, and shell to solid coupling.

8.4.5.1   Tie Constraint

To activate the "Interaction module", select "Interaction" in the module list as shown in Figure 8.21.

To define "Tie" constraint between all instances, click "Find contact pair" in the toolbox to open its dialog box and then click "Find contact pairs". The software tries to find all near faces and list them in the dialog box (see Figure 8.22)

Click on column "Type", then "Edit" to open the "Edit multiple cells" dialog box. In the box, select "Tie" constraint as type and click "Ok" to accept and close it (see Figure 8.23)

Finally, click "Ok" to close the "Find contact pairs" dialog box and define all tie constraints between part instances as shown in Figure 8.24.

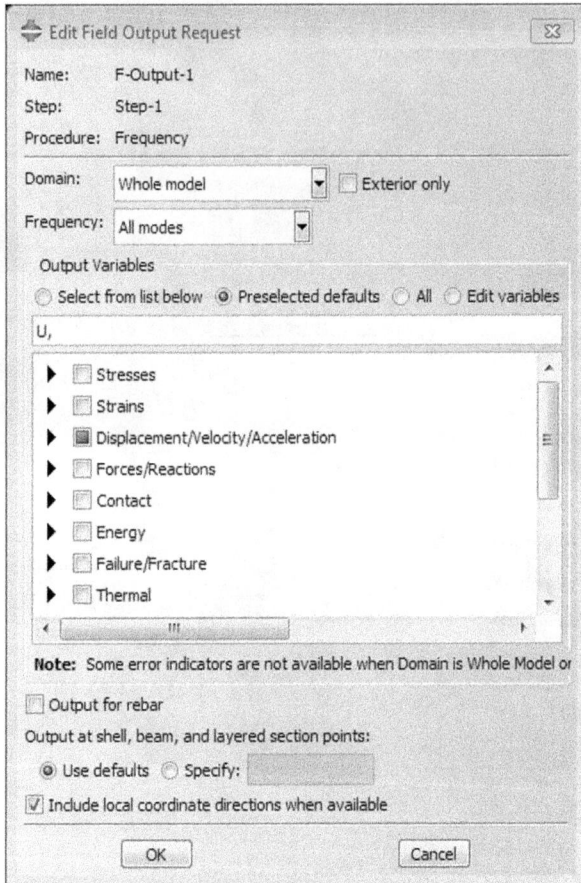

FIGURE 8.20   Checking preselected default variable as field output.

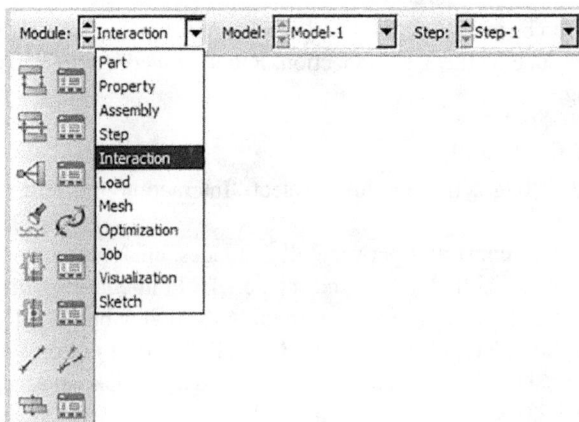

FIGURE 8.21   Interaction module.

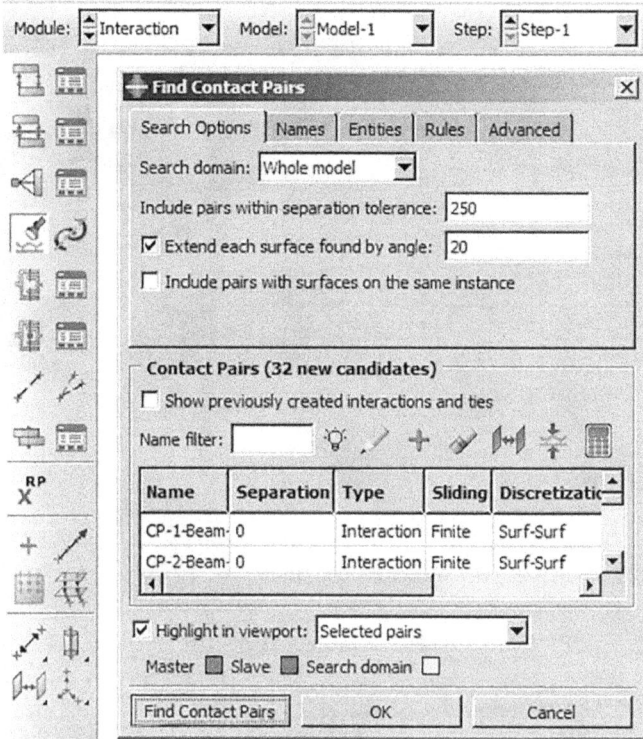

FIGURE 8.22   Create tie constraint using the find contact pair tool.

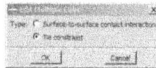

FIGURE 8.23   Edit contact pair to tie constraint.

8.4.6 LOAD MODULE

In this module, the user can create various loads, boundary conditions, prescribed conditions, etc.

In this example, it is not required to use these loads because this example aims to get natural frequency only and just boundary conditions should be defined.

To apply boundary conditions to the structure, first the user should activate the "Load" module.

In the module list located under the toolbar, click "Load" to activate the "Load" module. From the main menu bar, select "BC → Create". The "Create boundary condition" dialog box appears.

Name the boundary condition "Fix". From the list of steps, select "Initial" as the step in which the boundary condition will be activated. In the "Category" list, accept "Mechanical" as the default category selection. In the "Types for selected

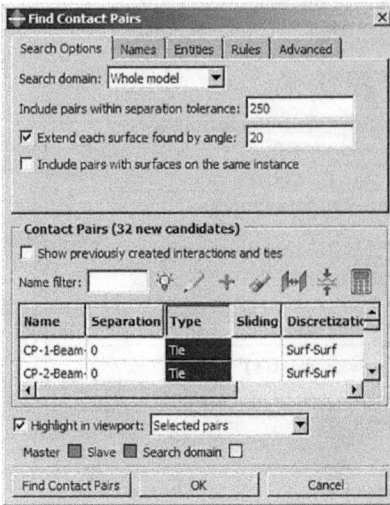

FIGURE 8.24 Finalizing tie constraints definition.

step" list, select "Symmetry/Axisymmetry/Encastre", and click "Continue" (see Figure 8.25).

Hold shift key on the keyboard and select all four bases of the structure, and once finished, click the middle mouse button to open the "Edit boundary condition" dialog box. Select "ENCASTRE" since all the rotational and translational degrees of freedom need to be constrained as a fixed support as shown in Figure 8.26.

Click "Ok" to create the boundary condition and to close the dialog box. The boundary condition at the column base is illustrated, as shown in Figure 8.27.

8.4.7 MESH MODULE

The mesh module is used to generate the finite element mesh. In the module list located under the toolbar, click on "Mesh" to open the "Mesh" module.

In the context bar, click on "Part" to unclick the assembly from the main menu bar. The part is recolored yellow.

Select "Mesh → Element type". In the viewport, select the entire frame as the region to be assigned to an element type. In the prompt area, click "Done". The "Element type" dialog box appears, as shown in Figure 8.28.

In the dialog box, select the following:

- Standard from the Element Library selection (the default).
- Linear from the Geometric Order (the default).
- 3D stress from the Family of elements.

Then, seeding should be defined. Seeding determines the approximate element size.

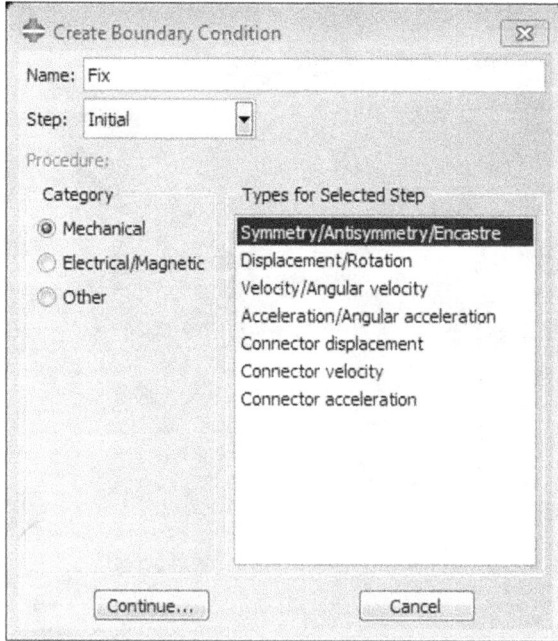

FIGURE 8.25 Create boundary condition.

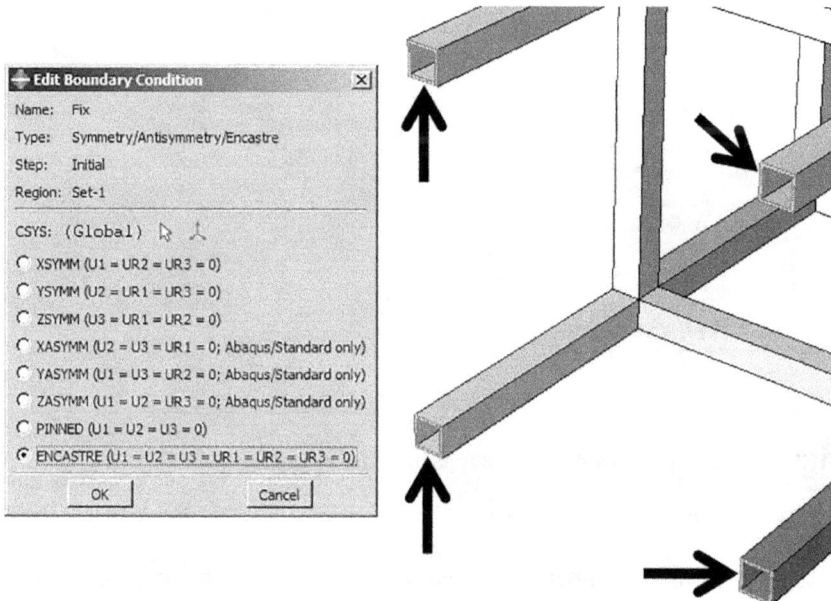

FIGURE 8.26 Define encastre boundary condition for bases.

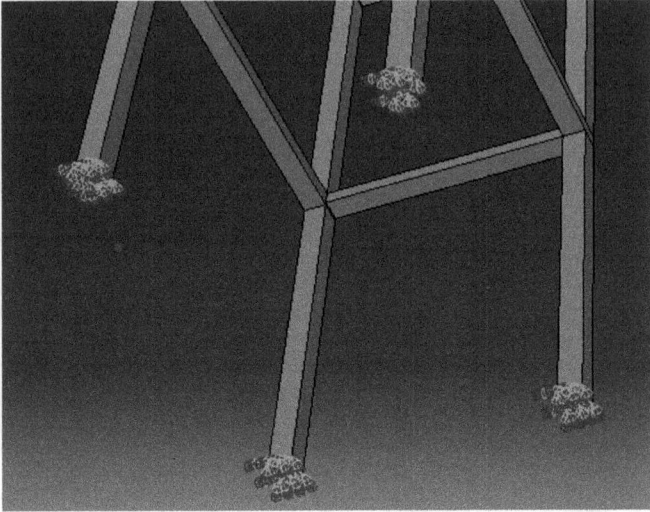

FIGURE 8.27   Encastred boundary conditions in the viewport.

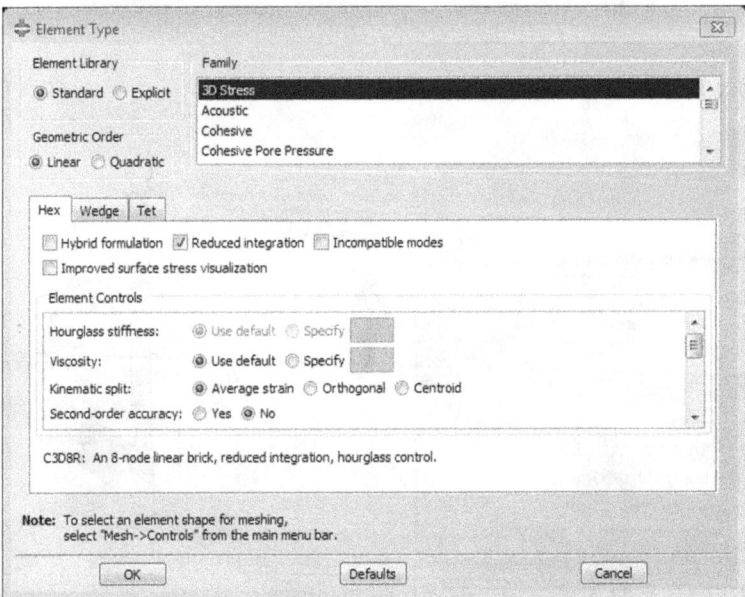

FIGURE 8.28   Selecting the element type in the mesh module.

From the main menu bar, select "Seed → Part" to open the General seeds dialog box.

Type 25 as the appropriate value for the "Approximate global size" of the mesh elements. Click "Ok" to accept seeding as shown in Figure 8.29.

To define finite element mesh for the part, select "Mesh → Part" and then click "Yes" in the prompt area to define the mesh (see Figure 8.30).

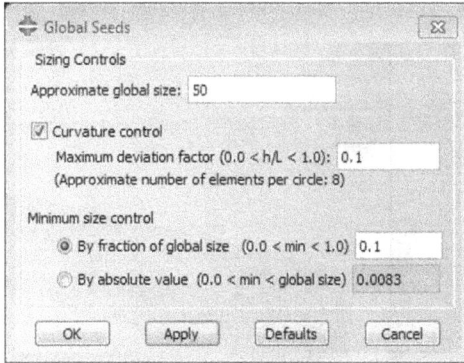

FIGURE 8.29 Assign the approximate global size as seed.

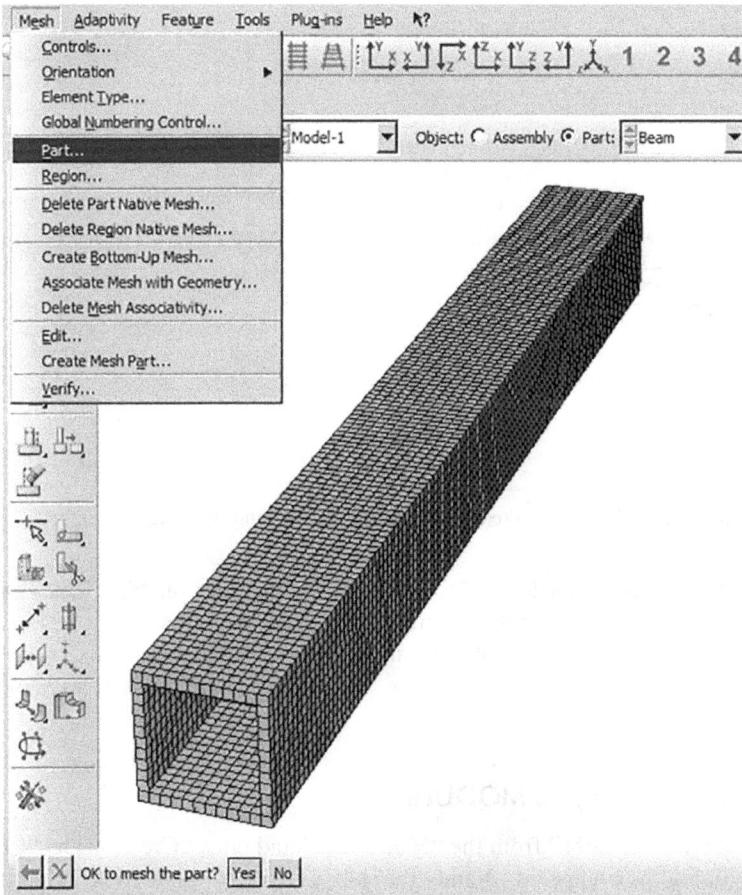

FIGURE 8.30 Meshing the part.

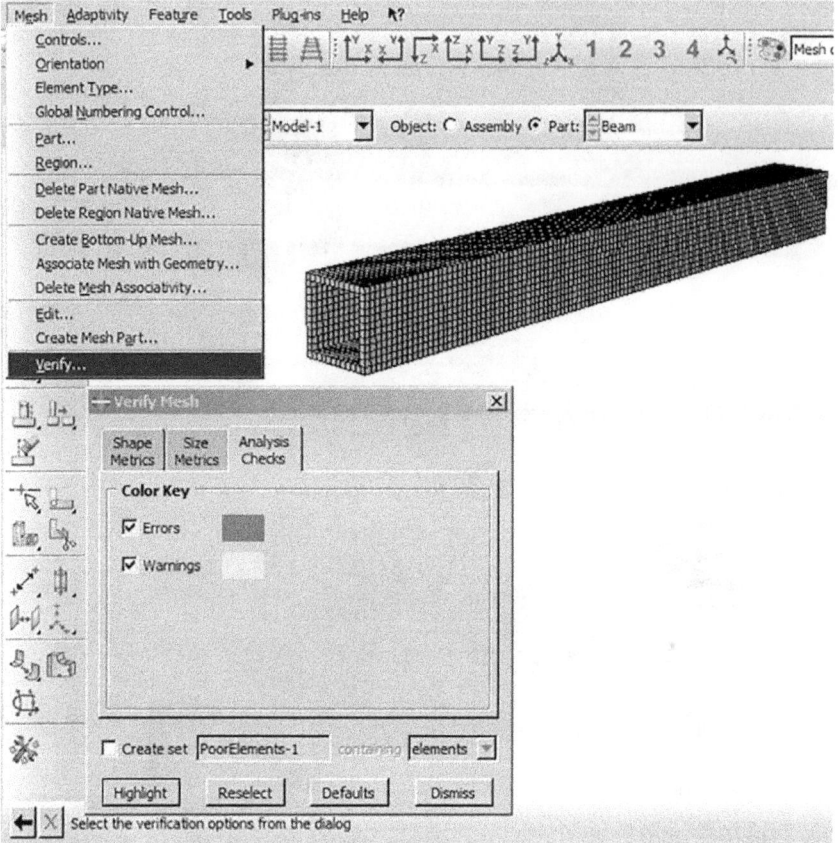

FIGURE 8.31 Verify meshing.

```
Part: Beam
   Number of elements :  5280,    Analysis errors:  0 (0%),  Analysis warnings:  0 (0%)
```

FIGURE 8.32 Checking finite element mesh for error and warning.

To check meshing, select "Mesh → Verify", then click on "Part" and "Done" in the prompt area to open the "Verify mesh" dialog box (see Figure 8.31).

Click "Highlight" to display elements that included errors and warnings. Because of the seed size, neither error nor warning was indicated as shown in Figure 8.32.

8.5   ANALYSIS: JOB MODULE

Double click on "Jobs" from the "Model tree" and click "Create". The "Create job" dialog box appears. Name the job as "Freq" and click "Continue" (see Figure 8.33).

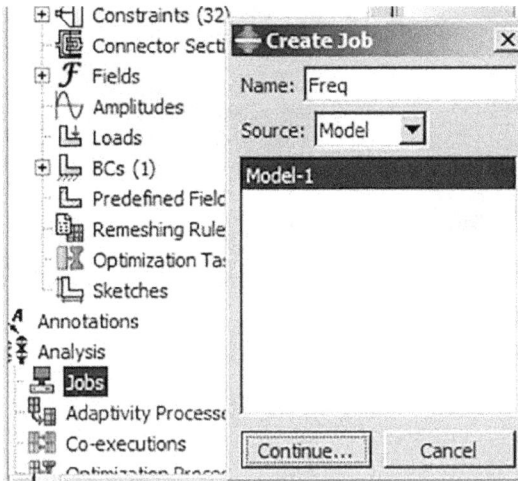

FIGURE 8.33   Create job.

The "Edit Job" dialog box appears. Click "Ok" to accept all other default job settings in the job editor and to close the dialog box as shown in Figure 8.34.

To begin the analysis, the job should be submitted. Right click on the job created recently, then click "Submit" to start checking the input file and running the analysis (see Figure 8.35).

To check the running progress, right click on the job and select "Monitor" to open the "Job monitor" dialog box (see Figure 6.36).

8.6   VISUALIZATION MODULE

Using this module, the user can extract and show all variables which are requested in the step module. When the job is completed, right click on the completed job and select "Results". The software is open the "Visualization module" and the "Output database Freq.odb" file (see Figure 8.37).

To see natural frequencies extracted by the software, select "Result → Step/frame" as shown in Figure 8.38.

A Step/Frame dialog box appears, as shown in Figure 8.39. The user can obtain the 10 natural frequencies for the considered model.

Select the first frequency to expand the corresponding mode shape. Then click on "Contour plot" from "Deformed shape" to display its mode shape as a contour plot (see Figure 8.40).

Again, open the "Step/frame" dialog box and double click on the fifth mode shape to update the contour plot in the viewport for the 5th mode shape as shown in Figure 8.41.

Perform the same for review the 9th mode shape in viewport (see Figure 8.42).

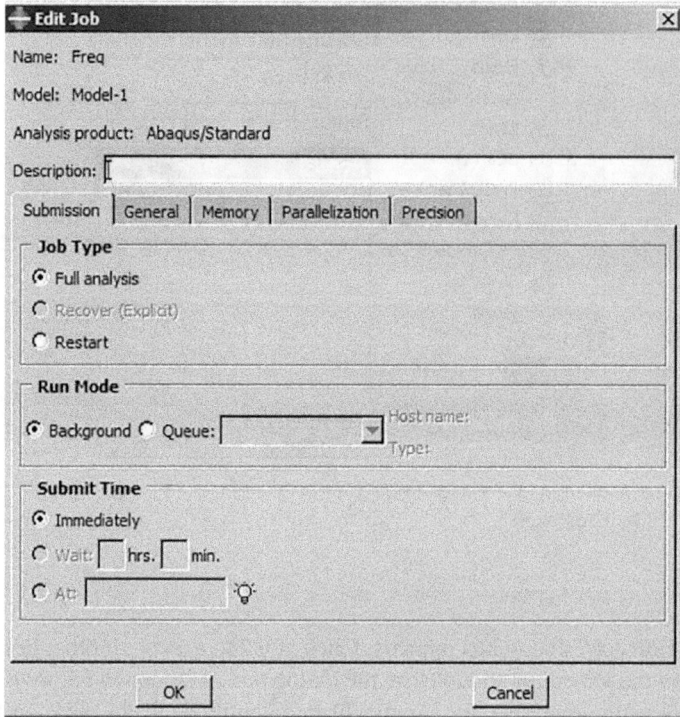

FIGURE 8.34 Job definition.

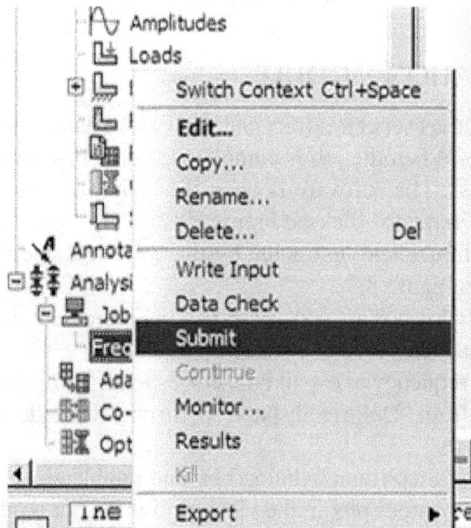

FIGURE 8.35 Job submission.

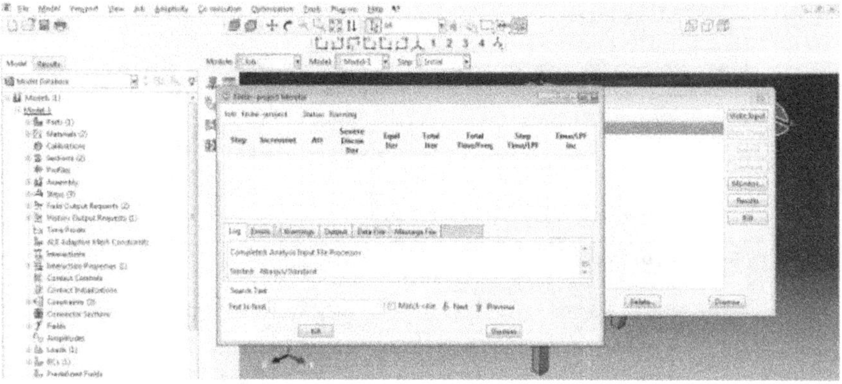

FIGURE 8.36 Monitor the job.

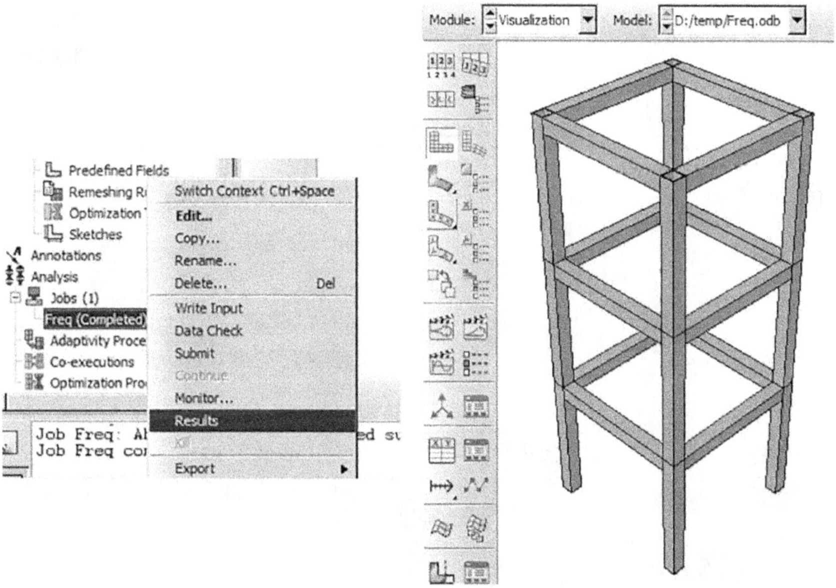

FIGURE 8.37 Visualization module.

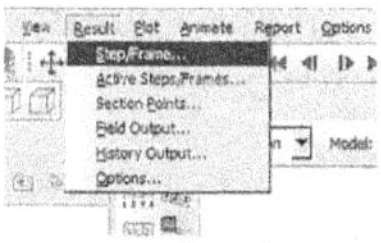

FIGURE 8.38 Accessing the extracted natural frequencies.

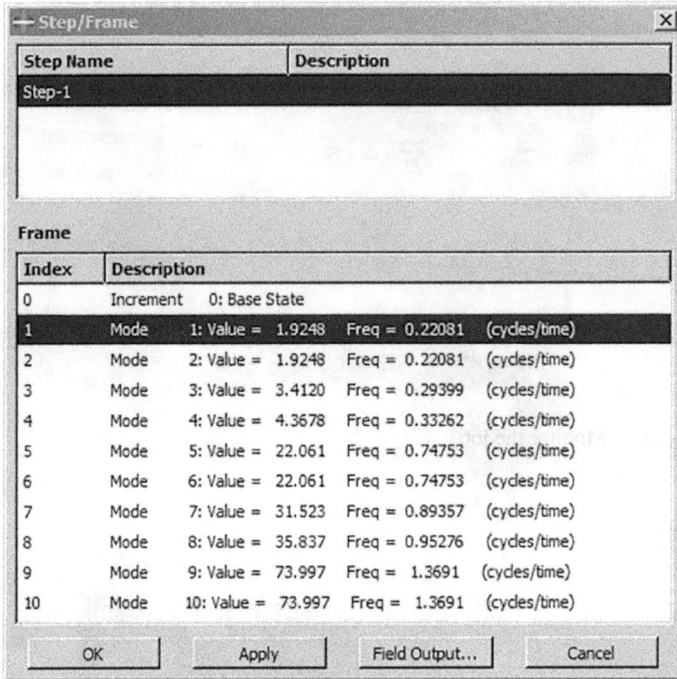

FIGURE 8.39   Result of natural frequency.

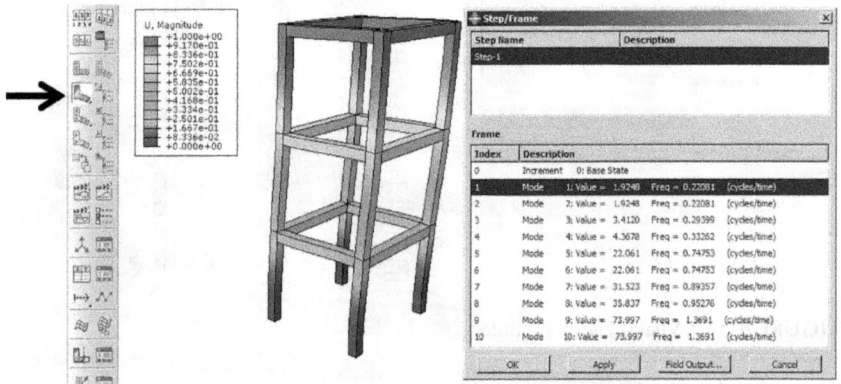

FIGURE 8.40   The first natural frequency and the corresponding mode shape of the structure.

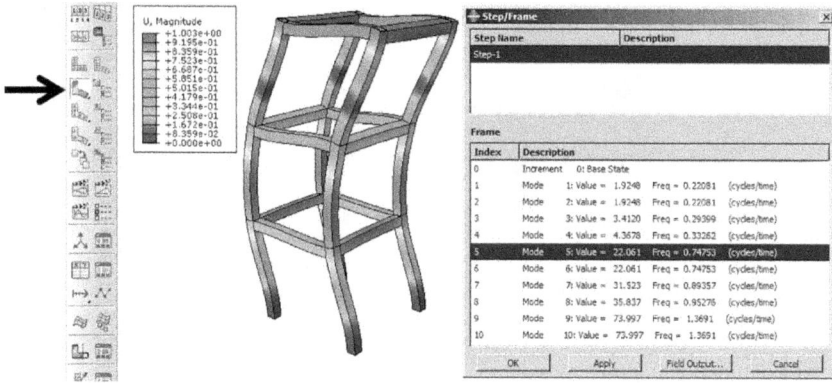

FIGURE 8.41 The fifth natural frequency and the corresponding mode shape of the structure.

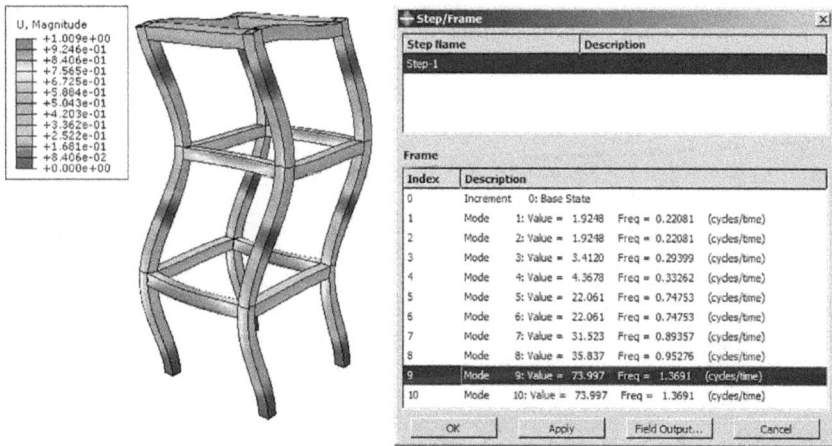

FIGURE 8.42 The ninth natural frequency and the corresponding mode shape of the structure.

Index

Actuator, 197
Analysis step, 38, 39, 77, 78, 138, 189, 253, 281

Boundary condition, 19, 20, 39, 42, 43, 78, 80,
    83, 84, 136, 137, 140, 194, 197, 227,
    260, 264, 265, 289, 290

CFRP, 59, 61, 65, 73, 92, 205, 206, 208, 212,
    216, 217, 218, 222, 226, 227, 228
Concrete Damaged Plasticity, 66, 117, 214, 244
Contact Pairs, 42, 75, 227, 287
Contour plot, 54, 87, 91, 98, 202, 271, 273, 295
Coupling, 42, 129, 192, 193, 287
Cracking Strain, 66, 118

Dowel bar, 101, 102, 105, 108, 114, 120

Earthquake, 2, 17, 19, 23, 27, 205
Embedded region, 42, 73, 75, 76, 77, 129, 132,
    134, 135, 226, 227, 259, 259, 287
Energy dissipation, 27, 159, 161, 206
Equivalent plastic strain, 271

Fractional exterior tolerance, 75, 77, 132,
    134, 259
Friction formulation, 130, 191, 259

Geometric Order, 47, 71, 143, 218, 265, 290

High Tensile Steel, 105, 117, 118
Hysteresis, 58, 88, 90, 91

Increment size, 17, 39, 79, 126, 190, 226, 253
Inelastic Strain, 66, 103, 118

Linear perturbation, 38, 282

Mode shape, 275, 276, 286, 295

Natural frequency, 275, 276, 282, 289, 298
Nonlinearities, 202

Plastic flow, 246
Plastic strain, 91, 105, 151, 156, 157, 200,
    245, 271
Poisson's ratio, 11, 32, 34, 61, 65, 117, 172, 213,
    244, 276, 278
Precast, 101, 102, 105, 106, 108, 110, 113, 114,
    117, 120, 121, 132, 135, 138, 140,
    142, 143, 146, 156, 158
Preloaded bolt connection, 159, 161

Rigid body, 42, 129, 132, 287
Round off, 77

Seed, 22, 46, 48, 49, 71, 143, 144, 184, 185, 188,
    218, 265, 290, 292, 294
Stiffness matrix, 4, 5, 6, 7
Subroutine, 2, 3, 4, 6, 7, 8, 11, 22

Testing protocol, 253
Tie constraint, 17, 42, 73, 129, 135, 287

Von Mises Stress, 22, 23, 157, 232

Wireframe, 113, 208

Yield stress, 61, 66, 105, 172, 245, 276
Young's modulus, 5, 6, 11, 19, 32, 34, 61, 65,
    104, 105, 117, 172, 213, 244, 276

For Product Safety Concerns and Information please contact our EU
representative GPSR@taylorandfrancis.com
Taylor & Francis Verlag GmbH, Kaufingerstraße 24, 80331 München, Germany

www.ingramcontent.com/pod-product-compliance
Lightning Source LLC
Chambersburg PA
CBHW060813220326
41598CB00022B/2606